Darney's Circuit Theory and Modelling

Updated and extended for EMC/EMI

Also available from IET by Ian B. Darney: Circuit Modeling for Electromagnetic Compatibility

Related titles on electromagnetic waves:

Dielectric Resonators, 2nd Edition Kajfez and Guillon
Electronic Applications of the Smith Chart Smith
Fiber Optic Technology Jha
Filtering in the Time and Frequency Domains Blinchikoff and Zverev
HF Filter Design and Computer Simulation Rhea
HF Radio Systems and Circuits Sabin
Microwave Field-Effect Transistors: Theory, design and application, 3rd Edition Pengelly
Microwave Semiconductor Engineering White
Microwave Transmission Line Impedance Data Gunston
Optical Fibers and RF: A natural combination Romeiser
Oscillator Design and Computer Simulation Rhea
Radio-Electronic Transmission Fundamentals, 2nd Edition Griffith, Jr
RF and Microwave Modeling and Measurement Techniques for Field Effect Transistors Jianjun Gao
RF Power Amplifiers Albulet
Small Signal Microwave Amplifier Design Grosch
Small Signal Microwave Amplifier Design: Solutions Grosch
2008+ Solved Problems in Electromagnetics Nasar
Antennas: Fundamentals, design, measurement, 3rd Edition Blake and Long
Designing Electronic Systems for EMC Duff
Electromagnetic Measurements in the Near Field, 2nd Edition Bienkowski and Trzaska
Fundamentals of Electromagnetics with MATLAB®, 2nd Edition Lonngren *et al.*
Fundamentals of Wave Phenomena, 2nd Edition Hirose and Lonngren
Integral Equation Methods for Electromagnetics Volakis and Sertel
Introduction to Adaptive Arrays, 2nd Edition Monzingo *et al.*
Microstrip and Printed Antenna Design, 2nd Edition Bancroft
Numerical Methods for Engineering: An introduction using MATLAB® and computational electromagnetics Warnick
Return of the Ether Deutsch
The Finite Difference Time Domain Method for Electromagnetics: With MATLAB® simulations Elsherbeni and Demir
Theory of Edge Diffraction in Electromagnetics Ufimtsev
Scattering of Wedges and Cones with Impedance Boundary Conditions Lyalinov and Zhu
Circuit Modeling for Electromagnetic Compatibility Darney
The Wiener–Hopf Method in Electromagnetics Daniele and Zich
Microwave and RF Design: A systems approach, 2nd Edition Steer
Spectrum and Network Measurements, 2nd Edition Witte
EMI Troubleshooting Cookbook for Product Designers Andre and Wyatt
Transmission Line Transformers Raymond Mack and Jerry Sevick
Electromagnetic Field Standards and Exposure Systems Grudzinski and Trzaska
Practical Communication Theory, 2nd Edition Adamy
Complex Space Source Theory of Spatially Localized Electromagnetic Waves Seshadri
Electromagnetic Compatibility Pocket Guide: Key EMC facts, equations and data Wyatt and Jost
Antenna Analysis and Design Using FEKO Electromagnetic Simulation Software Elsherbeni, Nayeri and Reddy
Scattering of Electromagnetic Waves by Obstacles Kristensson

Adjoint Sensitivity Analysis of High Frequency Structures with MATLAB® Bakr, Elsherbeni and Demir
Developments in Antenna Analysis and Synthesis, Volume 1 and Volume 2 Mittra
Advances in Planar Filters Design Hong
Post-processing Techniques in Antenna Measurement Castañer and Foged
Nano-Electromagnetic Communication at Terahertz and Optical Frequencies, Principles and Applications Alomainy, Yang, Imran, Yao and Abbasi
Nanoantennas and Plasmonics: Modelling, design and fabrication Werner, Campbell and Kang
Electromagnetic Reverberation Chambers: Recent advances and innovative applications Guillaume Andrieu
Radio Wave Propagation in Vehicular Environments Azpilicueta, Vargas-Rosales, Falcone and Alejos
Advances in Mathematical Methods for Electromagnetics Kobayashi and Smith
Theory and Practice of Modern Antenna Range Measurements 2nd Expanded Edition, Volume 1 Clive Parini, Stuart Gregson, John McCormick, Daniël Janse van Rensburg and Thomas Eibert
Theory and Practice of Modern Antenna Range Measurements 2nd Expanded Edition, Volume 2 Clive Parini, Stuart Gregson, John McCormick, Daniël Janse van Rensburg and Thomas Eibert
Emerging Evolutionary Algorithms for Antennas and Wireless Communications Sotirios Goudos
Bioelectromagnetics in Healthcare: Advanced sensing and communication applications W. Wittow
Applications of Deep Learning in Electromagnetics: Teaching Maxwell's equations to machines Maokun Li

Darney's Circuit Theory and Modelling

Updated and extended for EMC/EMI

Ian B. Darney and Chris M. Hewitt

The Institution of Engineering and Technology

Published by SciTech Publishing, an imprint of The Institution of Engineering and Technology, London, United Kingdom

The Institution of Engineering and Technology is registered as a Charity in England & Wales (no. 211014) and Scotland (no. SC038698).

First published 2023

The Institution of Engineering and Technology
Futures Place
Kings Way, Stevenage
Hertfordshire, SG1 2UA, United Kingdom

www.theiet.org

British Library Cataloguing in Publication Data
A catalogue record for this product is available from the British Library

ISBN 978-1-83953-953-4 (hardback)
ISBN 978-1-83953-954-1 (PDF)

Typeset in India by MPS Limited
Printed in the UK by CPI Group (UK) Ltd, Eastbourne

Cover Image Credit: Original by Alberto Menendez Cervero
modified and enhanced by Chris M. Hewitt

Contents

About the authors

Ian B. Darney (deceased) was a graduate of Glasgow University, UK. A highly knowledgeable electrical engineer, Mr Darney had extensive experience of EMC, particularly in the aircraft industry. He was a member of the IET and the European Organisation for Civil Aviation Equipment (EUROCAE).

Chris M. Hewitt is an autonomous agent design engineer with 9 years' HMRAF experience installing, repairing and calibrating radar and missile systems, alongside test and communication equipment. He is a recipient of the Support Command Certificate of Merit for 'Initiative and Inventiveness towards the improvement of engineering services of the RAF', with extensive knowledge of EMC/EMI. His current focus is on developing the AE42 concept; a gravity centric and empirical data approach to grand unification.

Foreword by Chris M. Hewitt

This book represents the life's work of Ian Brook Darney who passed away on the 26 December 2021 after a short illness and expands upon and continues his previous publication 'Circuit Modelling for Electromagnetic Compatibility'. Ian had already begun this book shortly before he became ill and his wish was for it to be completed and published. Thanks to the dedication and support of his wife Frances, his daughter Susan White and son Peter Darney, his wish has become reality.

It was through his son, who I became friends with whilst we were studying at Bristol University, that I first met Ian in the summer of 1993. Shortly after Ian's death I was approached by his son Peter to ascertain the feasibility of using all of the available material and the framework Ian had set out to fulfil his wish. Whilst evaluating the material it became apparent that Ian's forte was most definitely mathematics, as such, he only made the top–down derivations available from his website, his assumption being that the mathematics involved were self-evident. Unfortunately, even highly competent people with first year degree level mathematics found the derivations difficult to follow, as such, after consulting with Peter, I modified the original transmission line model section into a step-by-step derivation of the model (Section 2.2) using a bottom–up approach and likewise introduced Appendix A, Inductance and Capacitance Primitives to show the 'Primitives' derivations from first principles. Any mathematical errors in these sections are my responsibility as are any errors in the images in this book, which have been 'upgraded' from the ones originally provided by Ian. Other than the aforementioned sections, very little of the main text body has been altered, with the exception of correcting 'typos' and syntax.

A both personal and professional opinion of the research conducted by Ian can best be described as visionary. Ian's research into EMC began well before EMC became a major issue in an increasingly technological dependent and connected world. The almost exponential increase in the environment of human-generated electromagnetic waves has subsequently increased the complexity of using the Maxwell equations for determining EMC and EMI susceptibility of devices, with the increased number of associated variables required to produce results consummate with real-world scenarios, larger and faster computers are required to manipulate the equations in a timely manner to facilitate a device's development and manufacture. An often overlooked consequence of this is not only that the more complex such equations become and thus more susceptible to induced errors but much more energy is required to generate the requisite results. The method employed by Ian throughout this book reduces the complexity of the equations needed to obtain not just satisfactory results but results backed by empirical data, requiring substantially less Energy and thus less impact on the environment. In fact, all the equations demonstrated by Ian have not only been verified by empirical testing but have been continually refined during the testing phases. This is how research and development should be conducted, creating a mathematical model based on the data rather than trying to reinterpret the data to fit an 'established' theoretical model, which is often the case in today's world!

Figure border colour key

A yellow border indicates the figure is a Mathcad worksheet.

A green border indicates the figure is a Mathcad or an Excel plot.

A blue border is used to indicate an illustration, a model, a block or a circuit diagram.

A purple border indicates a table.

A red border indicates a photographic-derived image.

A red and green border indicates that the figure is a photo/plot combination.

Internet link identification

Internet links are indicated in the text body by This Colour Text. The URLs of the links can be found in Appendix C.

Chapter 1
Introduction

1.1 Introduction

Circuit theory provides all the analytical tools necessary to predict the behaviour of any electronic circuit. Central to teaching on this theory is the concept of the 'Equipotential Ground'. Invoking this concept prevents any attempt to use any circuit network to analyse electromagnetic (EM) interference (EMI).

A review of the relationship of circuit theory with EM theory identifies a way of augmenting these analytical tools, and this enables circuit models to be developed to simulate all forms of EMI, thus effectively updating and extending circuit theory.

Details of how this objective can be achieved were initially provided as a set of articles already available at the electronics world website in PDF format and can be reviewed by entering 'Updating Circuit Theory' in the EW search panel. This book is a compendium of those articles grouped by chapters and referenced within following sections and describes the approach adopted.

1.2 Background

All the mechanisms involved in the propagation of EMI are defined by the relationships derived in EM theory. Circuit theory is a development and simplification of the mathematics involved in the derivation process. It has proved to be a reliable and accurate method of designing electrical and electronic systems. Over the past century, circuit theory has provided all the analytical tools necessary to predict the functional behaviour of these systems. Even so, the unwanted coupling associated with EMI has become increasingly troublesome, to such an extent that it became necessary to develop regulatory requirements to ensure that different systems do not interfere with each other.

It is impossible to avoid some level of EM coupling. So the requirements for EM compatibility (EMC) have been formulated in a way which ensures that the level of interference of the equipment-under-review is within acceptable limits. At the end of the manufacturing process, the equipment is subjected to a series of defined tests in an EMC test house. If it passes the tests, it can be claimed that the equipment has met a defined set of formal regulations. These regulations can then be included in the specification which defines the performance of the equipment.

A whole industry has developed to provide advice and guidelines to equipment manufacturers as to how to design and develop their product in a way which ensures that it has a good chance of meeting the formal EMC requirements. But if the equipment fails to meet these requirements, the consultants will have long gone.

1.3 Assessment

As yet, there is no universally accepted method of designing electronic equipment to meet the formal EMC requirements. This is due to several factors.

The mathematics required to analyse the behaviour of three dimensional EM fields is very complicated. When the author was a student of electrical engineering in Glasgow in the late 1950s, he never even heard of the operations 'div', 'del', and 'curl', let alone how to manipulate them. He learnt of the existence of the Maxwell Equations several years later when he was talking to a colleague at British Aerospace. The professors who set the curriculum he followed at university probably decided that the manipulations of such concepts were well beyond the ability of the average student.

The complexity of electronic systems in vehicles, aircraft, spacecraft, and ships has created the development of various specialities. Each speciality creates its own jargon and abbreviations. It is often the case that one specialist has difficulty in understanding the reasoning of another specialist. Many abbreviations mean different things to different people.

An almost impenetrable barrier to the electronic system designer who hopes to glean information on the topic is the language and jargon of any paper on computational EMs.

The problem and the solution can both be found in the analytical tools of circuit theory: a theory which is dramatically easier to understand.

1.4 Problem

Circuit theory is a development and simplification of EM theory. As far as EMC is concerned, it contains a simplification too far. This simplification is the concept of the zero-volt ground-plane, represented as a 'ground' symbol or an 'earth' symbol on circuit diagrams.

SPICE (*Simulation Programs with Integrated Circuit Emphasis*) analysis uses this concept to good effect, by calculating the voltage on every node of the network with respect to a single node at which the voltage is defined as zero. Such a representation allows components such as resistors, inductors, capacitors, diodes, and transistors to be mounted on a circuit board and represented as a circuit in which there is one-to-one correlation between the symbol on the diagram and the component on the board. This allows extremely complex assemblies to be analysed. But, by its very nature, it cannot simulate EMI.

Circuit theory has proved to be so useful and so reliable that many engineers have been led to believe that its rules are the Laws of Physics. In attempting to correlate these rules with the observed phenomenon of EMI, a whole new philosophy has been developed; 'Grounding Philosophy'. This concept has been endorsed by several eminent researchers. Many authors have integrated this approach into books on EMC. The inclusion of many and varied relationships of EM theory (all of which are undisputable) in these books lends credibility to this false philosophy.

It has been recommended that a terminal be installed on the conducting structure and that all the return conductors of a subsystem be connected to that terminal. That is, all voltages in the subsystem can be referred to this zero-volt terminal. There are signal grounds, logic grounds, power grounds, and many other variations on the theme. Some engineers have advocated such a system with fervour worthy of any religious zealot.

Any system using these guidelines to define the design requirements for any system can be guaranteed to cause problems. These can be inconsequential, annoying, dangerous, or mission critical. The consequences can also be lethal, as with the in-flight breakup over the Atlantic Ocean of Trans World Airlines Flight 800. Figure 46 of the Accident Report clearly shows that the wiring of the fuel system is based on the concept of the single point ground.

1.5 Solution

Any link carrying signals or power between two electronic units mounted on a conducting structure is carried by three conductors; the send conductor, the return conductor, and the structure. In this representation, the 'structure' represents the effect of every other conductor in the system.

A single conductor can be represented as a T-network with inductors and resistors in the horizontal branches and a capacitor in the vertical branch. Three conductors in parallel are represented as a Triple-T network with the capacitors connected together at a single point. This point represents a location in the far distance where the energy level of the EM field is zero. Such a representation is closer to reality, where the energy is carried by photons propagating back and forth between the conductors; that is, carried by the EM field.

Any electronic system can be represented as a block diagram, where the contents of each block define the function of that block and the lines between blocks represent the signals transmitted between them. Instead of focussing attention on the contents of any block, the EMC analysis could treat each link as a separate entity.

The conductors of any signal link can be represented as a Triple-T network. By adding the interface circuitry of the blocks at each end, the cross-coupling between the signal loop and the common-mode loop can be analysed. The effect of an external EM field can be simulated by adding a voltage source in series with the structure.

By treating each signal link as a separate entity, the analysis of the EMC of the system can be broken down into a number of discrete problems, each of which can be analysed using the tools of circuit theory. The technique can be extended to simulate all the mechanisms associated with EMI coupling; conducted emission, conducted susceptibility, radiated emission, and radiation susceptibility.

1.6 Mesh analysis

The nature of the relationship between the two theories makes it natural to use loop equations to analyse the behaviour of the signal link.

When a voltage is applied between two conductors, current flows down the send conductor, creating an EM field. This propagates outwards at the speed of light and is reflected by the return conductor. This field arrives back at the send conductor and induces a voltage which enhances the forward current flow. Positive feedback ensures that current flow in the return conductor is in the opposite direction to that in the send conductor. The driving force is the EM field.

This mechanism is basic to both frequency analysis and transient analysis.

1.7 Approach

Since this approach involved the development of circuit models which could not be found in the literature on EMC, the only way of checking the accuracy of the simulation was to set up a bench test, record the results, and then compare those results with the model. Puzzling over any deviation between test results and the simulation involved opening old textbooks on EM theory and relating that theory to the model. New experiments could be devised and the process repeated.

Since the accuracy of each new model could be checked by comparing it with the actual performance of the hardware, and since the construction of that model was derived from the relationships of EM theory, sufficient confidence was acquired to proceed to the next stage.

This is the pattern followed throughout the book.

It was necessary to develop a set of equations for each model and to create a program which solved those equations. Mathcad software was used for this purpose since these was no need to translate these equations into machine language. Each equation could be copied onto the worksheet in the same form that it was derived.

To enhance understanding of the process, as well as to confirm the validity of the computation used to create response curves, every worksheet is replicated as a figure in the text. A significant feature of Mathcad is that the test results can be included in the worksheet, allowing test and model results to be displayed on a single graph. Every definition and every computation are visible on the same worksheet. Errors are not difficult to locate.

This approach means that each section can carry theory, details of a test setup, the test procedure, analysis, computation, correlation, and the assessment. This is different from that covered in traditional textbooks where the theory is introduced step-by-step. Each section in this book is self-standing. This allows the set of sections to be treated as a 'handbook' with easy comparison and validation between sections.

1.8 Frequency analysis

An experiment is described which compares the frequency response of a signal link with that of a Triple-T model. The two responses correlate closely over a range which includes that of full-wave resonance. This demonstrates the reliability of the model.

Any conducting surface of any cross-section can be simulated as an array of parallel conductors in which the current in any conductor will induce a voltage in every other conductor. By treating the array as a set of three composite conductors, a Triple-T model can be constructed of any cable assembly of any cross section. It is shown that a shield does not behave as a barrier to EM fields; it acts to neutralise the effect of those fields.

A dipole antenna can be represented as a series Inductor, Capacitor, Resistor (LCR) circuit in which the resistor simulates the effect of the environment. With a two-conductor cable, the environment is represented as a virtual conductor.

With EMC analysis, it is necessary to demonstrate that the radiated emission from the assembly-under-review is less than a defined limit and that the assembly is not susceptible to an external field of defined intensity. That is, worst-case conditions are assumed. It transpires that the equations for worst-case conditions are relatively simple. The magnetic field at a distance is proportional to the current in the structure multiplied by the distance. The voltage induced in a conductor is proportional to the strength of the threat field and the length of the conductor.

An example is provided of the method of analysing the susceptibility of a twin-conductor cable used to initiate an electro-explosive device. It goes on to show how to design a simple filter which ensures that the system is safe in the presence of the threat field.

The screens of coaxial cable normally consist of multiple strands of copper wire twisted round the solid central core in a braided pattern. This means that the length of any conductor of the shield is significantly longer than the core. Since charges are constrained to flow along conductors, the effect can be visualised as an extra length of conductor routed alongside the core. This extra conductor possesses the properties of inductance and capacitance. A test is described which provides data on the frequency responses of the admittance and the transfer admittance. A circuit model is created which replicates that response.

1.9 Transient analysis

Transmission line theory shows that any line can be defined in terms of its characteristic impedance and the time it takes a signal to traverse its length. If it is assumed that there are no losses, that impedance can be defined as a resistance. A shift register can be used to simulate the time delay. Reflections at each termination are defined by the reflection equations. A program can be created to simulate the transient response of a twin-conductor cable, whatever the interface circuitry.

With power supplies, it is not possible to avoid reflections, and these transients are the greatest cause of radiated emissions. The only way of reducing the level of these emissions is to absorb the unwanted energy in resistors. The design of a simple power line filter which absorbs transient energy is described.

This filter is significantly different from conventional filters which protect the load by sending unwanted energy back towards the source. The only place for that energy to go is into the environment; where it re-appears as unwanted radiation.

A circuit model is described which simulates the transient response of a signal link. This caters for the fact that the common-mode current propagates at a higher velocity than the differential-mode current.

Power supplies on aircraft use the conducting structure as a return conductor. It is reasoned that the inclusion of a return conductor would increase the weight of the wiring. Weight is a significant factor in the design of aircraft and spacecraft. However, if a small diameter conductor was routed alongside the send conductor and connected to structure at both ends, this would carry all the high-frequency components, without compromising the ability of the structure to carry the low frequency energy. A review of this effect could form part of a cost–benefit analysis.

A method of matching interface circuitry to the signal link is described. This minimises reflections in both the differential-mode and common-mode loops.

Ground bounce is due to the energy stored in the cables being released into the environment when power is disconnected. An experiment is used to illustrate this phenomenon.

An experiment is described which identifies the relationship between charges and photons. The energy required to emit photons from the surface of a conductor is equal to the energy needed to decelerate the charges which arrive radially at that surface.

1.10 Conclusion

The technique described in this book can be used to analyse all the mechanisms involved in the propagation of EMI. This approach avoids the need for endless debate on the topic of 'Ground Philosophy'. Equally, it avoids the need to invoke the complexities of EM computation.

Chapter 2
Cross coupling models

2.1 Introduction

A reprise of transmission line theory leads to the derivation of a pair of hybrid equations. These relate the current and voltage at one end of the line to voltage and current at the other. The line can also be represented as a T-network of impedance parameters. Mesh analysis of this network leads to a pair of Loop Equations.

Manipulating the loop equation into the same form as those derived from transmission line theory leads to two pairs of hybrid equations; one set derived from transmission line theory, the other set derived from circuit theory. Correlating the parameters leads to a set of transformation equations which define the parameters of the T-network in terms of the resistance, inductance, capacitance, and conductance of the line. Since the R, L, C, and G components are distributed along the line, the resultant circuit can be described as a distributed parameter model.

The simplest circuit model is also a T-network of lumped parameters, where values can be assigned to the components. This makes it possible to define the characteristics of a two-conductor line in terms of a simple lumped parameter model but to analyse it using the distributed parameter model.

A three-conductor line can be simulated as a Triple-T network, with inductors and resistors in the horizontal branches and capacitors in the vertical branches. Each T-network represents the properties of a single conductor. Transforming each T-network into a similar network of distributed parameters allows the response of the three-conductor to be simulated using the analytical tools of circuit theory.

Cross coupling between differential-mode signals and common-mode signals can be simulated using distributed parameter model, over a frequency range which extends well beyond that of half-wave resonance. Conducted emission and conducted susceptibility of a signal link or a power link can be defined during the initial design stage of a project.

Requirements for electromagnetic compatibility (EMC) dealt with in the design process in the same way as those of any other design requirement.

Establishing confidence in such an approach calls for some demonstration that it can be done. This calls for the use of transformers to inject signals into cable assemblies and transformers to monitor currents in those assemblies. Such equipment is usually associated with the highly expensive hardware to be found in any facility used to check EMC. It is entirely possible to manufacture the necessary equipment using components available from any supplier of electronic parts.

Energy can be induced in a cable using a voltage injection transformer. Such a device includes a monitor turn which allows the induced voltage to be monitored. The resultant current can be monitored by a current transformer. The construction and calibration of these transducers are described. These have a useable frequency range from 20 kHz to 20 MHz.

A test rig is described. This can be used to mount the cable under review. Since it is over 11 m long, it can be used to measure the response of the assembly over a frequency range well beyond that of half-wave resonance.

The first experiment was to measure the response of a twin conductor cable mounted on the rig. By short-circuiting the cable terminals to the conducting structure of the rig, it was possible to measure the amplitude of the currents over a range which included resonant conditions.

Assigning guess-values to the components of the general circuit model of the assembly and then modifying them until the response of the model coincided with that of the test results enabled a definitive circuit model of the assembly-under-review to be created. Given the existence of such a model, it becomes possible to predict the performance of a similar cable used to carry signals from one equipment unit to another when mounted on an actual system.

The experiment was then repeated to characterise the properties of a co-axial cable when mounted on the rig. Theoretical analysis of such an assembly results in a model which simulates the cylindrical screen as a conductor with zero inductance. In practice, the screen presents a significant inductance. It was reasoned that the extra inductance was due to the fact that the strands of the braid were longer than the core, because they were wound in a helix. So the circuit model was modified to take this into account. The end result was a general circuit model of a co-axial cable and a definitive circuit model of the actual assembly.

2.2 Step-by-step derivation of the transmission line model

2.2.1 Definition of the model

Figure 2.1 illustrates a schematic diagram of a twin-conductor transmission line. It is assumed that the cross-section of the conductor assembly is constant along the entire length. The length of the line is *l*, there is a voltage source *Vn* at the near end, and this source delivers a current *In* to the line. The current delivered to the load at the far end of the line is *If*, and the voltage between the terminals at the far end is *Vf*.

It is assumed that this line can be simulated as a number of LCR networks connected in series, and that one such segment is as shown in Figure 2.2. This represents the properties of the length *dx* at a distance *x* from the near end. Since each conductor possesses the properties of resistance and inductance, it is necessary to assign these properties to both.

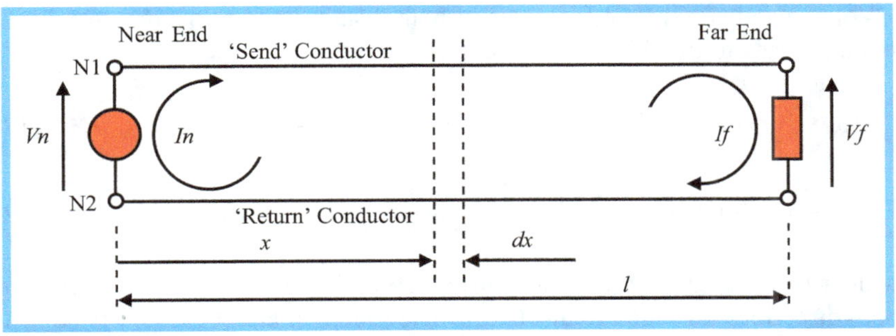

Figure 2.1 Schematic diagram of a twin-conductor transmission line

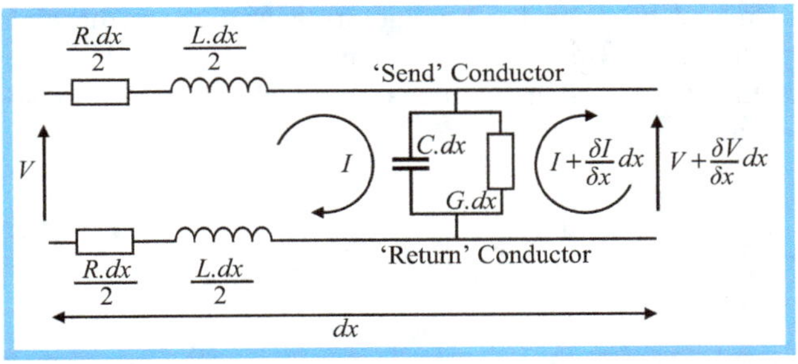

Figure 2.2 Voltages and currents in a single segment of the line

It is also assumed that the propagation of energy is from the near end to the far end and that, initially, there is no energy in the line. This means that both the voltage V and the current I increase as the distance x increases.

2.2.2 *Forward propagation*

It is assumed that this line can be simulated as a number of LCR networks connected in series, and that one such segment is as shown in Figure 2.2. This represents the properties of the length dx at a distance x from the near end. Since each conductor possesses the properties of resistance and inductance, it is necessary to assign these properties to both.

For the purpose of deriving the transmission line equations, the following definitions apply:

R = loop resistance per unit length: Ω/m
L = loop inductance per unit length: H/m
C = capacitance between conductors per unit length: F/m
G = conductance between conductors per unit length: S/m

Voltage and current are functions of both distance and frequency. Analysis of Figure 2.2 leads to the following equations:

$$(R + j \cdot \omega \cdot L) \cdot dx \cdot I = V - \left(V + \frac{\delta V}{\delta x} \cdot dx\right) = -\frac{\delta V}{\delta x} \cdot dx \tag{2.1}$$

$$(G + j \cdot \omega \cdot C) \cdot dx \cdot V = I - \left(I + \frac{\delta I}{\delta x} \cdot dx\right) = -\frac{\delta I}{\delta x} \cdot dx \tag{2.2}$$

It is clear from Figure 2.2 that each current flows in a loop. That is, current flows in both directions through the insulation. The send conductor can be visualised as a transmitting antenna. Charges are deposited on the return conductor. This creates current flow in the return conductor. It behaves as a receiving antenna. Current flowing in the opposite direction to that in the send conductor creates a magnetic field. This enhances the forward flow of current in the send conductor. The field so created between the conductors acts to steer the electromagnetic energy along the path defined by the routing of the cable. This means that the inductance of the return conductor plays a crucial role in the behaviour of any electronic system.

Hence,

$$\frac{\delta V}{\delta x} = -(R + j \cdot \omega \cdot L) \cdot I \tag{2.3}$$

$$\frac{\delta I}{\delta x} = -(G + j \cdot \omega \cdot C) \cdot V \tag{2.4}$$

Differentiating (2.3) gives

$$\frac{\delta^2 V}{\delta x^2} = -(R + j \cdot \omega \cdot L) \cdot \frac{\delta I}{\delta x} = (R + j \cdot \omega \cdot L) \cdot (G + j \cdot \omega \cdot C) \cdot V \tag{2.5}$$

Differentiating (2.4) gives

$$\frac{\delta^2 I}{\delta x^2} = -(G + j \cdot \omega \cdot C) \cdot \frac{\delta V}{\delta x} = (R + j \cdot \omega \cdot L) \cdot (G + j \cdot \omega \cdot C) \cdot I \tag{2.6}$$

If γ the propagation constant is defined as

$$\gamma = \sqrt{(R + j \cdot \omega \cdot L) \cdot (G + j \cdot \omega \cdot C)} \tag{2.7}$$

then substituting λ^2 for $(R + j \cdot \omega \cdot L) \cdot (G + j \cdot \omega \cdot C)$ in (2.5) and (2.6) gives

$$\frac{\delta^2 V}{\delta x^2} = \gamma^2 \cdot V \tag{2.8}$$

$$\frac{\delta^2 I}{\delta x^2} = \gamma^2 \cdot I \tag{2.9}$$

These are second order differential equations. One solution for (2.8) is as follows:

$$V = Vn \cdot e^{-\gamma \cdot x} \tag{2.10}$$

Differentiating (2.10) gives

$$\frac{\delta V}{\delta x} = -\gamma \cdot Vn \cdot e^{-\gamma \cdot x} \tag{2.11}$$

Differentiating (2.11) gives

$$\frac{\delta^2 V}{\delta x^2} = \gamma^2 \cdot Vn \cdot e^{-\gamma \cdot x}$$

Invoking (2.10) leads to (2.8):

$$\frac{\delta^2 V}{\delta x^2} = \gamma^2 \cdot V$$

This confirms that (2.10) is one solution to (2.8). Similarly, one solution to (2.9) is as follows:

$$I = In \cdot e^{-\gamma \cdot x} \tag{2.12}$$

From (2.3)

$$I = -\frac{1}{R + j \cdot \omega \cdot L} \cdot \frac{\delta V}{\delta x} \tag{2.13}$$

Using (2.11) to substitute for $\frac{\delta V}{\delta x}$ gives

$$I = \frac{\gamma}{R + j \cdot \omega \cdot L} \cdot Vn \cdot e^{-\gamma \cdot x} = \frac{\sqrt{(R + j \cdot \omega \cdot L) \cdot (G + j \cdot \omega \cdot C)}}{R + j \cdot \omega \cdot L} \cdot Vn \cdot e^{-\gamma \cdot x}$$

Giving

$$I = \sqrt{\frac{G + j \cdot \omega \cdot C}{R + j \cdot \omega \cdot L}} \cdot Vn \cdot e^{-\gamma \cdot x} \tag{2.14}$$

The characteristic impedance *Zo* is defined as

$$Zo = \sqrt{\frac{R + j \cdot \omega \cdot L}{G + j \cdot \omega \cdot C}} \tag{2.15}$$

So

$$I = \frac{1}{Zo} \cdot Vn \cdot e^{-\gamma \cdot x} \tag{2.16}$$

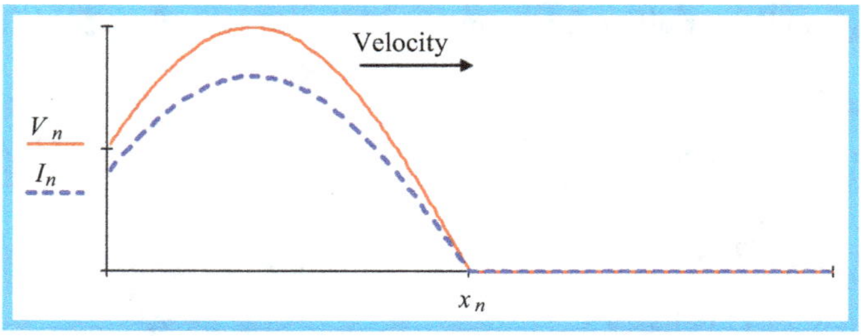

Figure 2.3 Relationship between voltage and current waveforms

And, invoking (2.10)

$$I = \frac{V}{Zo} \tag{2.17}$$

This means that, at any point, the ratio of voltage to current is equal to *Zo*, the characteristic impedance.

2.2.3 Rates of change

Invoking (2.10) to substitute for $Vn \cdot e^{-\gamma \cdot x}$ in (2.11) gives

$$\frac{\delta V}{\delta x} = -\gamma \cdot V \tag{2.18}$$

Differentiating (2.16) gives $\dfrac{\delta I}{\delta x} = -\dfrac{\gamma}{Zo} \cdot Vn \cdot e^{-\gamma \cdot x}$

Invoking (2.10) gives $\dfrac{\delta I}{\delta x} = -\dfrac{\gamma}{Zo} \cdot V$

Re-arranging: $-\gamma \cdot V = Zo \cdot \dfrac{\delta I}{\delta x}$

Invoking (2.18) to substitute for $-\gamma \cdot V$ gives rise to the significant relationship:

$$\frac{\delta V}{\delta x} = Zo \cdot \frac{\delta I}{\delta x} \tag{2.19}$$

This means that the rate of change of current is proportional to the rate of change of voltage. Since the current is also proportional to the voltage, there can be no phase difference between the current and voltage waveforms during forward propagation. Figure 2.3 illustrates this. A further deduction is that the charge propagates forward at the same velocity as the voltage and current.

2.2.4 Line losses

For a lossless line, $R = 0$ and $G = 0$. So (2.15) simplifies to

$$Ro = \sqrt{\frac{L}{C}} \tag{2.20}$$

Since (2.20) is a pure resistance, it is appropriate to define this parameter as the characteristic resistance. A significant feature of this relationship is that *Ro* is the independent of frequency. Also, as (2.19) indicates, for a lossless line, there is no phase difference between the voltage and current waveforms at any point. Figure 2.3 illustrates this.

With a lossy line, the parameters *R* and *G* become increasingly significant.

Each conductor can be regarded as an antenna. One of the characteristics of a transmitting dipole is that maximum efficiency is reached at the half-wave frequency, when it behaves as a resistor of approximately 73 Ω to the radio frequency amplifier. So if the frequency is doubled, a conductor the same length would behave as two dipoles in series and the load would be doubled. Since *G* represents the effect of losses, it is reasonable to surmise that this parameter would also double. That is, that the conductance is proportional to frequency.

The other parameter associated with line loss is the series resistor *R*. This is inversely proportional to the square root of frequency, due to skin effect.

2.2.5 Reflections

If the load is not precisely equal to *Zo*, there will be reflections at the far end. These manifest themselves as a voltage source at that end. Such a source will deliver current which flows from the far end to the near end. That is, a backward travel line wave is created. Current will flow in both directions in each conductor. Such a condition is catered for in circuit theory by the principle of superposition. To fully simulate the response of the line, it is necessary to include the backward travelling wave in the calculations.

2.2.6 Two-way propagation

When the signal reaches the termination at the far end, some of it is absorbed by the load, and some if it is reflected. This reflected signal flows from the far end to the near end. Energy is flowing in both directions along the line. So, a more comprehensive solution to (2.8) is, as the total voltage at any point is given by the equation:

$$V = H \cdot e^{\gamma \cdot x} + K \cdot e^{-\gamma \cdot x} \tag{2.21}$$

It is also possible to define the relationship as

$$V = A \cdot \cosh(\gamma \cdot x) + B \cdot \sinh(\gamma \cdot x) \tag{2.22}$$

That is

$$\begin{aligned} V &= A \cdot \left(\frac{e^{\gamma \cdot x} + e^{-\gamma \cdot x}}{2}\right) + B \cdot \left(\frac{e^{\gamma \cdot x} - e^{-\gamma \cdot x}}{2}\right) \\ &= \frac{A+B}{2} \cdot e^{\gamma \cdot x} + \frac{A-B}{2} \cdot e^{-\gamma \cdot x} \\ &= H \cdot e^{\gamma \cdot x} + K \cdot e^{-\gamma \cdot x} \end{aligned}$$

Equation (2.22) is used here because the final equations are amenable to conversion to equations which can be handled by matrix analysis; that is, by a circuit model.

From (2.13) and (2.22)

$$I = \frac{-1}{R + j \cdot \omega \cdot L} \cdot \frac{\delta V}{\delta x} = \frac{-\gamma}{R + j \cdot \omega \cdot L} \cdot [A \cdot \sinh(\gamma \cdot x) + B \cdot \cosh(\gamma \cdot x)]$$

Now, since

$$\frac{(R + j \cdot \omega \cdot L)}{\gamma} = \frac{R + j \cdot \omega \cdot L}{\sqrt{(R + j \cdot \omega \cdot L) \cdot (G + j \cdot \omega \cdot C)}} = \sqrt{\frac{R + j \cdot \omega \cdot L}{G + j \cdot \omega \cdot C}} = Zo$$

Then

$$I = -\frac{1}{Zo} \cdot [A \cdot \sinh(\gamma \cdot x) + B \cdot \cosh(\gamma \cdot x)] \tag{2.23}$$

2.2.7 Boundary conditions

The next task is to assign values to the parameters A and B. From an engineering point of view, the only voltages and currents of any interest are those which exist at the interfaces at the near and far ends as defined in Figure 2.1.

Thus from Figure 2.1, boundary conditions at the far end are as follows:

$V = Vf$ at $x = l$
$I = If$ at $x = l$

At the receiving end, x can be replaced by l the length of the cable between the near end and the far end in (2.22) and (2.23)

$$Vf = A \cdot \cosh(\gamma \cdot l) + B \cdot \sinh(\gamma \cdot l) \tag{2.24}$$

$$-Zo \cdot If = A \cdot \sinh(\gamma \cdot l) + B \cdot \cosh(\gamma \cdot l) \tag{2.25}$$

Multiplying (2.24) by $\sinh(\gamma \cdot l)$ and (2.25) by $\cosh(\gamma \cdot l)$

$$Vf \cdot \sinh(\gamma \cdot l) = A \cdot \cosh(\gamma \cdot l) \cdot \sinh(\gamma \cdot l) + B \cdot \sinh(\gamma \cdot l) \cdot \sinh(\gamma \cdot l) \tag{2.26}$$

$$-Zo \cdot If \cdot \cosh(\gamma \cdot l) = A \cdot \sinh(\gamma \cdot l) \cdot \cosh(\gamma \cdot l) + B \cdot \cosh(\gamma \cdot l) \cdot \cosh(\gamma \cdot l) \tag{2.27}$$

Subtracting (2.27) from (2.26) gives

$$Vf \cdot \sinh(\gamma \cdot l) + Zo \cdot If \cdot \cosh(\gamma \cdot l) = B \cdot \left(\sinh^2(\gamma \cdot l) - \cosh^2(\gamma \cdot l)\right) = -B$$

Hence

$$B = -(Vf \cdot \sinh(\gamma \cdot l) + Zo \cdot If \cdot \cosh(\gamma \cdot l)) \tag{2.28}$$

Multiplying (2.24) by $\cosh(\gamma \cdot l)$ and (2.25) by $\sinh(\gamma \cdot l)$

$$Vf \cdot \cosh(\gamma \cdot l) = A \cdot \cosh(\gamma \cdot l) \cdot \cosh(\gamma \cdot l) + B \cdot \sinh(\gamma \cdot l) \cdot \cosh(\gamma \cdot l) \tag{2.29}$$

$$-Zo \cdot If \cdot \sinh(\gamma \cdot l) = A \cdot \sinh(\gamma \cdot l) \cdot \sinh(\gamma \cdot l) + B \cdot \cosh(\gamma \cdot l) \cdot \sinh(\gamma \cdot l) \tag{2.30}$$

Subtracting (2.30) from (2.29)

$$Vf \cdot \cosh(\gamma \cdot l) + Zo \cdot If \cdot \sinh(\gamma \cdot l) = A \cdot \left(\cosh^2(\gamma \cdot l) - \sinh^2(\gamma \cdot l)\right)$$

Giving

$$A = Vf \cdot \cosh(\gamma \cdot l) + Zo \cdot If \cdot \sinh(\gamma \cdot l) \tag{2.31}$$

Substituting A and B in (2.22) gives

$$\begin{aligned} V = {} & [Vf \cdot \cosh(\gamma \cdot l) + Zo \cdot If \cdot \sinh(\gamma \cdot l)] \cdot \cosh(\gamma \cdot x) \\ & - [Vf \cdot \sinh(\gamma \cdot l) + Zo \cdot If \cdot \cosh(\gamma \cdot l)] \cdot \sinh(\gamma \cdot x) \\ = {} & Vf \cdot [\cosh(\gamma \cdot l) \cdot \cosh(\gamma \cdot x) - \sinh(\gamma \cdot l) \cdot \sinh(\gamma \cdot x)] \\ & + Zo \cdot If \cdot [\sinh(\gamma \cdot l) \cdot \cosh(\gamma \cdot x) - \cosh(\gamma \cdot l) \cdot \sinh(\gamma \cdot x)] \end{aligned}$$

Hence

$$V = Vf \cdot \cosh[\gamma(l - x)] + Zo \cdot If \cdot \sinh[\gamma(l - x)] \tag{2.32}$$

Similarly, substituting A and B in (2.23) gives

$$I = -\frac{1}{Zo} \cdot \left\{ \begin{array}{l} [Vf \cdot \cosh(\gamma \cdot l) + Zo \cdot If \cdot \sinh(\gamma \cdot l)] \cdot \sinh(\gamma \cdot x) \\ -[(Vf \cdot \sinh(\gamma \cdot l) + Zo \cdot If \cdot \cosh(\gamma \cdot l))] \cdot \cosh(\gamma \cdot x) \end{array} \right\}$$

$$= \frac{1}{Zo} \cdot \left\{ \begin{array}{l} Vf \cdot \sinh(\gamma \cdot l) \cdot \cosh(\gamma \cdot x) - Vf \cosh(\gamma \cdot l) \cdot \sinh(\gamma \cdot x) + \\ Zo \cdot If \cdot \cosh(\gamma \cdot l) \cdot \cosh(\gamma \cdot x) - Zo \cdot If \cdot \sinh(\gamma \cdot l) \cdot \sinh(\gamma \cdot x) \end{array} \right\}$$

And reducing the hyperbolic products gives

$$I = If \cdot \cosh[\gamma \cdot (l - x)] + \frac{Vf}{Zo} \cdot \sinh[\gamma \cdot (l - x)] \tag{2.33}$$

At the sending end, $x = 0$, $V = Vn$, $I = In$. Substituting these values in (2.32) and (2.33) gives

$$Vn = Vf \cdot \cosh(\gamma \cdot l) + Zo \cdot If \cdot \sinh(\gamma \cdot l) \tag{2.34}$$

And

$$In = \frac{Vf}{Zo} \cdot \sinh(\gamma \cdot l) + If \cdot \cosh(\gamma \cdot l) \tag{2.35}$$

Equation (2.34) defines the voltage at the near end in terms of the voltage and current at the far end. Equation (2.35) defines the current at the near end in terms of the voltage and current at the far end. These are called hybrid equations since the parameters are mixed. With nodal analysis, all the current variables are on the left-hand side of the set of equations. With mesh analysis, the left-hand side contains voltage variables. These hybrid parameters cannot represent the components of any circuit model, but they do form a stepping stone towards the creation of just such a model.

2.2.8 The transformation equations

Since every conductor possesses the properties of inductance, capacitance, resistance, and conductance, any two-conductor transmission line can be defined by the lumped-parameter circuit model as shown in Figure 2.4.

Invoking the concept of equivalent circuits, this can be simplified to that shown on Figure 2.5(a), where

$$\begin{aligned} Rd &= Rc_1 + Rc_2 \\ Ld &= Lc_1 + Lc_2 \\ \frac{1}{Cd} &= \frac{1}{Cc_1} + \frac{1}{Cc_2} \\ \frac{1}{Gd} &= \frac{1}{Gc_1} + \frac{1}{Gc_2} \end{aligned} \tag{2.36}$$

This simplification does not mean that the line at the bottom of the diagram actually represents an equipotential conductor. The relationship between the per-unit-length parameters and the lumped parameters is as follows:

$$\begin{aligned} Rd &= R \cdot l \\ Ld &= L \cdot l \\ Cd &= C \cdot l \\ Gd &= G \cdot l \end{aligned} \tag{2.37}$$

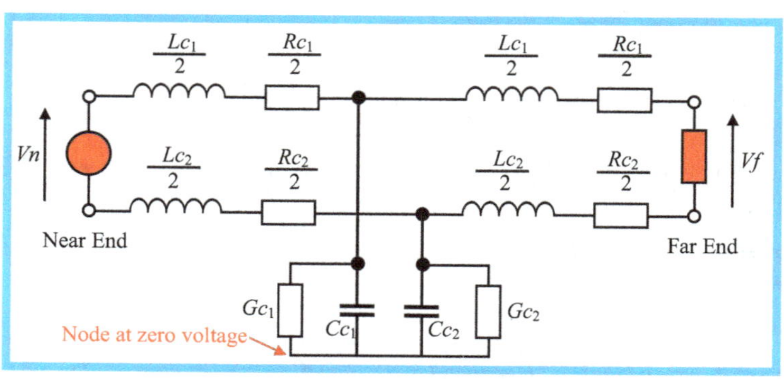

Figure 2.4 Lumped-parameter model

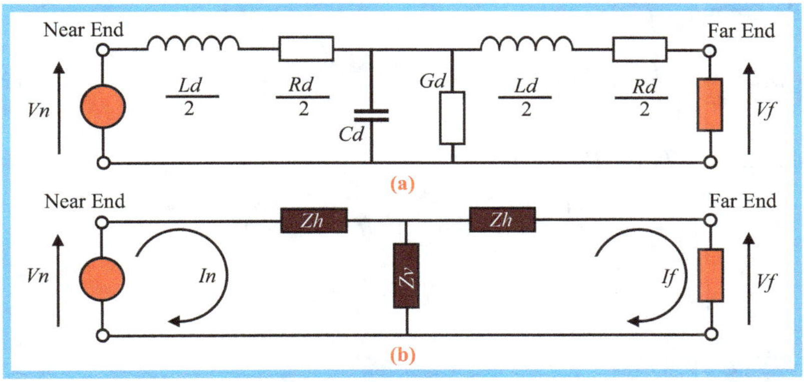

Figure 2.5 Single-T network models of a signal link: (a) lumped-parameter model and (b) distributed-parameter model

Substituting for *R, L, G,* and *C* in (2.7)

$$\gamma = \sqrt{\left(\frac{Rd}{l} + j \cdot \omega \cdot \frac{Ld}{l}\right) \cdot \left(\frac{Gd}{l} + j \cdot \omega \cdot \frac{Cd}{l}\right)} \tag{2.38}$$

If the parameter θ is defined as

$$\theta = \lambda \cdot l \tag{2.39}$$

Then the characteristic impedance and the propagation constant can be related to parameters which can be measured using electronic test equipment. There is now no need to invoke the use of per-unit-length parameters during the analysis of transmission lines.

Then (2.7) and (2.15) can be defined in terms of the lumped parameters:

$$\theta = \sqrt{(Rd + j \cdot \omega \cdot Ld) \cdot (Gd + j \cdot \omega \cdot Cd)} \tag{2.40}$$

$$Zo = \sqrt{\frac{Rd + j \cdot \omega \cdot Ld}{Gd + j \cdot \omega \cdot Cd}} \tag{2.41}$$

If it is hypothesised that the circuit model of Figure 2.5(b) is capable of simulating the relationships between the voltages and currents defined in Figure 2.1, the equations relating these parameters must correlate with those of (2.34) and (2.35). The mesh equations for Figure 2.5(b) are as follows:

$$Vn = (Zh + Zv) \cdot In - Zv \cdot If \tag{2.42}$$

$$Vf = Zv \cdot In - (Zh + Zv) \cdot If \tag{2.43}$$

where *Zh* are the horizontal impedances, and *Zv* are the components aligned vertically.

Re-arranging (2.43) gives

$$In = \frac{Vf}{Zv} + \left(1 + \frac{Zh}{Zv}\right) \cdot If \tag{2.44}$$

Substituting for *In* in (2.42)

$$\begin{aligned} Vn &= (Zh + Zv) \cdot \left[\frac{Vf}{Zv} + \left(1 + \frac{Zh}{Zv}\right) \cdot If\right] - Zv \cdot If \\ &= \left(1 + \frac{Zh}{Zv}\right) \cdot Vf + \left[Zh + Zv + \frac{Zh^2}{Zv} + Zh - Zv\right] \cdot If \end{aligned}$$

Simplifying further

$$Vn = \left(1 + \frac{Zh}{Zv}\right) \cdot Vf + Zh \cdot \left(2 + \frac{Zh}{Zv}\right) \cdot If \tag{2.45}$$

Equations (2.44) and (2.45) are hybrid equations derived from circuit theory which have been derived from the mesh Equations (2.42) and (2.43). They can be correlated with the hybrid Equations (2.35) and (2.34) which have been derived from electromagnetic theory. **This is the step which relates the two theories**.

Comparing (2.45) with (2.34) gives

$$1 + \frac{Zh}{Zv} = \cosh(\gamma \cdot l) \tag{2.46}$$

And

$$Zh \cdot \left(2 + \frac{Zh}{Zv}\right) = Zo \cdot \sinh(\gamma \cdot l) \tag{2.47}$$

Comparing (2.44) with (2.35)

$$\frac{1}{Zv} = \frac{1}{Zo} \cdot \sinh(\gamma \cdot l) \tag{2.48}$$

This comparison also replicates (2.46)

$$1 + \frac{Zh}{Zv} = \cosh(\gamma \cdot l)$$

Re-arranging (2.46) and invoking (2.39) gives

$$\frac{Zh}{Zv} = \cosh\theta - 1 \tag{2.49}$$

Substituting $\cosh\theta - 1$ for $\frac{Zh}{Zv}$ in (2.47) gives

$$Zh \cdot (1 + \cosh\theta) = Zo \cdot \sinh\theta$$

This leads to

$$Zh = \frac{\sinh \theta}{1 + \cosh \theta} \cdot Zo \tag{2.50}$$

Invoking the half angle formula

$$Zh = \frac{2 \cdot \sinh \frac{\theta}{2} \cdot \cosh \frac{\theta}{2}}{\cosh^2 \left(\frac{\theta}{2}\right) + \sinh^2 \left(\frac{\theta}{2}\right) + \cosh^2 \left(\frac{\theta}{2}\right) - \sinh^2 \left(\frac{\theta}{2}\right)} \cdot Zo$$

This can be simplified to

$$Zh = Zo \cdot \tanh \left(\frac{\theta}{2}\right) \tag{2.51}$$

Re-arranging (2.48)

$$Zv = Zo \cdot \operatorname{cosech} \theta \tag{2.52}$$

This means that the impedances *Zh* and *Zv* of Figure 2.5(b) can be derived directly from the lumped-parameter components of Figure 2.5(a), since, from (2.40) and (2.41)

$$\theta = \sqrt{(Rd + j \cdot \omega \cdot Ld) \cdot (Gd + j \cdot \omega \cdot Cd)}$$

And

$$Zo = \sqrt{\frac{Rd + j \cdot \omega \cdot Ld}{Gd + j \cdot \omega \cdot Cd}}$$

There is now a clear correlation between Figure 2.5(a) and (b). That is, the impedance of each branch of the T-network of Figure 2.5(a) can be transformed into an impedance derived from the transmission line equations. The transformations are as follows:

$$\frac{(Rd + j \cdot \omega \cdot Ld)}{2} \rightarrow Zo \cdot \tanh \left(\frac{\theta}{2}\right) \tag{2.53}$$

$$\frac{1}{Gd + j \cdot \omega \cdot Cd} \rightarrow Zo \cdot \operatorname{cosech} \theta \tag{2.54}$$

The impedances *Zh* and *Zv* can be described as distributed impedances, since they take account of the fact that the parameters *Rd, Ld, Cd,* and *Gd* are distributed along the length of the line. This means that it is possible to analyse the response of the line over the entire range of frequencies covered by the transmission line equations. The distributed-parameter model of Figure 2.5(b) can be analysed using the analytical tools of circuit theory to simulate a response.

2.2.9 Propagation velocity

The forward propagation of the signal is defined by

$$V = Vn \cdot e^{-\gamma \cdot x} \tag{2.55}$$

Since γ is a complex number, it can be defined as

$$\gamma = \alpha + j \cdot \beta \tag{2.56}$$

where α represents the loss constant, and β represents the phase constant.

Up till now, all the voltage and current parameters have been defined as phasors. If Vt is the voltage at any point x at time t, then, from (2.55)

$$Vt = Vpeak \cdot e^{j \cdot \omega \cdot t} \cdot e^{-(\alpha + j \cdot \beta) \cdot x} \tag{2.57}$$

where $Vpeak$ is the peak value of the waveform, and Vt is a function of time and distance. Re-arranging this gives

$$Vt = Vpeak \cdot e^{-\alpha \cdot x} \cdot e^{j \cdot \omega \cdot \left(t - \frac{\beta \cdot x}{\omega}\right)}$$

Let

$$v = \frac{\omega}{\beta} \tag{2.58}$$

Then

$$Vt = Vpeak \cdot e^{-\alpha \cdot x} \cdot e^{j \cdot \omega \cdot \left(t - \frac{x}{v}\right)} \tag{2.59}$$

If

$$\omega \cdot \left(t - \frac{x}{v}\right) = A \tag{2.60}$$

where A is a constant angle, then

$$t - \frac{x}{v} = \frac{A}{\omega}$$
$$x = v \cdot t - \frac{A \cdot v}{\omega}$$

where v is the velocity of propagation. Giving

$$\frac{\delta x}{\delta t} = v \tag{2.61}$$

With a lossless line, $Rd = 0$, $Gd = 0$ and $\alpha = 0$. Also

$$\gamma = \alpha + j \cdot \beta = \frac{1}{l} \cdot \sqrt{j \cdot \omega \cdot Ld \cdot j \cdot \omega \cdot Ld} = j \cdot \frac{\omega}{l} \cdot \sqrt{Ld \cdot Cd}$$

That is

$$\beta = \frac{\omega}{l} \cdot \sqrt{Ld \cdot Cd} \tag{2.62}$$

This leads to

$$l = \tfrac{\omega}{\beta} \cdot \sqrt{Ld \cdot Cd} = v \cdot \sqrt{Ld \cdot Cd}$$

Giving

$$\sqrt{Ld \cdot Cd} = \frac{l}{v} = T \tag{2.63}$$

where T is the time taken for the leading edge of the waveform to propagate from one end of the line to the other. If there is an open circuit at the far end of the line, the assembly will behave as though there was a short circuit at load at the near end. The line will resonate and the first peak in the response will occur at

$$T = \frac{fq}{4} \tag{2.64}$$

This means that if the value of Ld has been measured previously, Equations (2.63) and (2.64) can be used to derive a value for Cd.

2.2.10 Conclusion

A circuit model has been derived to simulate the behaviour of a two-conductor cable over the entire frequency range covered by transmission line theory. During the derivation process, two sets of equations have been formulated; one based on the relationships defined by electromagnetic theory and the other created from mesh analysis of a hypothetical circuit model. Correlating the parameters of the two sets of equations allows the component values of the circuit model to be defined in terms of the relationships of electromagnetic theory. This is the key step in the process.

Another, equally important, step can be identified. Each conductor possesses the properties of inductance, resistance, capacitance, and conductance, which has been shown to play an essential role in propagating the electromagnetic energy along the path defined by the routing of the cable. The basic model for a two-conductor model is a Twin-T circuit, with resistors and inductors in the horizontal branches, capacitors in the vertical branches, and the only point of interconnection being the node joining the capacitors. By invoking the concept of equivalent circuits, the properties of the return conductor can be subsumed into a single T-network. It can then be assumed that the return conductor can be represented by an equipotential surface. This simplifies the mathematics.

This leads to the generally held belief that the conducting structure of an electrical system is actually an equipotential surface. (Witness the ubiquitous ground symbol in electronic circuit diagrams.) Such an assumption is manifestly false and is the most probable cause of the vast majority of EMI problems.

2.3 The three-conductor model

2.3.1 Primitive capacitance

Figure 2.6 illustrates the relationship between a length of charged conductor and the electric field at a point. It is assumed that the length of the section is l metres, the line charge density is ρ coulombs/metre and that the point $P(r,l)$ is located a radial distance r from the centre of the conductor. Then the radial component of the electric field is as follows:

$$E = \frac{\rho}{4 \cdot \pi \cdot \varepsilon} \cdot \frac{1}{r} \cdot [\sin(\alpha_1) + \sin(\alpha_2)] \tag{2.65}$$

The work done on this section of conductor due to unit charge moving from infinity to the surface of the conductor can be calculated by integrating the value of the electric field over the distance between $r = \infty$ and $r = 0$. This leads to

$$Vp_{1,1} = \frac{\rho}{2 \cdot \pi \cdot \varepsilon} \cdot \ln\left(\frac{l}{r_{1,1}}\right) \text{V} \tag{2.66}$$

The capacitance of the conductor is the ratio of charge to voltage. So

$$Cp_{1,1} = \frac{\rho \cdot l}{Vp_{1,1}} = \frac{2 \cdot \pi \cdot \varepsilon \cdot l}{\ln\left(\frac{l}{r_{1,1}}\right)} \text{F} \tag{2.67}$$

If a second conductor is installed parallel to conductor 1, as shown on Figure 2.7, the voltage on conductor 2 due to the charge on conductor 1 would be

$$Vp_{2,1} = \frac{\rho}{2 \cdot \pi \cdot \varepsilon} \cdot \ln\left(\frac{l}{r_{2,1}}\right) \text{V} \tag{2.68}$$

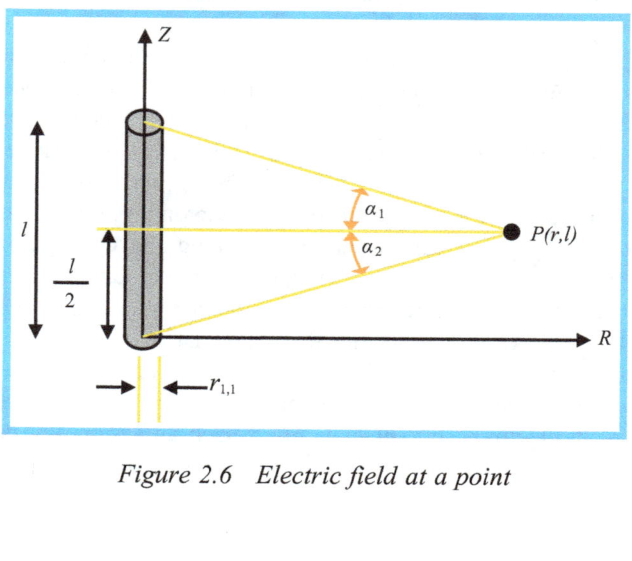

Figure 2.6 Electric field at a point

And the primitive capacitance would be

$$Cp_{2,1} = \frac{2 \cdot \pi \cdot \varepsilon \cdot l}{\ln\left(\frac{l}{r_{2,1}}\right)} \mathrm{F} \tag{2.69}$$

2.3.2 Primitive inductance

Figure 2.8 illustrates a length of conductor with a constant current *Ip* flowing in the z direction. The magnetic field at the point *P* is as follows:

$$H = \frac{Ip}{4 \cdot \pi \cdot r} \cdot [\sin(\alpha_1) + \sin(\alpha_2)] \mathrm{A/m} \tag{2.70}$$

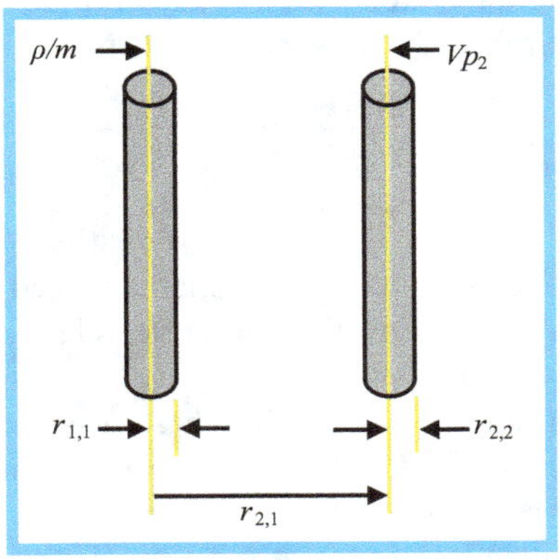

Figure 2.7 Voltage on conductor 2 due to charge on conductor 1

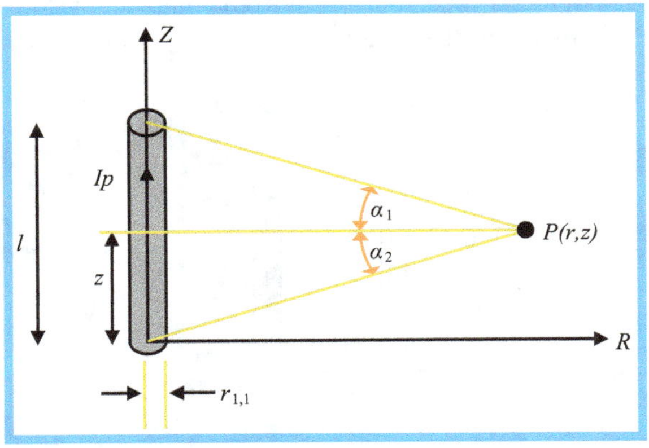

Figure 2.8 Magnetic field at a point

The flux density B at the point P is as follows:

$$B = \mu \cdot H \tag{2.71}$$

The total flux φ which passes through a rectangular strip extending from the surface of the conductor to a very large distance is as follows:

$$\varphi = \int B \cdot ds \tag{2.72}$$

where s is the surface of the rectangular strip.

Dividing this flux by the current Ip gives the primitive inductance:

$$Lp_{1,1} = \frac{\varphi}{Ip} = \frac{\mu \cdot l}{2 \cdot \pi} \cdot \ln\left(\frac{l}{r_{1,1}}\right) \text{H} \tag{2.73}$$

If a second conductor is routed alongside conductor 1, as illustrated in Figure 2.9, the primitive inductance of conductor 2 can be defined as

$$Lp_{2,1} = \frac{\mu \cdot l}{2 \cdot \pi} \cdot \ln\left(\frac{l}{r_{2,1}}\right) \text{H} \tag{2.74}$$

There are a lot of integrations between Equations (2.65) and (2.66), and between Equations (2.70) and (2.74). For anyone interested in checking the validity of these relationships. A full derivation of these relationships can be found in Appendix A.

It is clear from Figures 2.6 and 2.8 that these primitive parameters are properties of the electromagnetic field in the region surrounding the conductors. They are not the properties of the conductors.

The primitive capacitance and the primitive inductance are two of the building blocks for all circuit models used to simulate cable coupling mechanisms. They are essential parameters in circuit models which relate currents and voltages in conductors to the electromagnetic field.

The term 'primitive' has been borrowed from computer systems technology. It means that the parameter is the simplest that the system can handle. Primitive

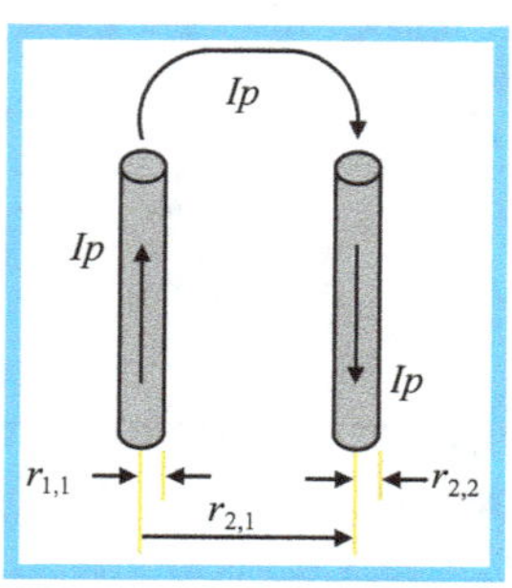

Figure 2.9 Current in a two-conductor assembly

capacitance and primitive inductance are the smallest entities that can be assigned the dimensions of Farad and Henry.

Some researchers have used the term 'partial' to define these entities. But this term creates confusion when partial currents and partial voltages in conductors are being analysed.

2.3.3 Primitive impedance

Transmission line theory derives an expression for the characteristic impedance of a zero-loss line. For a single, isolated, conductor, this would be

$$Zp_{1,1} = \sqrt{\frac{Lp_{1,1}}{Cp_{1,1}}} \tag{2.75}$$

Invoking (2.73) and (2.67) to substitute for $Lp_{1,1}$ and $Cp_{1,1}$

So that:

$$Zp_{1,1} = \frac{1}{2 \cdot \pi} \cdot \sqrt{\frac{\mu}{\varepsilon}} \cdot \ln\left(\frac{l}{r_{1,1}}\right) \tag{2.76}$$

2.3.4 Primitive equations

For a three-conductor cable where the cross section is as defined in Figure 2.10, the primitive equations are as follows:

$$\begin{aligned} Vp_1 &= Zp_{1,1} \cdot Ip_1 + Zp_{1,2} \cdot Ip_2 + Zp_{1,3} \cdot Ip_3 \\ Vp_2 &= Zp_{2,1} \cdot Ip_1 + Zp_{2,2} \cdot Ip_2 + Zp_{2,3} \cdot Ip_3 \\ Vp_3 &= Zp_{3,1} \cdot Ip_1 + Zp_{3,2} \cdot Ip_2 + Zp_{3,3} \cdot Ip_3 \end{aligned} \tag{2.77}$$

where

$$Zp_{i,j} = \frac{1}{2 \cdot \pi} \cdot \sqrt{\frac{\mu}{\varepsilon}} \cdot \ln\left(\frac{l}{r_{i,j}}\right) \tag{2.78}$$

$r_{i,j}$ is the separation between conductors i and j. $\sqrt{\frac{\mu}{\varepsilon}}$ is the intrinsic impedance

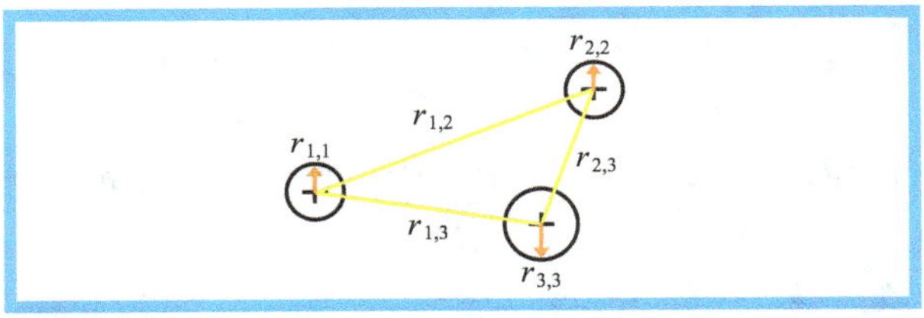

Figure 2.10 Cross section of conductors in a three-conductor assembly

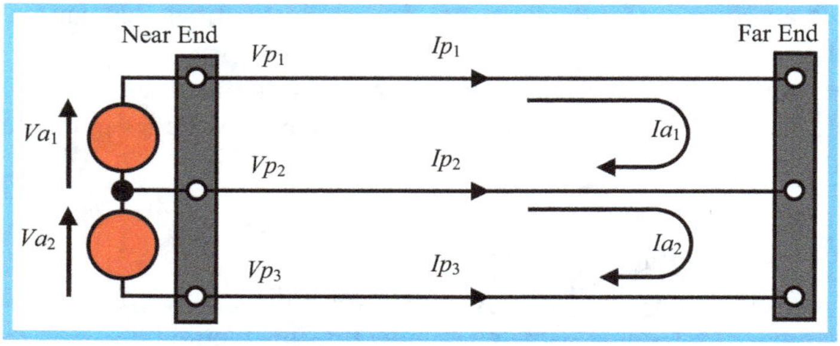

Figure 2.11 Relationship between loop parameters and primitive parameters

2.3.5 Loop equations

In practical situations, measurements using test equipment involve loops. That is, voltages Va_1 and Va_2 are measured between two adjacent terminals and currents Ia_1 and Ia_2 flow out of one terminal and back via the other. Figure 2.11 defines the relationship between loop parameters and primitive parameters.

Relationships between primitive currents and loop currents are as follows:

$$\begin{aligned} Ip_1 &= Ia_1 \\ Ip_2 &= Ia_2 - Ia_1 \\ Ip_3 &= -Ia_2 \end{aligned} \tag{2.79}$$

Relationships between loop voltages and primitive voltages are as follows:

$$\begin{aligned} Va_1 &= Vp_1 - Vp_2 \\ Va_2 &= Vp_2 - Vp_3 \end{aligned} \tag{2.80}$$

This leads to the loop equations

$$\begin{aligned} Va_1 &= Za_{1,1} \cdot Ia_1 + Za_{1,2} \cdot Ia_2 \\ Va_2 &= Za_{2,1} \cdot Ia_1 + Za_{2,2} \cdot Ia_2 \end{aligned} \tag{2.81}$$

where

$$\begin{aligned} Za_{1,1} &= Zp_{1,1} - Zp_{1,2} - Zp_{2,1} + Zp_{2,2} \\ Za_{1,2} &= Zp_{1,2} - Zp_{1,3} - Zp_{2,2} + Zp_{2,3} \\ Za_{2,1} &= Zp_{2,1} - Zp_{2,2} - Zp_{3,1} + Zp_{3,2} \\ Za_{2,2} &= Zp_{2,2} - Zp_{2,3} - Zp_{3,2} + Zp_{3,3} \end{aligned} \tag{2.82}$$

2.3.6 Circuit equations

Figure 2.12 illustrates a circuit model which can replicate the response of the loop equations. The circuit equations are as follows:

$$\begin{aligned} Va_1 &= (Zc_1 + Zc_2) \cdot Ia_1 - Zc_2 \cdot Ia_2 \\ Va_2 &= -Zc_2 \cdot Ia_1 + (Zc_2 + Zc_3) \cdot Ia_2 \end{aligned} \tag{2.83}$$

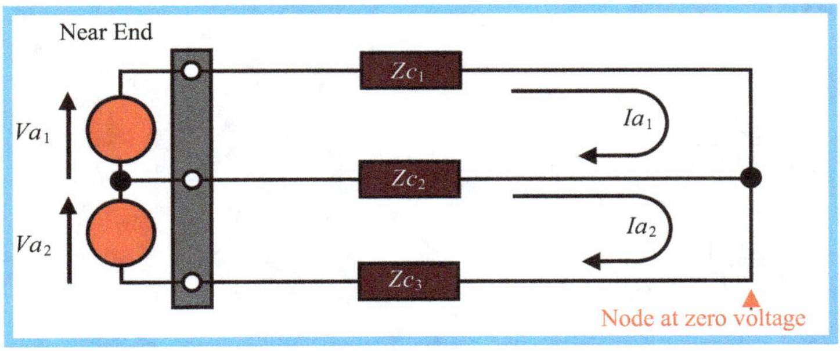

Figure 2.12 Circuit model which replicates the loop equations

Correlating the Z-parameters of (2.81) and (2.83) gives

$$\begin{aligned} Za_{1,1} &= Zc_1 + Zc_2 \\ Za_{1,2} &= -Zc_2 = Za_{2,1} \\ Za_{2,2} &= Zc_2 + Zc_3 \end{aligned} \tag{2.84}$$

The circuit parameters can now be defined in terms of the loop parameters

$$\begin{aligned} Zc_1 &= Za_{1,1} + Za_{1,2} \\ Zc_2 &= -Za_{1,2} \\ Zc_3 &= Za_{2,2} + Za_{1,2} \end{aligned} \tag{2.85}$$

Using (2.82) to replace the loop parameters with primitive parameters

$$\begin{aligned} Zc_1 &= Zp_{1,1} - Zp_{2,1} - Zp_{1,3} + Zp_{2,3} \\ Zc_2 &= Zp_{2,2} - Zp_{1,2} - Zp_{2,3} + Zp_{1,3} \\ Zc_3 &= Zp_{3,3} - Zp_{3,1} - Zp_{2,3} + Zp_{2,1} \end{aligned} \tag{2.86}$$

Capacitor values for the capacitors and the inductors can be derived from

$$\begin{aligned} Cc_k &= \frac{T}{Zc_k} \\ Lc_k &= T \cdot Zc_k \end{aligned} \tag{2.87}$$

Where the number k identifies the conductor and

$$T = \frac{l}{v} = l \cdot \sqrt{\mu \cdot \varepsilon} \tag{2.88}$$

T is the time taken for the front edge of a signal to propagate from one end of the line to the other, and v is the velocity of propagation of the signal.

2.3.7 Definition of parameters

This leads to:

$$Cc_1 = \frac{2 \cdot \pi \cdot \varepsilon \cdot l}{\ln\left(\frac{r_{1,2} \cdot r_{1,3}}{r_{1,1} \cdot r_{2,3}}\right)}$$
$$Cc_2 = \frac{2 \cdot \pi \cdot \varepsilon \cdot l}{\ln\left(\frac{r_{1,2} \cdot r_{2,3}}{r_{2,2} \cdot r_{1,3}}\right)} \qquad (2.89)$$
$$Cc_3 = \frac{2 \cdot \pi \cdot \varepsilon \cdot l}{\ln\left(\frac{r_{1,3} \cdot r_{2,3}}{r_{3,3} \cdot r_{1,2}}\right)}$$

And

$$Lc_1 = \frac{\mu \cdot l}{2 \cdot \pi} \cdot \ln\left(\frac{r_{1,2} \cdot r_{1,3}}{r_{1,1} \cdot r_{2,3}}\right)$$
$$Lc_2 = \frac{\mu \cdot l}{2 \cdot \pi} \cdot \ln\left(\frac{r_{1,2} \cdot r_{2,3}}{r_{2,2} \cdot r_{1,3}}\right) \qquad (2.90)$$
$$Lc_3 = \frac{\mu \cdot l}{2 \cdot \pi} \cdot \ln\left(\frac{r_{1,3} \cdot r_{2,3}}{r_{3,3} \cdot r_{1,2}}\right)$$

2.3.8 General circuit model

Since the input impedance will be the same when the source voltage is applied to the far end, the circuit model must be symmetrical. This leads to Figure 2.13.

This model allows all forms of intra-system coupling to be analysed. One of the conductors can simulate the effect of the conducting structure, plus the effect of all other conductors in the system. If the structure is used as a general-purpose

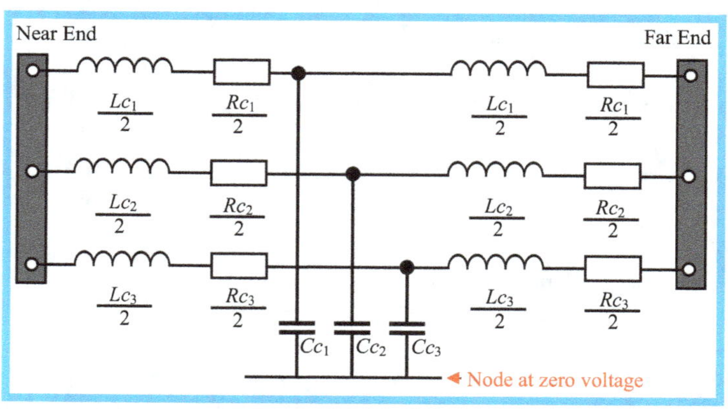

Figure 2.13 Circuit model of three-conductor assembly

return conductor, the model can simulate the coupling between the culprit and victim loops. In this case, the structure acts as a common-impedance.

If a dedicated return conductor is used to carry the signal or power, the model will simulate the coupling between the differential-mode and the common-mode loop. In this configuration, the structure acts as a shield.

2.3.9 Relating the theories

The formulae of Electromagnetic Theory lead inexorably to the conclusion that current in every conductor will create a voltage in every other conductor in the vicinity.

Circuit theory assumes that the current flowing in any branch of a network is solely dependent on the voltage across that branch.

The only connection between the two theories is the fact that a set of loop equations derived from electromagnetic theory can be correlated precisely with a set of loop equations derived from the analysis of a circuit model.

Analysing the mechanisms involved in the propagation of electromagnetic interference involves the use of time-variant, three-dimensional electromagnetic fields. Computational electromagnetics requires skills which can only be acquired by extra training; training which is not available to the average electrical engineer.

It is entirely possible for any electrical engineer to analyse the performance of a circuit model.

Circuit theory is a development and simplification of electromagnetic theory. The latter defines the equations; the rules of circuit theory provide the analytical tools which can solve those equations.

The problem at the moment is that, from the point of view of electromagnetic interference, the teaching of circuit theory is oversimplified. It assumes that the electromagnetic field associated with current flowing along a conductor bears a one-to-one relationship with the current in that conductor. The return conductor can be represented as an equipotential surface. These assumptions manifest themselves in the use of the ubiquitous 'ground' or 'earth' symbols in circuit diagrams.

It should be possible for academia to include these simple relationships in any course of study of circuit theory. Engineers will then be provided with the skills necessary to analyse any problem caused by electromagnetic interference.

2.4 The voltage injection transformer

2.4.1 Design

Transformer coupling is essential when carrying out measurements of the mechanisms involved in cable coupling, since any direct connection with the test equipment will inevitably change the configuration of the wiring being tested. It is important that the impedance being reflected into the cable is much lower than that of the cable, over the entire range of frequencies being applied. The device should operate over as wide a range of frequencies as is practicable.

Figure 2.14 Voltage transformer

A practical requirement is that the transformer should be able to couple into a cable-form without disconnecting that cable. So, it should be a split-core assembly.

A test requirement is that it should be possible to monitor the voltage being induced in the cable at the same time as the response of that assembly is being monitored. It is essential that the output of the system is measured at exactly the same time as the input.

A transformer meeting this specification is shown in Figure 2.14. Although it is not as robust as any device bought commercially, it does the job at a fraction of the cost and has proved to be reliable over several years.

The core itself is a 'cable suppression core assembly', manufactured by Fair Rite Products Corp. The primary winding consists of 10 turns of 22 Standard Wire Gauge (SWG) enamelled copper wire wound round one of the split cores.

The monitor winding is a single turn of the same type of wire wound round the other core. This configuration ensures that the voltage induced in the monitor turn is exactly the same as that induced in the cable-under-test.

The windings are terminated in a standard terminal block, which provides a mount for the other components. A 68 Ω 2 W resistor is connected in parallel with a 240 Ω 0.6 W resistor and the primary winding. This provides a 53 Ω resistor in parallel with the primary.

Ideally, the co-axial cable would have been terminated by a 50 Ω resistor to match its characteristic impedance. However, it is inevitable that reflected impedance of the loop-under-test will act as a load in parallel with the termination resistor. To compensate for this loading effect, a value of 53 Ω was chosen.

Two BNC connectors are connected to the terminal block to provide an interface with 50 Ω co-axial cable. A section of a plastic box was used to provide more rigidity to the assembly.

2.4.2 Characterisation test

Testing was carried out using the set-up of Figure 2.15. A splitter box was used to provide an input to channel 1 of the oscilloscope. This allowed the input to the voltage transformer to be monitored. The output of the monitor turn on the transformer was monitored by channel 2. Each 16.5 Ω resistor is a parallel combination of two 33 Ω resistors.

The signal generator was set to provide a sinusoidal output at a particular frequency, and the amplitudes of the peak-to-peak signals were measured on the oscilloscope. This process was carried out over a number of defined frequencies, and the results were tabulated.

For each spot frequency, the ratio of output voltage *Vch2* to input voltage *Vch1* was calculated, and this ratio gave the value of the transfer function *TFt* at that frequency. Plotting this parameter against the frequency, as shown in Figure 2.16, gives an indication of the bandwidth.

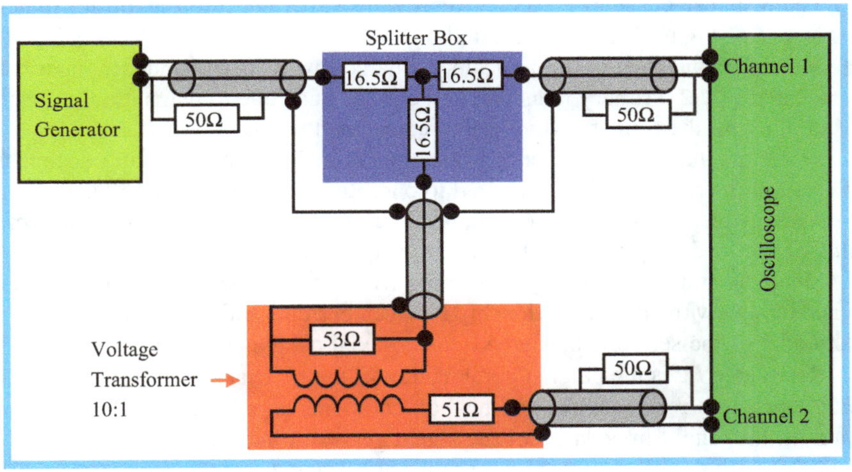

Figure 2.15 Set-up to characterise the voltage transformer

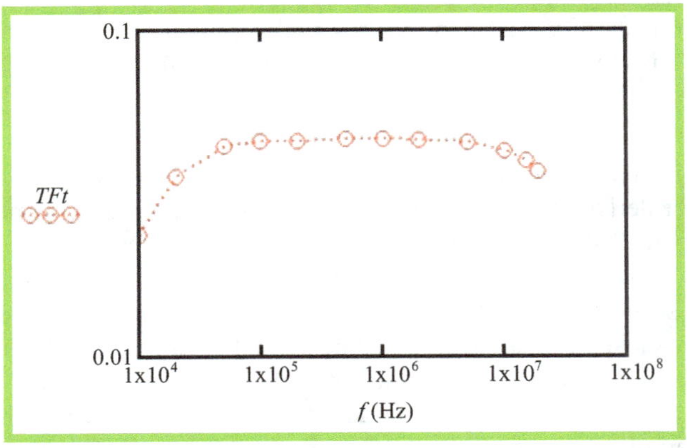

Figure 2.16 Transfer function of voltage transformer

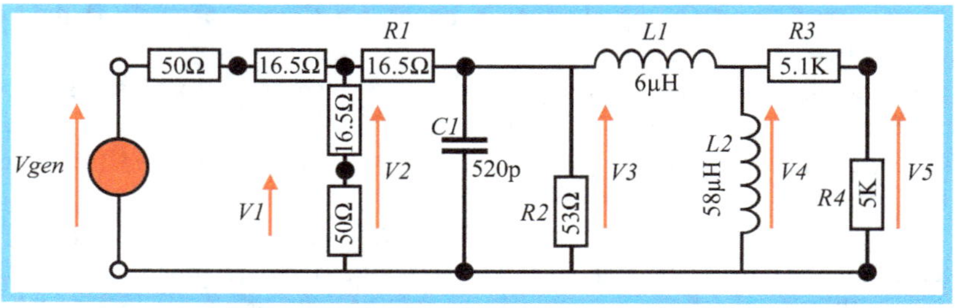

Figure 2.17 Circuit model of characterisation test

The response curve of Figure 2.16 provides sufficient information for the circuit model of Figure 2.17 to be created.

Mutual inductance (the inductance shared by primary and secondary) is represented by *L2*. Primary inductance (the inductance due to magnetic flux which links only with the primary winding) is represented by *L1*.

Since the turns-ratio is ten to one, the impedance ratio of the secondary reflected in the primary is one hundred to one. Resistance *R3* is the reflected value of the series resistance of the secondary circuit, and resistance *R4* is the reflected value of the load resistance at channel 2 input terminals.

Since the simulation is only concerned with amplitudes of the signals, there is no need to know the phase relationship between currents and voltages. So, as far as this model of the set-up is concerned, the co-axial cable is transparent.

Resistance *R2* is the parallel combination of the 68 Ω and 240 Ω resistors. The 16.5 Ω resistors represent the splitter-box components, whilst the 50 Ω resistors represent the output impedance of the signal generator and the input impedance of channel 2 of the oscilloscope. The capacitance *C1* represents the capacitance of the primary winding.

The voltage *V1* represents the voltage as monitored by channel 1 of the scope, *V2* is the voltage at the junction of the 16.5 Ω resistors, *V3* is the voltage applied to the input terminals of the transformer, *V4* is the voltage across the mutual inductance, and *V5* is ten times the voltage monitored by channel 2.

2.4.3 Worksheet

Since the objective of the circuit model is the creation of a curve which correlates with the curve derived from the test results, it is necessary for the same software to be used. Figure 2.18 shows part of a Mathcad worksheet which calculates the response of the model. Input data is specified at the top of the worksheet. Running the program several times with different values assigned to the components of the model allows close correlation to be achieved between test results and the simulation of the model. Figure 2.19 illustrates the final set of curves. The values assigned to the components *C1*, *L1*, and *L2* were those displayed on the worksheet when correlation had been achieved. The process is similar to that used in successive approximation programs for non-linear equations.

$$R1 := 16.5 \qquad R2 := \frac{68\cdot 240}{68+240} = 52.987$$

$$R3 := 5100 \qquad R4 := 5000$$

$$Lp := 64\cdot 10^{-6} \qquad L1 := 6.0\cdot 10^{-6}$$

$$L2 := Lp - L1 = 5.8\times 10^{-5} \qquad V2 := \frac{50+R1}{50}\cdot V1 = 1.33$$

$$V1 := 1$$

$$Turns := 10 \qquad C1 := 520\cdot 10^{-12}$$

$$TFm_i := \left|\begin{array}{l} \omega \leftarrow 2\cdot\pi\cdot F_i \\ Y2 \leftarrow \dfrac{1}{j\cdot\omega\cdot L2} + \dfrac{1}{R3+R4} \\ Z2 \leftarrow \dfrac{1}{Y2} \\ Y1 \leftarrow \dfrac{1}{R2} + \dfrac{1}{Z2 + j\cdot\omega\cdot L2} + j\cdot\omega\cdot C1 \\ Z1 \leftarrow \dfrac{1}{Y1} \\ V3 \leftarrow \dfrac{Z1}{Z1+R1}\cdot V2 \\ V4 \leftarrow \dfrac{Z2}{Z2 + j\cdot\omega\cdot L1}\cdot V3 \\ V5 \leftarrow \dfrac{R4}{R3+R4}\cdot V4 \\ \dfrac{|V5|}{Turns} \end{array}\right.$$

Figure 2.18 Part of Mathcad worksheet which calculates the response of the model

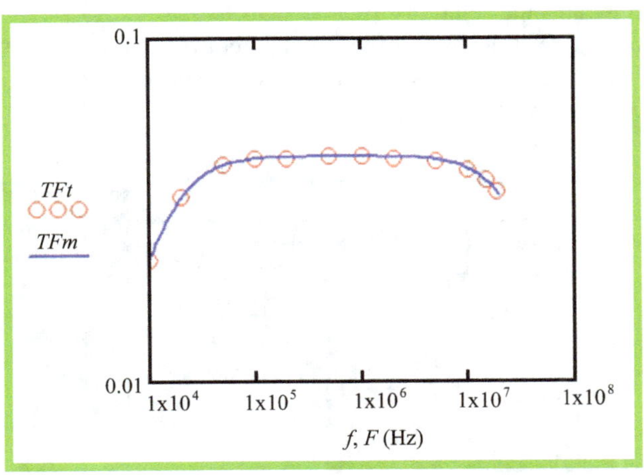

Figure 2.19 Transfer functions derived from testing TFt and modelling TFm

2.4.4 Application

The fact that there is close correlation means that the model can now be used to calculate the response of the transformer at any given frequency. It can also be used to interpret the mechanisms involved and relate them to the electromagnetic field. The useable range of this transformer is from 20 kHz to 20 MHz. It is reasonable to expect that transformers can be constructed to match the operating frequency range of any particular system-under-review.

The design, build, and characterisation processes are quite simple and can be replicated by anyone wishing to investigate the characteristics of his or her assembly-under-review.

2.5 The current monitor transformer

2.5.1 Design

Bench testing can be used to confirm that any electronic system meets its EMC requirements, and this can be done during the development stage. One essential unit of equipment needed for this is the Current Monitor Transformer. The construction of a low-cost current transformer, its function, purpose, calibration and characterisation are described.

Figure 2.20 illustrates the method of coupling the test equipment to the loop-under-test. With such a configuration, the load reflected into the primary circuit (the loop-under-test) is the load presented to the secondary divided by the square of the number of turns. Hence, the greater the number of turns, the less the value of the reflected load, and the less will be the effect of the test equipment on the system-under-review.

The load 'seen' by the secondary winding is the 51 Ω resistor *R1* in parallel with the 50 Ω impedance of the load at the input terminals of the oscilloscope. Figure 2.21 illustrates the relationship.

The circuit model relevant to the input circuitry of the oscilloscope is illustrated in Figure 2.22. This shows that, as far as this configuration is concerned, the co-axial cable is transparent.

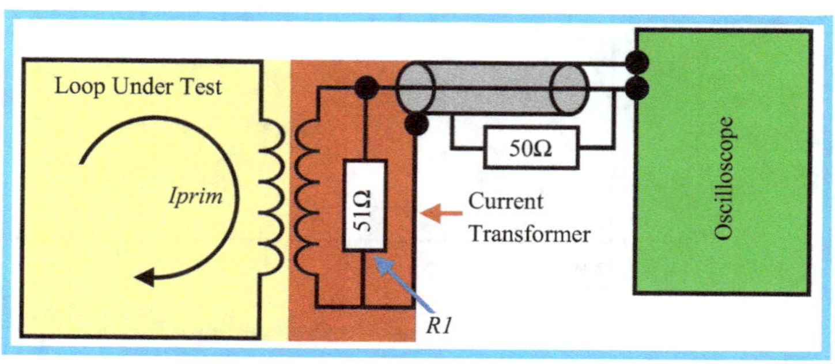

Figure 2.20 Use of current transformer

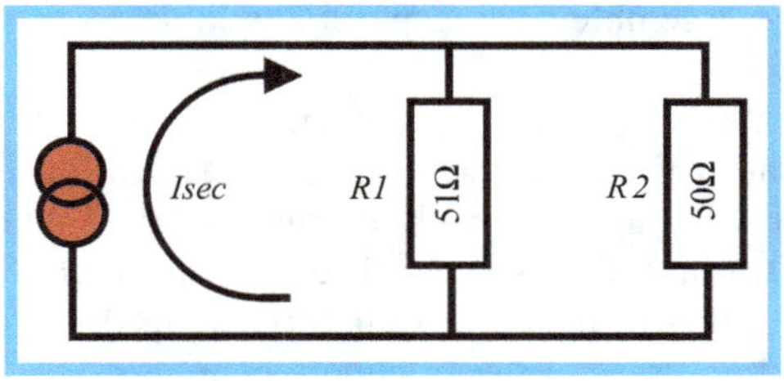

Figure 2.21 Transformer as a current source

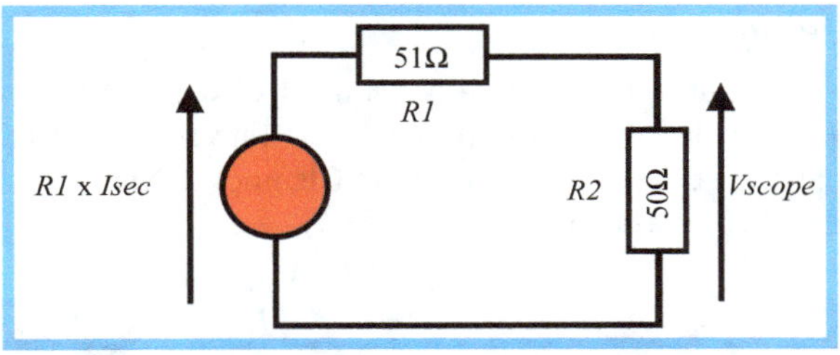

Figure 2.22 Transformer as a voltage source

From Figure 2.20, the voltage appearing at the input terminals of the oscilloscope is as follows:

$$Vscope = \mathrm{Rsec} \cdot \mathrm{Isec} \tag{2.91}$$

where

$$\mathrm{Rsec} = \frac{\mathrm{R1} \cdot \mathrm{R2}}{\mathrm{R1} + \mathrm{R2}} \tag{2.92}$$

The current in the loop-under-test is as follows:

$$Iprim = Turns \cdot \mathrm{Isec} \tag{2.93}$$

where *Turns* is the number of turns on the secondary winding. This means that the voltage monitored by the oscilloscope is directly proportional to the current in the loop under test. It also means that the load *Rprim* reflected into the loop-under-test is as follows:

$$Rprim = \frac{\mathrm{Rsec}}{Turns^2} \tag{2.94}$$

2.5.2 Assembly

Since the author did not have the financial resources necessary to purchase a set of off-the-shelf instruments, a current transformer meeting these design considerations were assembled using parts available from a supplier of electronic components.

The core is a 'cable suppression core assembly' manufactured by the Fair Rite Products Corporation. It is exactly the same as that used for the voltage transformer. The secondary winding comprises ten turns of 22 SWG enamelled copper wire. Figure 2.23 illustrates the assembly. The tie wrap is removable and is used to ensure that the two halves of the core are tightly clamped together.

To characterise this transformer, a simple coupling jig was assembled. This delivered a primary current of known amplitude and known frequency and was designed to ensure that the loop-under-test was tightly coupled to the transformer.

2.5.3 Test set-up

The test set-up is shown in Figure 2.24. Channel 1 of the oscilloscope was used to measure the current delivered to the transformer; that is, the current in the 50 Ω resistor placed at the scope input connector. Channel 2 was used to monitor the output of the transformer assembly.

Figure 2.23 Current transformer assembly

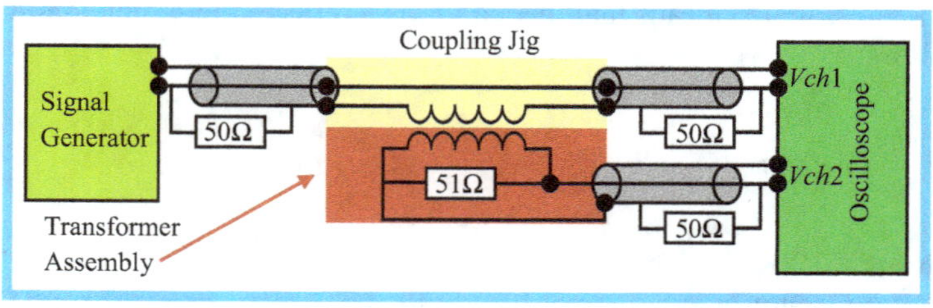

Figure 2.24 Set-up for calibration of current transformer

The signal generator was used to deliver a constant voltage at a set of spot frequencies f_s. The peak-to-peak voltage *Vch1* at channel 1 of the oscilloscope gave a measure of the peak-to-peak current *Iprim* flowing in the primary of the transformer:

$$Iprim = \frac{Vch1}{50} \tag{2.95}$$

The output voltage of the transformer *Vch2* was monitored at channel 2, and the transfer impedance *ZTt* calculated. This was repeated at each spot frequency:

$$ZTt = \frac{Vch2}{Iprim} \tag{2.96}$$

A Mathcad worksheet was used to calculate and display the relationship between the transfer impedance and the frequency. The resulting graph is shown in Figure 2.25.

2.5.4 Circuit model

The task then was to create a circuit model which replicated this relationship.

Figure 2.26 is a development of the simple model of Figure 2.21. *R2* and *R3* represent the 51 Ω resistor in the transformer assembly and the 50 Ω resistor at

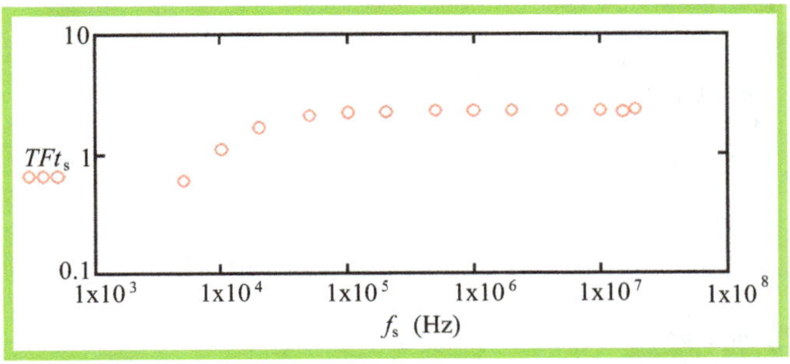

Figure 2.25 Transfer impedance of current transformer

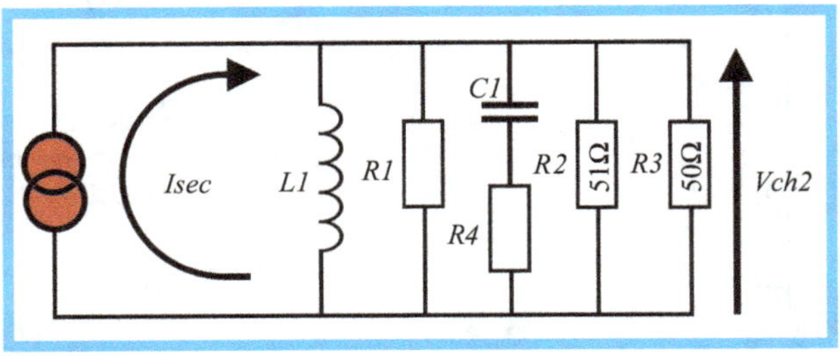

Figure 2.26 Circuit model of current transformer

channel 2 input connector, respectively. *L1* represents the inductance of the transformer winding, whilst *R1* represents transformer losses. These losses could be due to the magnetic field from the loop-under-test which does not link with the transformer core. Another cause of losses is the eddy current in the core. Capacitance *C1* and resistor *R4* were added to the model to simulate additional losses at frequencies over 2 MHz.

2.5.5 Simulation

A copy of the Mathcad worksheet used to carry out a frequency response analysis of this circuit model is provided in Figure 2.27. This worksheet first calculates a set of 100 frequencies, on a logarithmic scale between 5 kHz and 20 MHz and stores them in the vector ***F***. It then defines the component values of the circuit model as well as the number of turns *Turns* on the secondary.

The equations used in the function *ZTm* are derived from an inspection of the circuit model. The impedance *Z2* is the impedance as 'seen' by the current generator. It defines the ratio of *Vch*2 to *Isec*. Dividing *Z2* by the number of turns gives the ratio of *Vch*2 to *Iprim*, that is the transfer impedance of the circuit model.

Figure 2.28 illustrates the correlation between the test results and the response of the model.

Although there was an initial discrepancy between the two curves, a few adjustments of *L1* and *R1* led to a curve which intersected the data points at the low

$$y1 := \log(5\cdot 10^3) \qquad y2 := \log(20\cdot 10^6) \qquad m := \frac{y2 - y1}{100}$$

$$i := 1..101$$

$$F_i := \left| \begin{array}{l} y \leftarrow m\cdot(i-1) + y1 \\ 10^y \end{array} \right.$$

Calculating a set of 100 frequencies, equally spaced between 5 kHz and 20 MHz

$$R1 := 300 \quad R2 := 51 \quad L1 := 200\cdot 10^{-6}$$

$$R3 := 50 \quad R4 := 850 \quad C1 := 60\cdot 10^{-12}$$

$$Turns := 10$$

$$ZTm_i := \left| \begin{array}{l} \omega \leftarrow 2\cdot\pi\cdot F_i \\ Z1 \leftarrow R4 + \dfrac{1}{j\cdot\omega\cdot C1} \\ Y2 \leftarrow \dfrac{1}{R1} + \dfrac{1}{R2} + \dfrac{1}{R3} + \dfrac{1}{j\cdot\omega\cdot L1} + \dfrac{1}{Z1} \\ Z2 \leftarrow \dfrac{1}{Y2} \\ ZT \leftarrow \dfrac{|Z2|}{Turns} \end{array} \right.$$

Calculating the value of *ZTm* at each frequency *F*

ZTm is the ratio *Vch2*/*Iprim* derived from the response of the circuit model

Figure 2.27 Calculating the transfer impedance of the circuit model

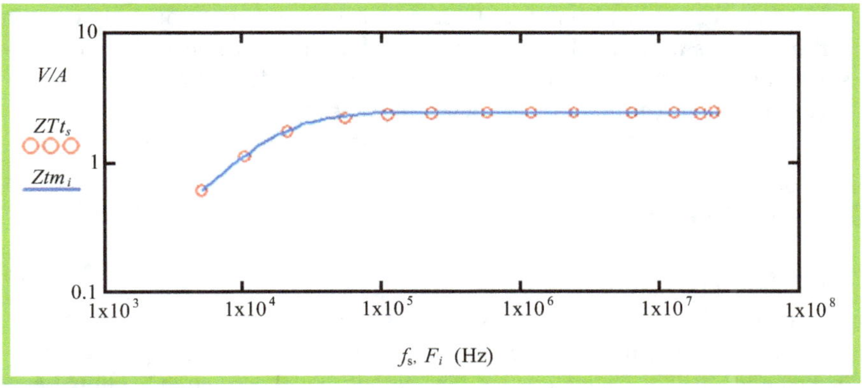

Figure 2.28 Correlating the circuit model with the test results

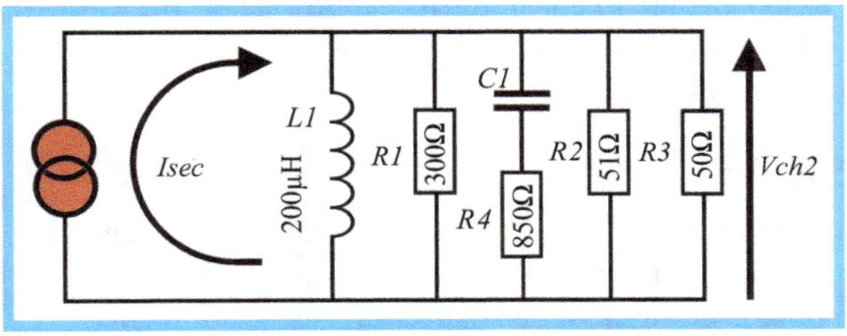

Figure 2.29 Definitive circuit model

frequencies. Varying the values of *C1* and *R4* led to a curve which intersected the data point above 2 MHz as well.

The process is akin to the use of successive approximations to solve non-linear equations.

ZTm is the ratio *Vch2/Iprim* derived from the response of the circuit model

2.5.6 *Definitive circuit model*

The end result is a model which effectively defines the relationship between the current in the loop under test and the voltage measured at the input terminals of the oscilloscope. That is, the measurements of *Vch2* can be used to calculate the value of *Iprim* at any frequency in the range 5 kHz to 20 MHz. This being so, Figure 2.29 can be described as a 'Definitive Circuit Model' for this particular transformer.

2.5.7 *Application*

Such a transformer can be used to measure the amplitude of any current in any cable routed through the core. In this particular example, the bandwidth is from 10 kHz to 20 MHz. Since the reflected impedance in the assembly under test is extremely low, it has minimal effect on the characteristics of that assembly.

The transformer has been used to carry out a wide variety of cable coupling experiments. It is well suited to the task of bench testing of cable assemblies and or in situ testing of system installations, prior to formal EMC testing.

The technique can be used to develop low-cost current transformers which operate over lower frequency ranges or higher frequency ranges.

2.6 The signal link

2.6.1 Test rig

The first task in any analysis of cross coupling is to build a test rig on which to lay the conductors. After installing a cable on the rig, the assembly can be subjected to tests which identify its characteristics. The circuit model derived in Section 2.4 can then be used to assign parameter values to the components.

The basic rig consists of a length of copper pipe installed round the walls of a room, with a wood batten fixed along its length. Then the cable-under-test was installed along the length of the batten. Figure 2.30 shows a cross section of the assembly.

2.6.2 Test set-up

The method of measurement is illustrated by the set-up of Figure 2.31. The copper pipe can represent the conducting structure of a vehicle, aircraft, spacecraft, ship, or the 'earth' conductor of any assembly of electronic equipment.

One of the cable conductors was used to send a signal, whilst the other was assigned to carry the return current. The terminals at the far end were shorted

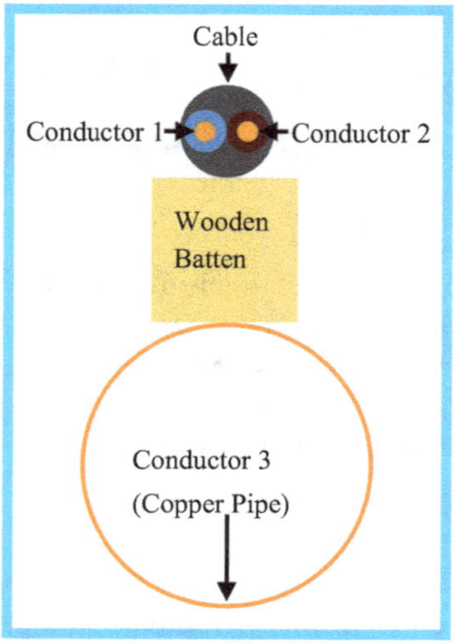

Figure 2.30 Cross-section of test rig

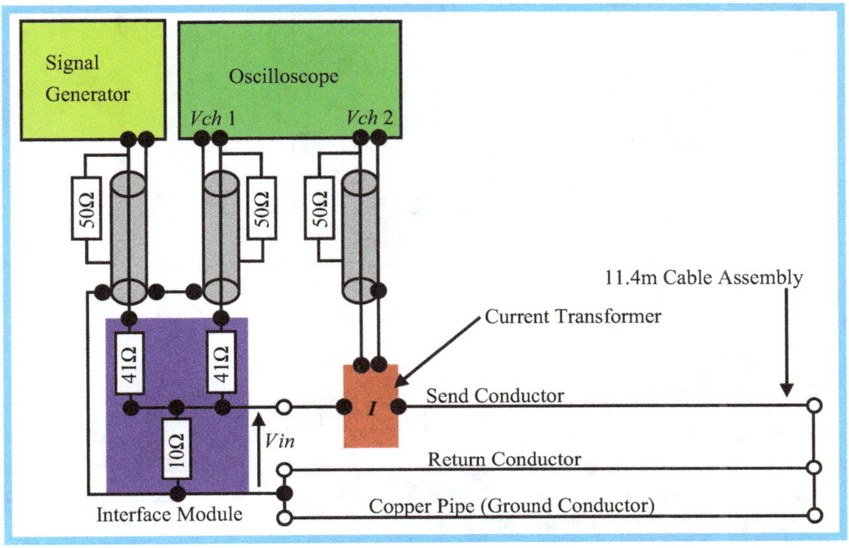

Figure 2.31 Set-up to characterise the test rig

together. A signal generator was used to provide a variable frequency voltage source *Vin* at the near end, and the peak-to-peak amplitude was measured by channel 1 of the oscilloscope.

From Figure 2.31

$$Vch1 = \frac{50}{50 + 41} \cdot Vin \tag{2.97}$$

Since the source impedance was less than 10 Ω, the loading effect of the assembly under review had minimal effect on the load presented to the signal generator.

2.6.3 Current measurement

Current in the send conductor *Iprim* flows through the core of the current transformer and the output of the ten-turn secondary is monitored by *Vch2* of the oscilloscope. The relationship between these two parameters is defined by the circuit model of Figure 2.32. The method of deriving this model is described in Section 2.5.

At any particular frequency there is a one-to-one correlation between the amplitude of the input current and the amplitude of the output voltage. Conversely, if the amplitude of *Vch2* is known, then the amplitude of *Iprim* can be calculated. It is simply a matter of invoking the equations of circuit theory.

Calculating the ratio of *Iprim* to *Vin* gives a value for the admittance:

$$Yt = \frac{Iprim}{Vch1} \tag{2.98}$$

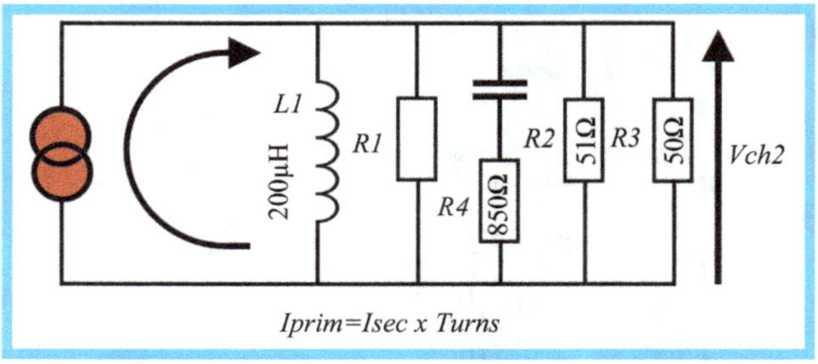

Figure 2.32 Circuit model of current transformer

2.6.4 Test results

When the configuration was as shown on Figure 2.31 and this measurement was carried out over a set of spot frequencies, the data was as recorded in the Mathcad worksheet shown on Figure 2.33. The worksheet calculates the frequency response of the admittance of *Yt1*, the admittance of the differential-mode loop.

Then the current transformer was clamped round both conductor and the test repeated. The data collected is shown on the worksheet of Figure 2.34. This calculated the frequency response of the transfer admittance *Yt2*, the ratio of the current in the common-mode loop to the voltage applied to the differential-mode loop.

The results of the calculations carried out by pages 1 and 2 of the worksheet are illustrated in Figure 2.35. (It is worth noting that the vertical scaling on the two graphs is different.)

2.6.5 Lumped-parameter model

The next step in the design process was to create a model that replicates these test results. The model of Figure 2.36(a) is derived in Section 2.3. Initial values for the components were derived from the geometry of the assembly.

From measurements on the assembly of Figure 2.30:

$r11 = r22 = 0.61$ mm	Radii of send and return conductors
$r33 = 7.5$ mm	Radius of copper pipe
$r12 = 2.45$ mm	Separation between send and return conductors
$r13 = r23 = 23$ mm	Separation of conductors from pipe

The length of the test rig was measured to be: $l = 11.4$ m

The initial values of the inductors were as follows:

$$
\begin{aligned}
Lc1 &= \frac{\mu_o \cdot \mu_r \cdot l}{2 \cdot \pi} \cdot \ln\left(\frac{r12 \cdot r13}{r11 \cdot r23}\right) = 3.17 \times 10^{-6}\ \text{H} \\
Lc2 &= \frac{\mu_o \cdot \mu_r \cdot l}{2 \cdot \pi} \cdot \ln\left(\frac{r12 \cdot r23}{r22 \cdot r13}\right) = 3.17 \times 10^{-6}\ \text{H} \\
\text{Lc3} &= \frac{\mu_o \cdot \mu_r \cdot l}{2 \cdot \pi} \cdot \ln\left(\frac{r13 \cdot r23}{r33 \cdot r12}\right) = 7.66 \times 10^{-6}\ \text{H}
\end{aligned}
\tag{2.99}
$$

$$data3 := \begin{pmatrix} 1 & 800 & 86 \\ 1.5 & 820 & 54 \\ 2 & 840 & 36 \\ 2.5 & 840 & 21 \\ 3 & 840 & 13 \\ 3.5 & 840 & 6 \\ 3.72 & 840 & 4 \\ 4 & 840 & 8.4 \\ 4.5 & 840 & 18 \\ 5 & 840 & 32 \\ 5.5 & 820 & 50 \\ 6 & 800 & 80 \\ 6.5 & 660 & 140 \\ 6.93 & 480 & 190 \\ 7.5 & 720 & 110 \\ 8 & 800 & 70 \\ 9 & 840 & 23 \\ 9.66 & 840 & 12 \\ 10 & 820 & 28 \\ 10.41 & 700 & 92 \\ 10.5 & 740 & 76 \\ 11 & 840 & 24 \\ 11.69 & 840 & 10 \\ 12 & 840 & 14 \end{pmatrix}$$

$$R1 := 100$$
$$R2 := 51$$
$$R3 := 50$$
$$R4 := 850$$
$$L1 := 200 \cdot 10^{-6}$$
$$C1 := 60 \cdot 10^{-12}$$
$$Turns := 10$$
$$K := \frac{50 + 41}{50} = 1.82$$
$$fh3 := 6.93 \cdot 10^{6}$$

$$s := 1..rows(data3) \qquad F1_s := data3_{s,1} \cdot 10^{6}$$

$$Yt1_s := \left| \begin{array}{l} Vch1 \leftarrow data3_{s,2} \cdot 10^{-3} \\ Vin \leftarrow K \cdot Vch1 \\ Vch2 \leftarrow data3_{s,3} \cdot 10^{-3} \\ \omega \leftarrow 2 \cdot \pi \cdot F1_s \\ Z1 \leftarrow R4 + \dfrac{1}{j \cdot \omega \cdot C1} \\ Y2 \leftarrow \dfrac{1}{R1} + \dfrac{1}{R2} + \dfrac{1}{R3} + \dfrac{1}{j \cdot \omega \cdot L1} + \dfrac{1}{Z1} \\ Isec \leftarrow |Y2| \cdot Vch2 \\ Iprim \leftarrow Isec \cdot Turns \\ \dfrac{Iprim}{Vin} \end{array} \right.$$

Figure 2.33 Mathcad worksheet page 1: data from differential-mode test

The initial values of the capacitors were as follows:

$$Cc1 = \frac{2 \cdot \pi \cdot \varepsilon_o \cdot \varepsilon_r \cdot l}{\ln\left(\dfrac{r12 \cdot r13}{r11 \cdot r23}\right)} = 4.56 \times 10^{-10} \text{ Farad}$$

$$Cc2 = \frac{2 \cdot \pi \cdot \varepsilon_o \cdot \varepsilon_r \cdot l}{\ln\left(\dfrac{r12 \cdot r23}{r22 \cdot r13}\right)} = 4.56 \times 10^{-10} \text{ Farad} \qquad (2.100)$$

$$Cc3 = \frac{2 \cdot \pi \cdot \varepsilon_o \cdot \varepsilon_r \cdot l}{\ln\left(\dfrac{r13 \cdot r23}{r33 \cdot r12}\right)} = 1.89 \times 10^{-10} \text{ Farad}$$

$$data4 := \begin{pmatrix} 1 & 800 & 26 \\ 1.5 & 800 & 16 \\ 2 & 800 & 11 \\ 2.5 & 820 & 5.8 \\ 3 & 820 & 5.6 \\ 4 & 820 & 2 \\ 5 & 840 & 2 \\ 6 & 800 & 3.8 \\ 7 & 500 & 4.8 \\ 8 & 780 & 13 \\ 9 & 840 & 25 \\ 10 & 780 & 110 \\ 10.26 & 680 & 170 \\ 10.5 & 780 & 100 \\ 11 & 840 & 40 \\ 12 & 840 & 15 \end{pmatrix}$$

$$s := 1..rows(data4) \qquad F2_s := data4_{s,1} \cdot 10^6$$

$$Yt2_s := \left| \begin{array}{l} Vch1 \leftarrow data4_{s,2} \cdot 10^{-3} \\ Vin \leftarrow K \cdot Vch1 \\ Vch2 \leftarrow data4_{s,3} \cdot 10^{-3} \\ \omega \leftarrow 2 \cdot \pi \cdot F2_s \\ Z1 \leftarrow R4 + \dfrac{1}{j \cdot \omega \cdot C1} \\ Y2 \leftarrow \dfrac{1}{R1} + \dfrac{1}{R2} + \dfrac{1}{R3} + \dfrac{1}{j \cdot \omega \cdot L1} + \dfrac{1}{Z1} \\ Isec \leftarrow |Y2| \cdot Vch2 \\ Iprim \leftarrow Isec \cdot Turns \\ \dfrac{Iprim}{Vin} \end{array} \right.$$

Figure 2.34 Mathcad worksheet page 2: data from transfer admittance test

Resistances of the conductors were assumed to be

$$\begin{aligned} Rc1 &= 0.17\,\Omega \\ Rc2 &= 0.17\,\Omega \\ Rc3 &= 0.005\,\Omega \end{aligned} \tag{2.101}$$

These equations are used to assign values to the parameters recorded on the fourth page of the worksheet. A copy of this page is provided in Figure 2.37.

2.6.6 Distributed-parameter model

Textbook analysis of transmission lines leads to the conclusion that, unless the line is terminated by its characteristic impedance, reflections will occur at that termination. This means that there will be partial currents flowing in both directions in all three-conductors.

If it is assumed that the signals are oscillatory, a circuit model can be created to simulate the action of the line. Each conductor is represented as a T-network of impedances, where the value of each *Z*-parameter is a function of the *R*, *L*, and *C*

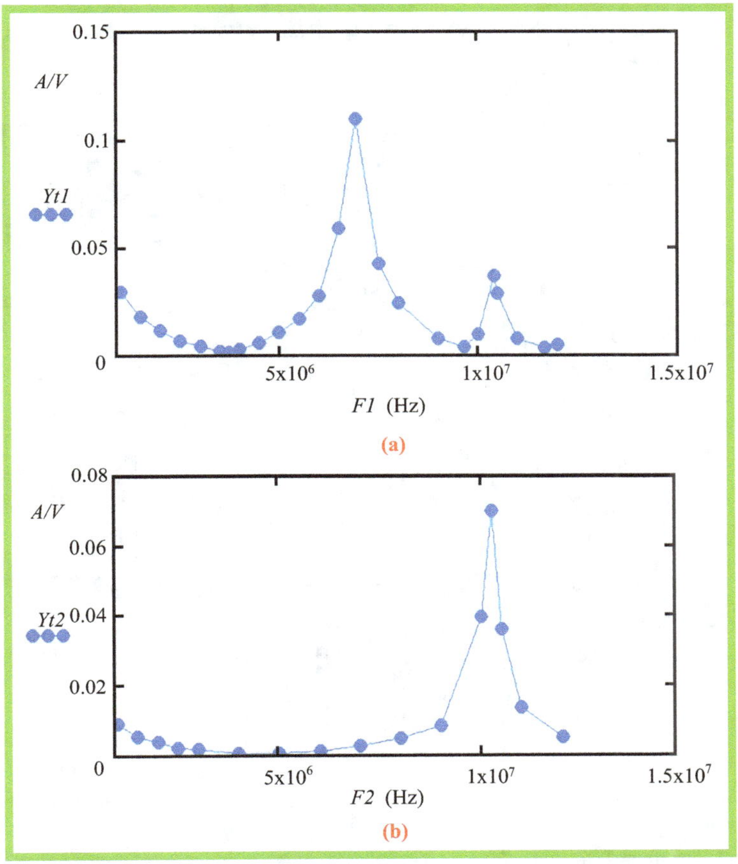

Figure 2.35 Page 3 of Mathcad worksheet: frequency response tests of test rig: (a) admittance of differential-mode loop, (b) transfer admittance of common-mode loop

values of the lumped-parameter model. The transformation equations, derived in Section 2.2.3, define the relationships.

This relationship can also be applied to the three-conductor configuration. This is illustrated by comparing Figure 2.36(b) with (a).

2.6.7 Computation

The function *Zbranch(f)* at the top of the Mathcad worksheet of Figure 2.38 invokes this relationship to assign values to the impedances Zh_i and Zv_i , where the character '*i*' identifies the associated conductor. The output of this function is a pair of vectors containing these impedance values at the frequency *f*.

The function *Zloop(f)* picks up these values and uses them to compile a set of loop impedances. Each loop impedance is the sum of the branch impedances along

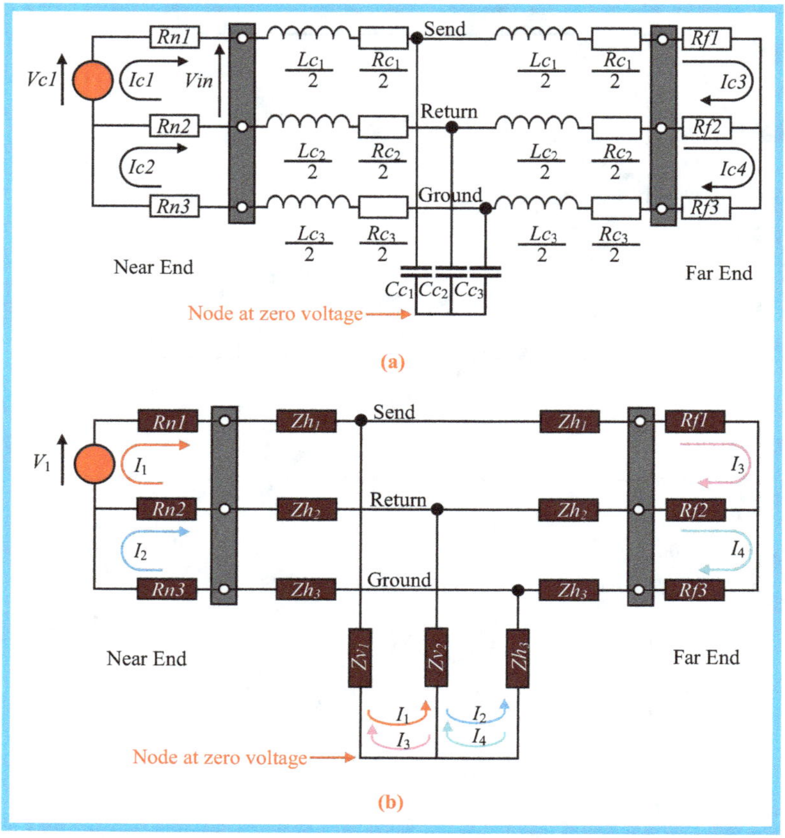

Figure 2.36 Triple-T models of the test rig: (a) Lumped-parameter model and (b) distributed-parameter model

the path of a particular current flow in the circuit of Figure 2.36(b). For example, *Z11* is the sum of Rn_1, Zh_1, Zv_1, Zv_2, Zh_2, and Rn_2, since that is the path followed by I_1. Since mesh analysis is used, it is assumed that many currents can flow in any particular impedance. Hence, *Z12* is the sum of Rn_2, Zv_1 and Zv_2, since these are the impedances shared by I_1 and I_2. The minus sign exists because I_1 and I_2 are flowing in opposite directions in these components.

The final action of the subroutine of *Zloop(f)* is to assemble these loop parameters into a four-by-four array. There are only ten independent loop parameters, since *Z12* = *Z21*, *Z13* = *Z31*, etc. This means that the array is symmetrical. This array is the output of the function Zloop(*f*).

The next action of the worksheet is to set-up the control variable *i* and use it to define the frequency vector F_i. The voltage vector *V* assumes that the voltage source is in loop 1, that its amplitude is 1 Volt and that there is no source in any of the other three loops.

$$Lc1 := 3.0\cdot 10^{-6} \qquad Lc3 := 6.7\cdot 10^{-6}$$

$$Cc1 := 1730\cdot 10^{-12} \qquad Cc3 := 300\cdot 10^{-12}$$

$$Rc := \begin{pmatrix} 0.17 \\ 0.17 \\ 4.5\cdot 10^{-3} \end{pmatrix} \qquad Gc := \begin{pmatrix} 0 \\ 0 \\ 0 \end{pmatrix}$$

$$Rn := \begin{pmatrix} 8 \\ 0 \\ 0 \end{pmatrix} \qquad Rf := \begin{pmatrix} 0 \\ 0 \\ 0 \end{pmatrix}$$

$$Lc := \begin{pmatrix} Lc1 \\ Lc1 \\ Lc3 \end{pmatrix} \qquad \frac{Lc}{2} = \begin{pmatrix} 1.5\times 10^{-6} \\ 1.5\times 10^{-6} \\ 3.35\times 10^{-6} \end{pmatrix}$$

$$Cc := \begin{pmatrix} Cc1 \\ Cc1 \\ Cc3 \end{pmatrix} \qquad Cc = \begin{pmatrix} 1.73\times 10^{-9} \\ 1.73\times 10^{-9} \\ 3\times 10^{-10} \end{pmatrix}$$

Figure 2.37 Page 4 of worksheet. Assigning values to the circuit parameters.

The function *Ym1* then calculates the magnitude of the current I_1 over the same range of frequencies employed in creating the graphs of Figure 2.35. The function *Ym2* does the same for the magnitude of the current I_2.

It was now possible to display the output of the simulation on the same pair of graphs as that used to record the test results. The results were encouraging, in that the shapes of the curves were similar. But there was a significant discrepancy in the values of the peak frequencies and in the amplitudes of the curves. However, by making incremental changes to the inductors and capacitors assigned to those of Figure 2.36(a) at the start of the worksheet and re-running the program, it became possible to establish the correlation illustrated in Figure 2.39.

Differential-mode current is defined as a flow along the send conductor and back via the return conductor. Figure 2.39(a) illustrates the response of this current at the near end of the cable. There are two peaks. The first is at the half-wave frequency, *fh* = 6.93 MHz. The second is at 10.41 MHz. This second peak is due to the interaction with the common-mode loop.

Current in the common-mode loop is defined as a flow along the send conductor and back via the ground conductor. This peaks at 10.26 MHz, the half-wave frequency. This frequency is much higher than that of the differential-mode loop since the relative permittivity is much less. The relative permittivity of the differential-mode loop is higher because the material between the send and return conductors is solid insulation, as illustrated in Figure 2.30. Most of the insulation in the common-mode loop is air.

2.6.8 Modelling the peaks

However, there was still a big difference between the amplitude of the peaks recorded by the test results and those derived from the simulation.

$$
\begin{aligned}
Zbranch(f) := \; & \omega \leftarrow 2\cdot\pi\cdot f \\
& \text{for } i \in 1..3 \\
& \quad \theta \leftarrow \sqrt{(Rc_i + j\cdot\omega\cdot Lc_i)\cdot(Gc_i + j\cdot\omega\cdot Cc_i)} \\
& \quad Zo \leftarrow \sqrt{\frac{Rc_i + j\cdot\omega\cdot Lc_i}{Gc_i + j\cdot\omega\cdot Cc_i}} \\
& \quad Zh_i \leftarrow Zo\cdot tanh\left(\frac{\theta}{2}\right) \\
& \quad Zv_i \leftarrow Zo\cdot csch(\theta) \\
& (Zh \;\; Zv)
\end{aligned}
$$

$$
\begin{aligned}
Zloop(f) := \; & (Zh \;\; Zv) \leftarrow Zbranch(f) \\
& Z11 \leftarrow Rn_1 + Zh_1 + Zv_1 + Zv_2 + Zh_2 + Rn_2 \\
& Z12 \leftarrow -(Zh_2 + Zv_2 + Rn_2) \\
& Z13 \leftarrow -(Zv_1 + Zv_2) \\
& Z14 \leftarrow Zv_2 \\
& Z22 \leftarrow Rn_2 + Zh_2 + Zv_2 + Zv_3 + Zh_3 + Rn_3 \\
& Z23 \leftarrow Zv_2 \\
& Z24 \leftarrow -(Zv_2 + Zv_3) \\
& Z33 \leftarrow Rf_2 + Zh_2 + Zv_2 + Zv_1 + Zh_1 + Rf_1 \\
& Z34 \leftarrow -(Zh_2 + Zv_2 + Rf_2) \\
& Z44 \leftarrow Rf_3 + Zh_3 + Zv_3 + Zv_2 + Zh_2 + Rf_2 \\
& \begin{pmatrix} Z11 & Z12 & Z13 & Z14 \\ Z12 & Z22 & Z23 & Z24 \\ Z13 & Z23 & Z33 & Z34 \\ Z14 & Z24 & Z34 & Z44 \end{pmatrix}
\end{aligned}
$$

$$i := 100..1200 \qquad F_i := i\cdot 10^4 \qquad V := \begin{pmatrix} 1 \\ 0 \\ 0 \\ 0 \end{pmatrix}$$

$$
\begin{aligned}
Ym1_i := \; & f \leftarrow F_i \\
& Z \leftarrow Zloop(f) \\
& I \leftarrow lsolve(Z, V) \\
& |I_1|
\end{aligned}
\qquad
\begin{aligned}
Ym2_i := \; & f \leftarrow F_i \\
& Z \leftarrow Zloop(f) \\
& I \leftarrow lsolve(Z, V) \\
& |I_2|
\end{aligned}
$$

Figure 2.38 Copy of page 5 of the Mathcad worksheet. The main program

The significant feature was that the actual amplitude of the peaks reduced as the frequency increased. So, instead of assuming that the values of the conductances were zero, it was hypothesised that they were significant; and proportional to frequency. Given this assumption, the circuit model of Figure 2.36(a) was modified to introduce a resistance value in parallel with each capacitor.

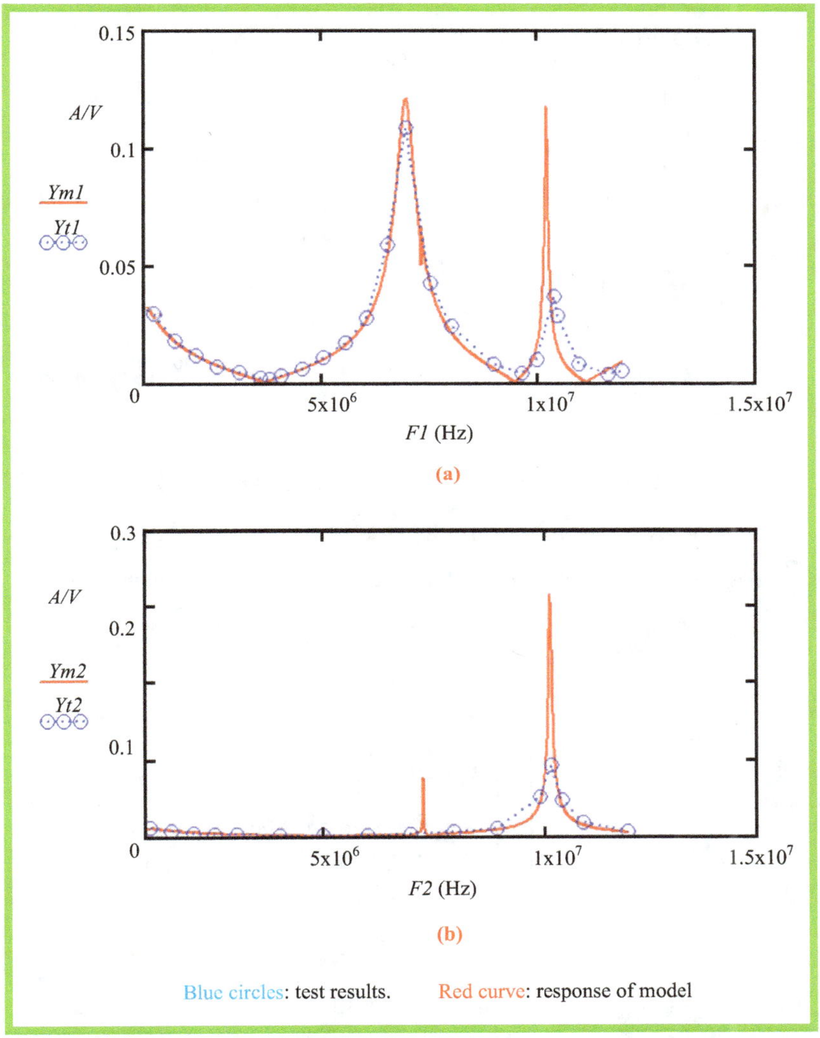

Figure 2.39 Initial correlation between test and model, page 5 of the worksheet: (a) current in the differential-mode loop conductor and (b) current in common-mode loop

It was also assumed that the value of each resistance was inversely proportional to the frequency. This led to the relationship:

$$Gc_i = \frac{1}{Re_i} \cdot \frac{f}{fh} \tag{2.102}$$

where Re_i is the leakage resistance assigned to conductor i, f is the frequency, and fh is the frequency at which the first peak occurs in Figure 2.35(a).

$$Lc1 := 3.0\cdot 10^{-6} \quad Lc3 := 6.7\cdot 10^{-6} \quad Cc1 := 1730\cdot 10^{-12} \quad Cc3 := 300\cdot 10^{-12}$$

$$Rss1 := 170\cdot 10^{-3} \quad Rss3 := 4.5\cdot 10^{-3} \quad Re1 := 2000 \quad Re3 := 4000$$

$$Rn := \begin{pmatrix} 8 \\ 0 \\ 0 \end{pmatrix} \quad Rf := \begin{pmatrix} 0 \\ 0 \\ 0 \end{pmatrix} \quad Ge1 := \frac{1}{Re1} \quad Ge3 := \frac{1}{Re3}$$

$$Lc := \begin{pmatrix} Lc1 \\ Lc1 \\ Lc3 \end{pmatrix} \quad \frac{Lc}{2} = \begin{pmatrix} 1.5 \times 10^{-6} \\ 1.5 \times 10^{-6} \\ 3.35 \times 10^{-6} \end{pmatrix}$$

$$Cc := \begin{pmatrix} Cc1 \\ Cc1 \\ Cc3 \end{pmatrix} \quad Cc = \begin{pmatrix} 1.73 \times 10^{-9} \\ 1.73 \times 10^{-9} \\ 3 \times 10^{-10} \end{pmatrix}$$

$$Rss := \begin{pmatrix} Rss1 \\ Rss1 \\ Rss3 \end{pmatrix} \quad \frac{Rss}{2} = \begin{pmatrix} 0.085 \\ 0.085 \\ 2.25 \times 10^{-3} \end{pmatrix}$$

$$Ge := \begin{pmatrix} Ge1 \\ Ge1 \\ Ge3 \end{pmatrix}$$

Figure 2.40 Page 4 of worksheet after inclusion of amendment

$$\begin{aligned} Zbranch(f) := \; & \omega \leftarrow 2\cdot\pi\cdot f \\ & \text{for } i \in 1..3 \\ & \quad Rc_i \leftarrow Rss_i \\ & \quad Gc_i \leftarrow Ge_i\cdot\frac{f}{fh3} \\ & \quad \theta \leftarrow \sqrt{(Rc_i + j\cdot\omega\cdot Lc_i)\cdot(Gc_i + j\cdot\omega\cdot Cc_i)} \\ & \quad Zo \leftarrow \sqrt{\frac{Rc_i + j\cdot\omega\cdot Lc_i}{Gc_i + j\cdot\omega\cdot Cc_i}} \\ & \quad Zh_i \leftarrow Zo\cdot\tanh\left(\frac{\theta}{2}\right) \\ & \quad Zv_i \leftarrow Zo\cdot\operatorname{csch}(\theta) \\ & (Zh \;\; Zv) \end{aligned}$$

Figure 2.41 Modified function

So, page 4 of the Mathcad worksheet was altered to include these parameters. The amended page is illustrated in Figure 2.40.

It was also necessary to modify the function *Zbranch(f)* on the main worksheet, to include the new parameters in the computation. A copy of the modified function is provided in Figure 2.41.

The end result was the graphs illustrated in Figure 2.42.

The values of the components of the circuit model in Figure 2.36(a) could then be defined by displaying the values that had finally been assigned to these variables

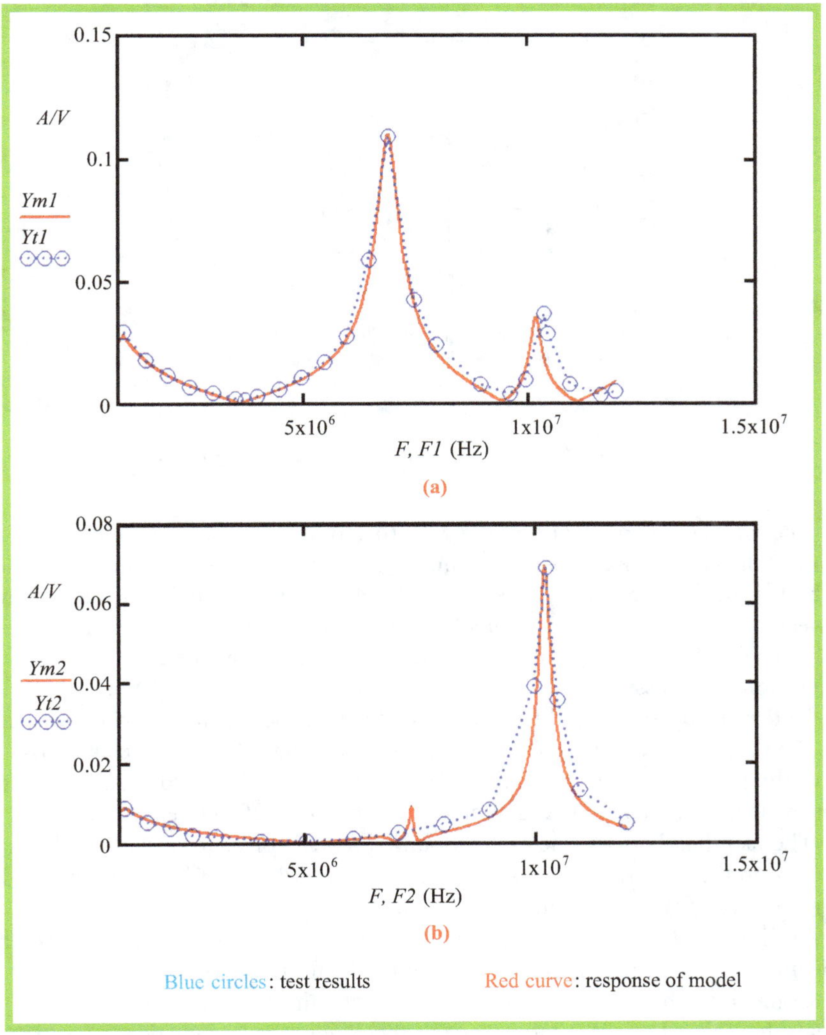

Figure 2.42 Final correlation between test and model: (a) current in send conductor and (b) current in ground conductor

in the worksheet. These are recorded in the circuit model in Figure 2.43. This model does not include the circuitry at either interface since the existence of that circuitry has no effect on the coupling between the conductors of the harness.

In the set-up of Figure 2.31, the interface at the far end is a short circuit between all three terminals. At the near end, the return conductor is shorted to the ground connection and the send conductor is supplied by a voltage source via a low value resistance. That is, the design of the test equipment is such that there is minimal interaction between the interface circuitry and the assembly-under-test.

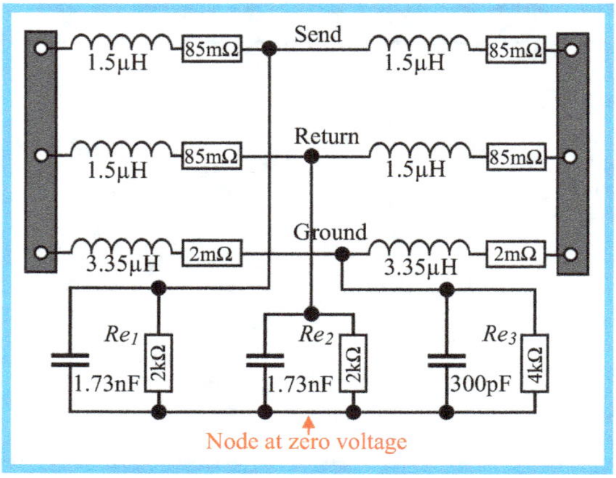

Figure 2.43 Definitive circuit model of the wiring of the test rig

The fact that the model of Figure 2.36(b) and its relationship with the model of Figure 2.36(a) is derived from the equations of electromagnetic theory provides credibility to the technique. The fact that close correlation has been achieved between test results and the simulation of the model provides a high degree of confidence in its reliability. The only feature of the model which could be subjected to question is the assignment of values to the resistors *Re* of (2.102).

Under steady state conditions measurement of the insulation of many cables has shown that the resistance is of the order of tens of MΩ. The only way of reconciling this fact with the existence of radiated power is to assume that any resistor which simulates the environment is a function of frequency.

The fact that the response of the modified model correlates closely with that of the tests on the hardware increases the level of confidence of its credibility and of the reliability of the measurement technique.

In Figure 2.43, the 2 kΩ resistors and the 4 kΩ resistor provide an effective indication of the level of radiation emanating from the assembly.

Comparing the initial values of (2.99) with the final values of Figure 2.43 reveals that the value of the inductance of the conductors has changed from 3.17 to 3.0 μH. This is probably due to the fact that transient current will tend to flow along the facing surfaces, thus reducing the inductance. The value of the inductance of the copper pipe has reduced from 7.66 to 6.7 μH, for the same reason.

Comparing the initial values of (2.100) with the capacitors of Figure 2.43 reveals that the values of Cc_1 and Cc_2 have increased from 456 to 1,730 pF. This is because the relative permittivity of the insulation is much higher than unity. The capacitor associated with the copper pipe has reduced from 1,890 to 300 pF. The probable reason for this is that less current emanating from the structure is arriving back at the cable, because significant current is flowing out into the environment in the form of radiated interference. The resistor Re_3 simulates this radiation.

2.6.9 Definitive circuit model

The fact that there is close correlation between the test results and those of the simulation means that the circuit diagram of Figure 2.43 provides a reliable model of the interference coupling between the differential-mode loop and the common-mode loop of the signal link.

It is possible to add the interface circuitry of the assembly to achieve a complete circuit model of the link. The completed model can then be used to predict the effect of intra-system interference on the performance of the configuration under review.

2.6.10 Applications

Circuitry can be added at the near end to simulate the interaction between the signal link and its interface with an equipment unit connected at the near end. Adding circuitry to simulate the effect of the signal receiver at the far end completes the model of the equipment-under-review. The predicted response of the equipment-under-test (EUT) can be defined over a band of frequencies which includes the resonant peaks. The current in the common-mode loop can be used to estimate the conducted emission of the EUT.

Similarly, the introduction of a voltage source in series with the ground conductor of Figure 2.43 can be used to analyse the conducted susceptibility of the EUT.

The model can also be used to simulate the cross-coupling between two signals which share the structure as the return conductor.

Since the R, L, and C values of the model are proportional to the length of the cable, the model can be used to predict the performance of a system in which the link-under-review is longer or shorter.

Chapter 3
Multi-conductor models

3.1 Composite conductors

3.1.1 Introduction

A method of simulating the distribution of current on the surfaces of conductors of a cable-form is described. This is dramatically more useful to designers than multi-coloured images of electromagnetic fields.

This is done by representing the surface of each conductor as an array of elemental conductors. The technique is introduced here by simulating the current distribution in two parallel conductors when these are short-circuited at one end and a voltage step of 1 V is applied at the other. Calculating the sum of the currents in the elemental conductors of each composite gives the total current in that conductor. There is then enough data to calculate the characteristic impedance, inductance, and capacitance of each conductor, and a circuit model of this simple assembly is derived.

The value of the capacitance between the conductors is shown to be precisely the same as that derived in books on electromagnetic theory. This provides confidence in the technique. The method can be developed to create a circuit model of any cable assembly.

The computations involve the use of a Mathcad worksheet; and sections of that worksheet are displayed at appropriate places in this section. The programs are easy to understand, since they are written in the same way as a set of equations in a textbook on mathematics. The illustrations of current distribution were created by transferring data from the worksheet to Microsoft Excel and invoking the 'Bubble Plot' option.

The formulae used in the equations are derived in Section 2.3.

The technique described in this section is a development of the method used by Culham Lightning Studies Unit to analyse the effect of lightning strikes on aircraft electronic systems.

3.1.2 Concept

The interaction between the external electromagnetic field and charges on any conductor takes place at the surface of that conductor. Since the charge intensity on that surface is proportional to the strength of the electromagnetic field at that location, one way of analysing the effect of that field is to simulate the way charges

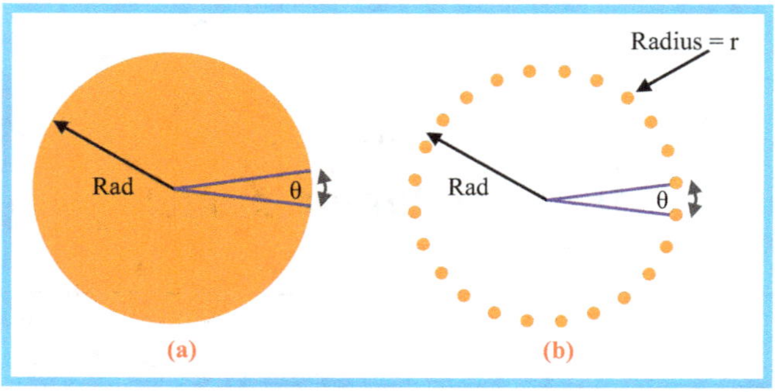

Figure 3.1 Concept of the composite conductor: (a) cross-section of the actual conductor and (b) array of elemental conductors

are distributed across the surface. This can be done by replacing each conductor with an array of elemental conductors, as illustrated in Figure 3.1.

Each elemental conductor represents a small segment, s, of the surface. Hence

$$s = 2 \cdot \pi \cdot r = \theta \cdot Rad \tag{3.1}$$

where r is the radius of the elemental conductor, and Rad is the radius of the composite conductor.

In this example, it is assumed that the conductor is isolated. So, the charge is distributed evenly round the surface. If n is the number of elemental conductors:

$$n \cdot \theta = 2 \cdot \pi$$

And

$$r = \frac{Rad}{n} \tag{3.2}$$

3.1.3 Twin conductor assembly

An array of 24 elemental conductors is illustrated in Figure 3.2. This represents the cross-section of two conductors spaced 4 mm apart. The radius of each conductor is 1 mm. This assembly can be defined using a table of three columns; recording the x co-ordinates, the y co-ordinates, and the radius of each elemental conductor.

In Section 2.3, a general formula is derived for primitive impedance:

$$Zp_{i,j} = \frac{1}{2 \cdot \pi} \cdot \sqrt{\frac{\mu}{\varepsilon}} \cdot \ln\left(\frac{l}{r_{i,j}}\right) \tag{3.3}$$

where $r_{i,j}$ is the separation between conductors i and j, $r_{i,i}$ is the radius of conductor i, μ is the permeability of the conductor, ε is the average permittivity of the insulation, and l is the length of the assembly.

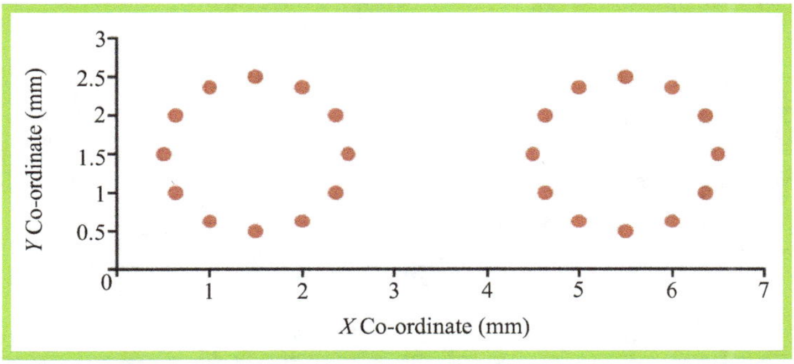

Figure 3.2 Using elemental conductors to simulate a twin-conductor assembly

$\mu := 4\cdot\pi\cdot 10^{-7}$ $\quad \varepsilon := 8.854\cdot 10^{-12}$	Fundamental constants
$n1 = 12 \quad n2 = 12$	Number of elemental conductors in each composite
$N = 24$	Total number of elemental conductors
$\boldsymbol{Zp} := \left\| H \leftarrow \frac{1}{2\cdot\pi}\cdot\sqrt{\frac{\mu}{\varepsilon}} \right.$	Calculating the primitive impedance of every elemental conductor
$for\ i \in 1..N$ $\quad for\ j \in 1..N$ $\qquad h \leftarrow \boldsymbol{x}_j - \boldsymbol{x}_i$ $\qquad v \leftarrow \boldsymbol{y}_j - \boldsymbol{y}_i$	The first action is to calculate the separation between every conductor and every other conductor, *rad*
$\qquad rad \leftarrow \sqrt{h^2 + v^2}$ $\qquad rad \leftarrow \boldsymbol{r}_i\ \ if\ \ rad = 0$ $\qquad \boldsymbol{Z}_{i,j} \leftarrow H\cdot ln\left(\frac{1}{rad}\right)$	Then this value is used to calculate the value of the primitive impedance
$\boldsymbol{Z}$	The output of this subroutine is the vector ***Z*** Its contents are transferred to ***Zp***
$i := 1..N$	
$\boldsymbol{Vp}_i := 1$	Setting a voltage of 1V along every elemental conductor
$\boldsymbol{Ip} := lsolve(\boldsymbol{Zp}, \boldsymbol{Vp})$	Calculating the value of the current along every element

Figure 3.3 Calculating values for the currents in the elemental conductors

It is assumed that a voltage of 1 V is applied along the length. Figure 3.3 is an extract from a Mathcad worksheet. This calculates the primitive impedances associated with each elemental conductor and records them in an array ***Zp***. It then defines the voltage vector ***Vp*** and calculates the value of all the currents in the

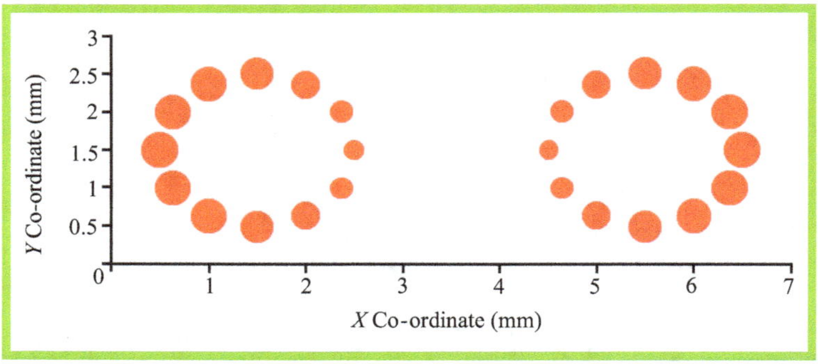

Figure 3.4 Distribution of antenna current in two parallel conductors

elemental conductors. The vector ***Ip*** records the value of the current in every elemental conductor.

The distribution of current is illustrated by the bubble plot of Figure 3.4. Since the same voltage is applied across each conductor, all the currents are flowing in the same direction. That is, current flow is concentrated on the outward facing surfaces.

This can be visualised in two ways:

- The current due to the magnetic field in the environment,

  or
- the charge distribution due to the electric field in the environment.

3.1.4 Differential mode current

To analyse differential mode current, it is assumed that the terminals at the far end of the cable are shorted together and that a voltage is applied between the two conductors at the near end. For the array of Figure 3.2, this voltage appears between elements 12 and 13 at the near end. Figure 3.5 illustrates this.

To calculate the values of the current *Ip* in each elemental conductor, it is first necessary to calculate the loop currents illustrated in Figure 3.5. The section of the Mathcad worksheet which performs this function is illustrated in Figure 3.6.

Each loop impedance is due to current flowing in each conductor of the loop, plus the mutual coupling between the two conductors of that loop. This results in the vector ***Zloop***. The voltage at the near end between conductors 12 and 13 is defined as 1 V. All the other voltages at the near end are zero. This is defined by the vector ***Vloop***. The third function of Figure 3.6 calculates the values of the loop currents and stores them in the vector ***Iloop***. The final action of this part of the worksheet is to calculate the value of the current in each elemental conductor and store it in the vector ***Ip***.

Figure 3.7 illustrates the current distribution in the assembly. The area of each red circle is proportional to the current flowing in the elemental conductor at that location. This current is flowing from left to right, as illustrated by the currents *Ip* in

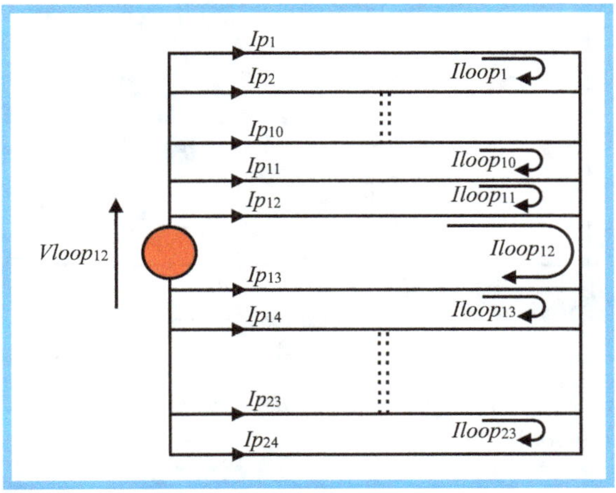

Figure 3.5 Definition of loop currents and voltages

$$\mathbf{Zloop} := \left| \begin{array}{l} for\ i \in 1..N-1 \\ \quad for\ j \in 1..N-1 \\ \quad\quad \mathbf{Z}_{i,j} \leftarrow \mathbf{Zp}_{i,j} - \mathbf{Zp}_{i,j+1} - \mathbf{Zp}_{i+1,j} + \mathbf{Zp}_{i+1,j+1} \\ \mathbf{Z} \end{array} \right.$$

Calculating the loop impedance of every pair of elemental conductors

$$\mathbf{Vloop} := \left| \begin{array}{l} for\ i \in 1..N-1 \\ \quad \left| \begin{array}{l} \mathbf{V}_i \leftarrow 0 \\ \mathbf{V}_i \leftarrow 1 \quad if\ i = n1 \end{array} \right. \\ \mathbf{V} \end{array} \right.$$

Locating the voltage source in loop $n1$

$$\mathbf{Iloop} := lsolve(\mathbf{Zloop}, \mathbf{Vloop})$$

Calculating the value of the current in every loop

$$\mathbf{Ip} := \left| \begin{array}{l} \mathbf{I}_1 \leftarrow \mathbf{Iloop}_1 \\ for\ i \in 2..N-1 \\ \quad \mathbf{I}_i \leftarrow \mathbf{Iloop}_i - \mathbf{Iloop}_{i-1} \\ \mathbf{I}_N \leftarrow -\mathbf{Iloop}_{N-1} \\ \mathbf{I} \end{array} \right.$$

Calculating the value of the current flowing along every elemental conductor

Figure 3.6 Calculating the values of the currents **Ip** *in the circuit of Figure 3.5*

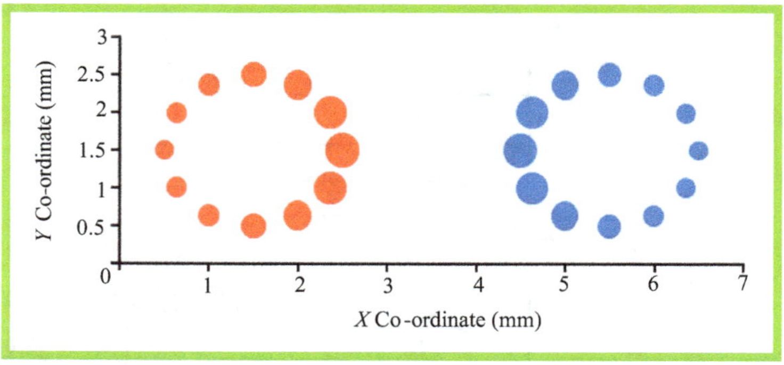

Figure 3.7 Distribution of differential-mode current in the two composite conductors

Figure 3.5. Each blue circle represents the magnitude of the current flowing from right to left in that diagram.

This depiction can be visualised as the current distribution in the send and return conductors when constant current is flowing in the loop, or as the distribution of charges when these conductors are isolated from each other.

3.1.5 Component values

The worksheet depicted in Figure 3.8 first defines the elemental conductors which simulate composite conductors 1 and 2 and then calculates the total current in each composite and stores them in the vector ***Ic***. This computation indicates that 6.33 mA flows along the send conductor and back along the return conductor when a voltage of 1 V is applied to the input terminals.

The next step in the computation is to calculate the sum of the voltages along every elemental conductor of the send conductor. Since the voltage along each elemental conductor is 0.5 V and there are 12 such conductors, it is necessary to divide the total voltage by 12. Similarly, the voltage along the return conductor is calculated to be −0.5 V. This data is recorded in the vector ***Vc***.

It is then a simple task to calculate the value of the impedances $\boldsymbol{Zc}_1$ and $\boldsymbol{Zc}_2$ presented by each composite conductor to the input voltage.

For a lossless line, the characteristic impedance is as follows:

$$Zc = \sqrt{\frac{Lc}{Cc}} \tag{3.4}$$

And the time taken for a step current to propagate from one end of the line to the other is as follows:

$$T = \sqrt{Lc \cdot Cc} \tag{3.5}$$

Both these relationships are derived in Section 2.2.

$$\mathbf{Start} := \begin{pmatrix} 1 \\ n_1 + 1 \end{pmatrix} = \begin{pmatrix} 1 \\ 13 \end{pmatrix}$$

Defining the first elemental conductor in each composite conductor

$$\mathbf{End} := \begin{pmatrix} n_1 \\ n_1 + n_2 \end{pmatrix} = \begin{pmatrix} 12 \\ 24 \end{pmatrix}$$

Defining the last elemental conductor in each composite conductor

```
Ic := | for h ∈ 1..2
      |   | S ← 0
      |   | for i ∈ Start_h..End_h
      |   |   S ← S + Ip_i
      |   | I_h ← S
      | I
```

Calculating the total current along each composite conductor

$$\mathbf{Ic} = \begin{pmatrix} 6.331 \times 10^{-3} \\ -6.331 \times 10^{-3} \end{pmatrix}$$

```
Vc := | for h ∈ 1..2
      |   | S ← 0
      |   | for i ∈ Start_h..End_h
      |   |   for j ∈ 1..N
      |   |     S ← S + Ip_j·Zp_i,j
      |   | V_h ← S
      | V / n_h
```

Selecting all the primitive impedances associated with composite conductor h multiplying each by every primitive current, adding up all these voltage increments and putting the total in the element $\boldsymbol{V_h}$

Since there are $\boldsymbol{n_h}$ elemental conductors in composite h, the sum total is divided by $\boldsymbol{n_h}$

This gives the voltage drop along each composite conductor

$$\mathbf{Vc} = \begin{pmatrix} 0.5 \\ -0.5 \end{pmatrix}$$

$$\mathbf{Zc} := \frac{\mathbf{Vc}}{\mathbf{Ic}} = \begin{pmatrix} 78.972 \\ 78.972 \end{pmatrix}$$

Characteristic impedance of each composite conductor

Figure 3.8 Computing the values for the circuit impedances

From (3.4),

$$Cc = \frac{Lc}{Zc^2} \tag{3.6}$$

Substituting for Cc in (3.5)

$$T = \frac{Lc}{Zc} \tag{3.7}$$

Giving

$$Lc = T \cdot Zc \tag{3.8}$$

$T := \sqrt{\mu \cdot \varepsilon} = 3.336 \times 10^{-9}$ Time taken for a step voltage to propagate along a length of 1 metre of the cable

$Lc := T \cdot Zc = \begin{pmatrix} 2.634 \times 10^{-7} \\ 2.634 \times 10^{-7} \end{pmatrix}$ Inductance of each composite conductor

$Cc := \frac{T}{Zc} = \begin{pmatrix} 4.224 \times 10^{-11} \\ 4.224 \times 10^{-11} \end{pmatrix}$ Capacitance of each composite conductor

"From the book Electromagnetic Concepts and applications by G.G. Skitek and S. V. Marshall, pages 208 to 209: Capacitance between two cylindrical conductors"

$b := 2 \cdot 10^{-3}$ $r := 1 \cdot 10^{-3}$ $l := 1$

$$Ctheory := \frac{\pi \cdot \varepsilon \cdot l}{ln\left(\frac{b + \sqrt{b^2 - r^2}}{r}\right)} = 2.112 \times 10^{-11}$$

Figure 3.9 Assigning values to the inductors and capacitors of the circuit model

Substituting for *Lc* in (3.6)

$$Cc = \frac{T}{Zc} \tag{3.9}$$

The section of the worksheet which assigns values to these parameters is illustrated in Figure 3.9.

The worksheet also compares the results with that derived from a book on electromagnetic theory. Since the capacitance of two 42.24 pF capacitors in series is 21.12 pF, this comparison validates the technique described above.

3.1.6 Circuit model

The circuit model for the assembly-under-review is illustrated in Figure 3.10. The inductors in this model are assigned values of 132 nH, since each conductor is represented as a T-network. The fictitious conductor at zero volts has no physical existence, but is an essential part of the model.

It is possible to simplify the model further, by invoking the concept of equivalent circuits, but this action would bring with it the assumption that the return conductor can be represented as a zero-volt surface.

3.1.7 Conclusion

The concept of the composite conductor has been introduced. The technique involves the use of elemental conductors to simulate the current distribution at the surface of each composite conductor they represent. It provides an analytical tool

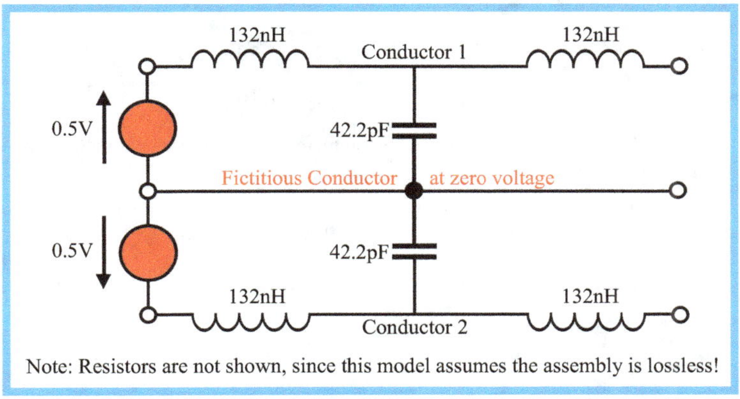

Figure 3.10 Circuit model of 1 m of twin conductor cable

which can be used to assign accurate values to the inductance and capacitance of those conductors. The value derived for the capacitance between the conductors has been shown to be identical to the value derived in textbooks on electromagnetic theory.

The technique can be developed to derive values for the components of the circuit model which simulates any conductor assembly, given information on the geometry of that assembly. There is no need to invoke the complexities of full-field analysis.

3.2 The multi-potential ground

3.2.1 Introduction

The structure can be used to propagate interference far and wide or to act as a shield. The choice lies with the designer.

A technique is described which defines the current and voltage distribution in the ground conductor of a cable assembly. It uses this data to create a circuit model which simulates cross-coupling of interference.

Full details are provided of the Mathcad worksheet which performs this task. The results are more meaningful than any full-field simulation. The technique can be adopted by any electronic system designer.

3.2.2 Assembly under review

The technique is based on the concept of composite conductors, where the conducting surfaces are simulated by an array of elemental conductors. The section under review is essentially that of a cable conduit on which two cable conductors are laid alongside each other. The graph near the bottom of Figure 3.11 illustrates this section. It is assumed that the section is constant along the entire length.

$X1 := 8\cdot10^{-3}$ $\quad Y1 := 3\cdot10^{-3}$ $\quad R1 := 1\cdot10^{-3}$ $\quad n_1 := 12$

$X2 := 12\cdot10^{-3}$ $\quad Y2 := 3\cdot10^{-3}$ $\quad R2 := 1\cdot10^{-3}$ $\quad n_2 := 12$

$i := 1..n_1$ Elemental conductors representing conductor 1

$\theta_i := \frac{2\pi}{n_1}\cdot(i)$

$x_i := X1 + R1\cdot\cos(\theta_i)$ $\quad y_i := Y1 + R1\cdot\sin(\theta_i)$ $\quad r_i := \frac{R1}{n_1}$

$i := n_1 + 1..n_1 + n_2$ Elemental conductors representing conductor 2

$\theta_i := \frac{2\pi}{n_2}\cdot(i - n_1)$

$x_i := X2 + R2\cdot\cos(\theta_i)$ $\quad y_i := Y2 + R2\cdot\sin(\theta_i)$ $\quad r_i := \frac{R2}{n_2}$

$X3 := 0.001$ $\quad Y3 := 0.006$ $\quad s := 0.0005$ $\quad n_3 := 79 - 24 = 55$

$X4 := 0.019$ $\quad Y4 := 0.001$

$i := 25..34$ $\quad p_i := i - 24$ $\quad x_i := X3$ $\quad y_i := Y3 - (p_i - 1)\cdot s$ $\quad r_i := \frac{s}{2\cdot\pi}$

$i := 35..69$ $\quad p_i := i - 34$ $\quad x_i := X3 + s\cdot p_i$ $\quad y_i := Y4$ $\quad r_i := \frac{s}{2\cdot\pi}$

$i := 70..79$ $\quad p_i := i - 69$ $\quad x_i := X4$ $\quad y_i := Y4 + s\cdot p_i$ $\quad r_i := \frac{s}{2\cdot\pi}$

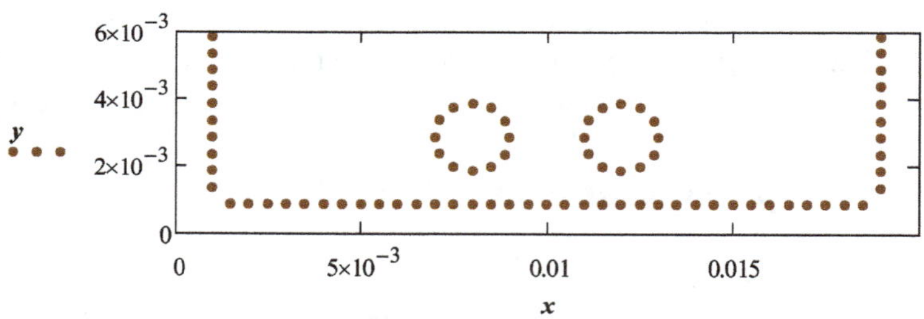

$n = \begin{pmatrix}12\\12\\55\end{pmatrix}$ $\quad N := n_1 + n_2 + n_3$ $\quad i := 1..N$

Figure 3.11 Defining the geometry of the cross-section

The X and Y co-ordinates and the radii of the conductors are defined by the first two lines of the worksheet. The number of elemental conductors in conductors is defined by $\boldsymbol{n}_1$ and $\boldsymbol{n}_2$, respectively.

The next three lines of the worksheet define the $\boldsymbol{x}$ and $\boldsymbol{y}$ co-ordinates and the radius $\boldsymbol{r}$ of each of the elemental conductors used to simulate conductor 1. The circumference of each element is equal to the path length around the surface it represents. This ensures that the strength of the magnetic field on the element is identical to that on the surface of the section of the conductor it simulates.

The co-ordinates of the top left of the conduit are defined as *X3* and *Y3*. The length of the surface of the section represented by the first elemental conductor representing the conduit is defined as *s*. The co-ordinates of the left-hand corner of the conduit are defined as *X3* and *Y4*. The co-ordinates of the right-hand corner are *X4* and *Y4*. The final elemental conductor of the simulation is located at the co-ordinates *X4* and *Y3*.

The next three lines in the worksheet define the x and y co-ordinates of the elemental conductors representing the conduit and the radius r of each.

The graph displays the location of all the elemental conductors. Data is recorded in the vectors $\boldsymbol{x}$, $\boldsymbol{y}$, and r. The vector $\boldsymbol{n}$ records the number of elements in each of the three composite conductors, and N is the total number of elements.

3.2.3 *Current distribution*

Figure 3.12 is a copy of the second page of the worksheet. This analysis assumes that all the terminations at the near end of the elemental conductors are shorted together, but that all the far-end terminations are left open-circuit. This ensures that the only connection between the conductors is capacitive. It is assumed that a voltage source of 1 V is inserted at the near end between conductors 1 and 2, that is, between elements $\boldsymbol{n}_1$ and $\boldsymbol{n}_1 + 1$.

The basic equation used in this computation is the expression for the primitive impedance:

$$Zp_{i,j} = \frac{1}{2 \cdot \pi} \cdot \sqrt{\frac{\mu}{\varepsilon}} \cdot \ln\left(\frac{l}{r_{i,j}}\right) \tag{3.10}$$

where μ is the permeability of the conductors, ε is the permittivity of the insulation, i and j identify the elemental conductors, l is the length of the assembly, and $r_{i,j}$ defines the radial separation between conductors i and j. This impedance is related to the primitive inductance and capacitance by

$$Zp_{i,j} = \sqrt{\frac{Lp_{i,j}}{Cp_{i,j}}} \tag{3.11}$$

This means that when the values of the primitive impedances have been calculated, the values of the primitive inductance and primitive capacitance of each conductor can also be determined.

There are three types of parameter: primitive, loop, and circuit. Primitive parameters define the relationship between currents and voltages in the conductors to the strength of the magnetic and electric field at the surfaces. Loop parameters are those due to current flowing down one conductor and back via the adjacent conductor; they can be measured by test equipment. Circuit parameters are those which are depicted on a circuit diagram.

$\mu := 4 \cdot \pi \cdot 10^{-7} \qquad \varepsilon := 8.854 \cdot 10^{-12}$ — Fundamental constants

$$
\begin{aligned}
\boldsymbol{Zp} := \; & H \leftarrow \frac{1}{2 \cdot \pi} \cdot \sqrt{\frac{\mu}{\varepsilon}} \\
& \text{for } i \in 1..N \\
& \quad \text{for } j \in 1..N \\
& \qquad h \leftarrow \boldsymbol{x}_j - \boldsymbol{x}_i \\
& \qquad v \leftarrow \boldsymbol{y}_j - \boldsymbol{y}_i \\
& \qquad rad \leftarrow \sqrt{h^2 + v^2} \\
& \qquad rad \leftarrow \boldsymbol{r}_i \;\; \text{if } rad = 0 \\
& \qquad \boldsymbol{Z}_{i,j} \leftarrow H \cdot ln\left(\frac{1}{rad}\right) \\
& \boldsymbol{Z}
\end{aligned}
$$

Calculating the primitive impedance of every elemental conductor

The first action is to calculate the separation between every conductor and every other conductor, *rad*. Then this value is used to calculate the value of the primitive inductance

Then the results are stored in an array of N rows by N columns

$$
\begin{aligned}
\boldsymbol{Zloop} := \; & \text{for } i \in 1..N-1 \\
& \quad \text{for } j \in 1..N-1 \\
& \qquad \boldsymbol{Z}_{i,j} \leftarrow \boldsymbol{Zp}_{i,j} - \boldsymbol{Zp}_{i,j+1} - \boldsymbol{Zp}_{i+1,j} + \boldsymbol{Zp}_{i+1,j+1} \\
& \boldsymbol{Z}
\end{aligned}
$$

Calculating the loop impedance of each adjacent pair of elemental conductors

$$
\begin{aligned}
\boldsymbol{Vloop} := \; & \text{for } i \in 1..N-1 \\
& \quad V_i \leftarrow 0 \\
& \quad V_i \leftarrow 1 \;\; \text{if } i = \boldsymbol{n}_1 \\
& \boldsymbol{V}
\end{aligned}
$$

Locating the voltage source in loop n_1

$\boldsymbol{Iloop} := lsolve(\boldsymbol{Zloop}, \boldsymbol{Vloop})$ — Calculating the values of all the loop currents

$$
\begin{aligned}
\boldsymbol{Ip} := \; & \boldsymbol{I}_1 \leftarrow \boldsymbol{Iloop}_1 \\
& \text{for } i \in 2..N-1 \\
& \quad \boldsymbol{I}_i \leftarrow \boldsymbol{Iloop}_i - \boldsymbol{Iloop}_{i-1} \\
& \boldsymbol{I}_N \leftarrow -\boldsymbol{Iloop}_{N-1} \\
& \boldsymbol{I}
\end{aligned}
$$

Using data on the values of the loop currents to derive values for the primitive currents

$\boldsymbol{A}_{i,1} := \boldsymbol{x}_i \cdot 10^4 \qquad \boldsymbol{A}_{i,2} := \boldsymbol{y}_i \cdot 10^4 \qquad \boldsymbol{A}_{i,3} := \boldsymbol{Ip}_i \cdot 10^4$ — Creating a table of values

Figure 3.12 Calculating the current in each elemental conductor

The first action of the worksheet is to calculate the values of all the primitive impedances and to record them in the N by N array ***Zp***. The contents of this array are then converted to loop impedances and stored in the Z***loop*** array. The ***Vloop*** vector places the voltage source between composite conductors 1 and 2.

The function *lsolve(**Zloop**, **Vloop**)* calculates the value of every loop current and stores the results in the vector ***Iloop***. The value of the current in every elemental conductor is calculated and the results stored in the vector ***Ip***. Finally, a table of values of x and y co-ordinates of each elemental conductor is listed alongside the current *Ip* in that conductor is assembled and recorded in the array ***A***.

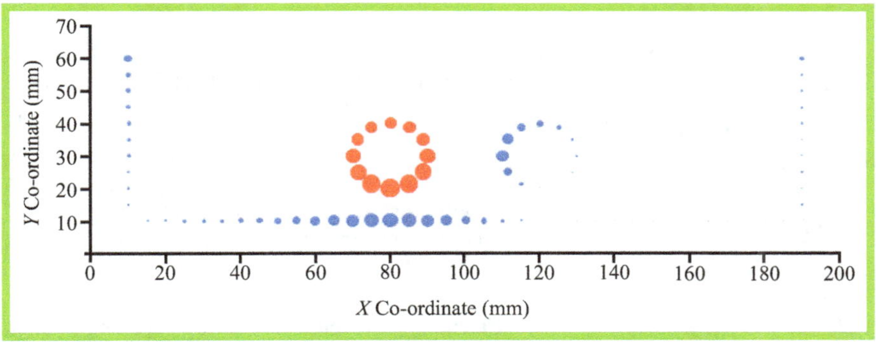

Figure 3.13 Distribution of current in the cross-section of the cable assembly red circles: current flowing forward along the conductor blue circles: current flowing back

This data can be displayed in tabular form by invoking the command '*A*='. The 'copy and paste' command can be used to transfer the table to Microsoft Excel. This software can then display the results as a bubble plot, where the diameter of each bubble is proportional to the amplitude of the current in the elemental conductor at that location. Figure 3.13 illustrates this.

This display shows that current is concentrated on adjacent surfaces and that a significant portion of the return current flows in the second conductor. The amplitude of this current is a measure of the level of interference picked up by any signal carried by this conductor. It can also be observed that the current density increases near the top edges of the conduit.

3.2.4 Circuit model

Since the amplitude of the current in every elemental conductor is now a known quantity, it becomes possible to calculate the voltage along the length of any one conductor. This task is performed by the section of the worksheet illustrated in Figure 3.14.

The ***Start*** and ***End*** vectors define the relationship between the identifiers for the elemental conductors and those for the composite conductors. This data is used to calculate the total current in each of the three composites and record the result in the vector ***Iq***. This establishes the fact that 13.7 mA flows forward along composite 1 and returns via the parallel combination of composite conductors 2 and 3.

The voltage along every elemental conductor is calculated and stored in the vector ***Vp***, and the table opposite verifies that the voltage along every elemental conductor of composite 1 is 0.941 V. The voltage along every other element is 0.059 V. That is, the total loop voltage is 1 V. The vector ***Vq*** records the voltages along the three composite conductors.

Given a known value for the current in each composite conductor and the voltage along its length, it is a simple matter to derive a value for the impedance.

$$\boldsymbol{Start} := \begin{pmatrix} 1 \\ n_1 + 1 \\ n_1 + n_2 + 1 \end{pmatrix} = \begin{pmatrix} 1 \\ 13 \\ 25 \end{pmatrix}$$

Defining the first elemental conductor in each composite conductor

$$\boldsymbol{End} := \begin{pmatrix} n_1 \\ n_1 + n_2 \\ N \end{pmatrix} = \begin{pmatrix} 12 \\ 24 \\ 79 \end{pmatrix}$$

Defining the last elemental conductor in each composite conductor

```
Iq :=  | for h ∈ 1..3
       |    | S ← 0
       |    | for i ∈ Start_h..End_h
       |    |    S ← S + Ip_i
       |    | I_h ← S
       | I
```

Calculating the total current flowing along each composite conductor

$$Iq \cdot 10^3 = \begin{pmatrix} 13.729 \\ -2.943 \\ -10.786 \end{pmatrix}$$

$$(Iq_2 + Iq_3) \cdot 10^3 = -13.729$$

```
Vp :=  | for i ∈ 1..N
       |    | S ← 0
       |    | for j ∈ 1..N
       |    |    | S ← S + Zp_i,j · Ip_j
       |    |    | V_i ← S
       | V
```

Calculating the voltage developed along every elemental conductor

$Vp =$

	0
0	0.941
1	0.941
2	0.941
3	0.941
4	0.941
5	0.941
6	0.941
7	0.941
8	0.941
9	0.941
10	0.941
11	0.941
12	-0.059
13	...

```
Vq :=  | for h ∈ 1..3
       |    | S ← 0
       |    | for i ∈ Start_h..End_h
       |    |    S ← S + Vp_i
       |    | V_h ← S / n_h
       | V
```

Calculating the voltage developed along every composite conductor

$$Vq = \begin{pmatrix} 0.941 \\ -0.059 \\ -0.059 \end{pmatrix}$$

Figure 3.14 Calculating the amplitudes of the currents and voltages for the composite conductors

This is done by the section of the worksheet depicted in Figure 3.15. Capacitance and inductance values can be calculated using the relationships derived in Section 2.3. The parameter T is the time taken for the front edge of a signal to propagate along a one metre length of the assembly.

$$Iq \cdot 10^3 = \begin{pmatrix} 13.729 \\ -2.943 \\ -10.786 \end{pmatrix} \qquad Vq = \begin{pmatrix} 0.941 \\ -0.059 \\ -0.059 \end{pmatrix} \qquad Zq := \frac{Vq}{Iq} = \begin{pmatrix} 68.525 \\ 20.121 \\ 5.491 \end{pmatrix}$$

$$T := \sqrt{\mu \cdot \varepsilon} = 3.336 \times 10^{-9}$$

$$Lq := T \cdot Zq \qquad \frac{Lq}{2} = \begin{pmatrix} 1.143 \times 10^{-7} \\ 3.356 \times 10^{-8} \\ 9.157 \times 10^{-9} \end{pmatrix}$$

$$Cq := \frac{T}{Zq} \qquad Cq = \begin{pmatrix} 4.868 \times 10^{-11} \\ 1.658 \times 10^{-10} \\ 6.075 \times 10^{-10} \end{pmatrix}$$

Figure 3.15 Calculating values for the circuit model

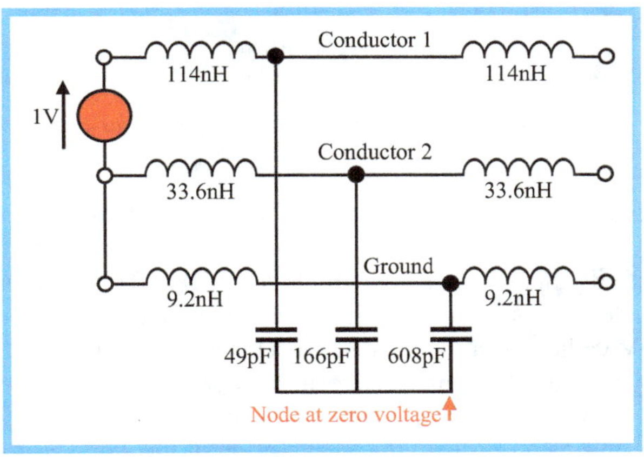

Figure 3.16 Circuit model of one metre length of the 'ground return' configuration

Figure 3.16 shows the circuit model for a one metre length of the assembly.

3.2.5 *Differential mode*

The circuit model of Figure 3.16 depicts a system configuration where the ground conductor is used to carry the return conductor and provides a method of calculating the level of interference picked up by conductor 2.

If it is decided to utilise conductor 2 as the return conductor, this new configuration can be modelled by applying the source voltage between conductors 1

```
Vloop :=  | for i ∈ 1..N − 1                    Locating the voltage sources
          |    | V_i ← 0
          |    | V_i ← 2  if i = n1
          |    | V_i ← −1  if i = n1 + n2
          | V
```

Figure 3.17 Redefining the loop voltages between the conductors

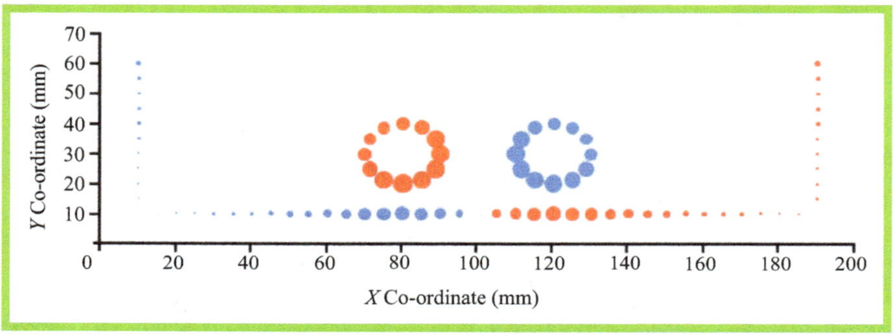

Figure 3.18 Current distribution for a differential-mode signal

and 2. This can be analysed simply by modifying the function used to define ***Vloop*** in Figure 3.12.

The modified function is shown in Figure 3.17. This inserts two voltage sources; conductor one is set at 2 V with respect to conductor 2 and conductor 2 is set at minus 1 V with respect to conductor 3. This configuration can be achieved in practice by applying a 2 V signal via a centre-tapped transformer.

Figure 3.18 illustrates the resultant current distribution.

Processing this data results in the final section of the worksheet displaying the L and C values shown in Figure 3.19. Since the value of the voltage developed along the structure is zero, there is no point in trying to assign an impedance value to that conductor.

Figure 3.20 provides a circuit model of the assembly when it is connected to a differential-mode supply.

3.2.6 Conclusion

The pictures say it all. Figures 3.13 and 3.18 show the voltage distribution in the conductors when a steady-state voltage is applied or the current distribution when the energy source is a step voltage.

If the structure is used as the return conductor, the electromagnetic energy used to carry the signal spreads far and wide.

$$Iq \cdot 10^3 = \begin{pmatrix} 16.672 \\ -16.672 \\ -1.492 \times 10^{-13} \end{pmatrix} \qquad Vq = \begin{pmatrix} 1 \\ -1 \\ 0 \end{pmatrix}$$

$$h := 1..2 \qquad Zq_h := \frac{Vq_h}{Iq_h} \qquad Zq = \begin{pmatrix} 59.98 \\ 59.98 \end{pmatrix}$$

$$T := \sqrt{\mu \cdot \varepsilon} = 3.336 \times 10^{-9} \qquad \frac{Lq}{2} = \begin{pmatrix} 1 \times 10^{-7} \\ 1 \times 10^{-7} \end{pmatrix}$$

$$Lq := T \cdot Zq$$

$$Cq := \frac{T}{Zq} \qquad Cq = \begin{pmatrix} 5.561 \times 10^{-11} \\ 5.561 \times 10^{-11} \end{pmatrix}$$

Figure 3.19 Circuit component values derived for differential-mode configuration

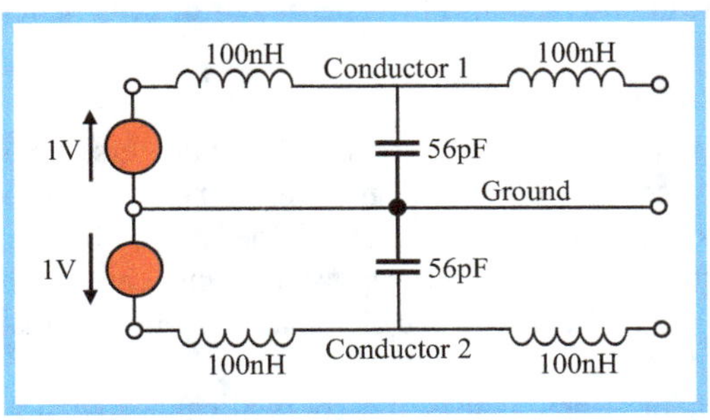

Figure 3.20 Circuit model of balanced configuration

If a dedicated conductor is used to carry the return current, the electromagnetic energy is effectively contained within the region close to the conductor pair. As well as that, the currents induced in the structure by that conductor pair balance out, and the total current reduces to zero.

The data used to create the pictures of current distribution can also be used to derive circuit models which can analyse the interference coupling over the entire frequency range covered by transmission line theory.

3.3 The nature of shielding

3.3.1 Introduction

The screen of a coaxial cable does not form a barrier to electromagnetic interference. It neutralises the effect of that interference.

Transient current flows along the core of a coaxial cable and creates an electromagnetic field which radiates out through the screen into the environment. In doing so, it induces a voltage in the screen which causes a current to flow in the opposite direction to that in the core. The screen current creates one field which radiates into the environment and another which radiates energy back to the core. The fields emanating from the core and the screen into the environment neutralise each other.

If a third conductor is routed alongside the cable and a step voltage applied between this conductor and the screen, current will flow along this conductor and back via the screen. The field emanating from the screen will induce a voltage along the core. But the field emanating from the third conductor will create an equal and opposite voltage.

Such a setup is analysed using the technique of composite conductors. The analysis shows that the shield behaves in the manner described above.

This type of analysis represents the surface of each conductor as a set of elemental conductors. Voltage induced in any conductor creates current in that conductor. Current in any conductor creates a voltage in every other conductor. It is assumed that the electromagnetic field penetrates conducting material.

The analysis is based on the relationships of electromagnetic theory, and the results are in line with observations of the behaviour of actual circuit configurations.

3.3.2 Section under review

This analysis invokes the concept of composite conductors, where each composite is represented by a set of elemental conductors arrayed round its periphery. The circumference of each elemental conductor is defined to be the same as the path length of the section of the surface it represents. This ensures that the strength of the magnetic field on the surface of the element is identical to that on the surface of the surface it simulates.

The section under review is that of a co-axial cable routed alongside a second conductor. The radius *R1* of the core of the coax is 0.33 mm. The radius *R2* of the screen is 1.9 mm, as is the radius *R3* of the third conductor. The separation *sep* between screen and third conductor is 6 mm. Each of these conductors is represented as an array of 12 elemental conductors. So, there are three composite conductors and 36 elemental conductors. Figure 3.21 is a copy of the first page of a Mathcad worksheet. The cross-section is illustrated at the bottom of Figure 3.21.

3.3.3 Primitive currents

Figure 3.22 is a copy of the second page of the worksheet. This calculates the amplitude of the current flowing in every elemental conductor when a 1 V step is applied between the core and screen at one end of the co-axial cable; that is, between composite conductors 1 and 2.

The basic equation used in this computation defines the impedance values assigned to the elemental conductors:

$$Zp_{i,j} = \frac{1}{2 \cdot \pi} \cdot \sqrt{\frac{\mu}{\varepsilon}} \cdot \ln\left(\frac{1}{r_{i,j}}\right) \ \Omega/\text{m} \tag{3.12}$$

$$X1 := 3\cdot10^{-3} \qquad Y1 := 3\cdot10^{-3} \qquad R1 := 0.33\cdot10^{-3} \qquad n_1 := 12$$

$$X2 := X1 \qquad Y2 := Y1 \qquad R2 := 1.9\cdot10^{-3} \qquad n_2 := 12$$

$$X3 := 9\cdot10^{-3} \qquad Y3 := 3\cdot10^{-3} \qquad R3 := 1.9\cdot10^{-3} \qquad n_3 := 12$$

$$N := n_1 + n_2 + n_3 = 36 \qquad sep := X3 - X1 = 6\times10^{-3}$$

$$i := 1..n_1$$

$$\theta_i := \frac{2\pi}{n_1}\cdot(i)$$

$$x_i := X1 + R1\cdot cos(\theta_i) \qquad y_i := Y1 + R1\cdot sin(\theta_i) \qquad r_i := \frac{R1}{n_1}$$

$$i := n_1 + 1..n_1 + n_2$$

$$\theta_i := \frac{2\pi}{n_2}\cdot(i - n_1)$$

$$x_i := X2 + R2\cdot cos(\theta_i) \qquad y_i := Y2 + R2\cdot sin(\theta_i) \qquad r_i := \frac{R2}{n_2}$$

$$i := n_1 + n_2 + 1..N$$

$$\theta_i := \frac{2\pi}{n_3}\cdot(i - n_1 - n_2)$$

$$x_i := X3 + R3\cdot cos(\theta_i) \qquad y_i := Y3 + R3\cdot sin(\theta_i) \qquad r_i := \frac{R3}{n_3}$$

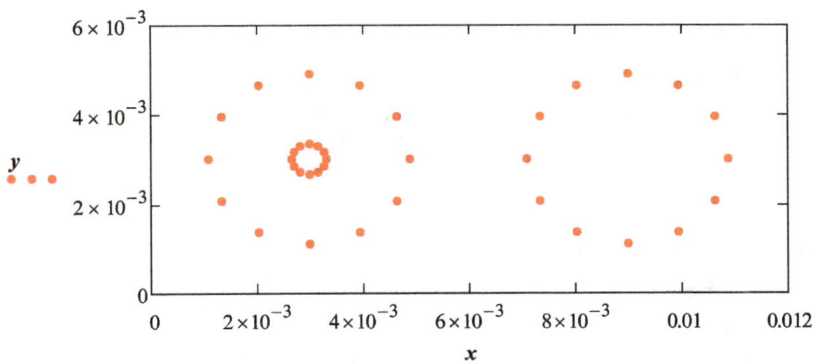

$$i := 1..N \qquad A_{i,1} := x_i\cdot10^3 \qquad A_{i,2} := y_i\cdot10^3 \qquad A_{i,3} := r_i\cdot10^3$$

Figure 3.21 Defining the cross section of the cable assembly

where μ is the permeability of the conductors, ε is the permittivity of the insulation, i and j identify the elemental conductors, and $r_{i,j}$ defines the radial separation between these conductors. If i is not equal to j, $Zp_{i,j}$ is the mutual impedance. If i equals j, $Zp_{i,i}$ defines the self-impedance of conductor i and $r_{i,i}$ is its radius. Equation (3.12) is derived in Section 2.3.

The loop impedance $Zloop_{i,j}$ is the impedance due to current flowing down one conductor and back via the adjacent conductor. The vector *Vloop* is defined as a set of voltage sources between each pair of elemental conductors, where the voltage

$\mu := 4\cdot\pi\cdot10^{-7}$ $\quad\varepsilon := 8.854\cdot10^{-12}$

Fundamental constants

```
Zp := | H ← (1/(2·π))·√(μ/ε)
      | for i ∈ 1..N
      |   for j ∈ 1..N
      |     | h ← x_j − x_i
      |     | v ← y_j − y_i
      |     | rad ← √(h² + v²)
      |     | rad ← r_i  if rad = 0
      |     | Z_i,j ← H·ln(1/rad)
      | Z
```

Calculating the primitive impedance of every elemental conductor

The first action is to calculate the separation between every conductor and every other conductor, *rad*. Then this value is used to calculate the value of the primitive inductance

Then the results are stored in an array of *N* rows by *N* columns

```
Zloop := | for i ∈ 1..N − 1
         |   for j ∈ 1..N − 1
         |     Z_i,j ← Zp_i,j − Zp_i,j+1 − Zp_i+1,j + Zp_i+1,j+1
         | Z
```

Calculating the loop impedance of each adjacent pair of elemental conductors

```
Vloop := | for i ∈ 1..N − 1
         |   | V_i ← 0
         |   | V_i ← 1  if i = n_1
         | V
```

Locating the voltage source in loop n_1

$$Iloop := lsolve(Zloop, Vloop)$$

Calculating the values of all the loop currents

```
Ip := | I_1 ← Iloop_1
      | for i ∈ 2..N − 1
      |   I_i ← Iloop_i − Iloop_i−1
      | I_N ← −Iloop_N−1
      | I
```

Using data on the values of the loop currents to derive values for the primitive currents

$B_{i,1} := x_i\cdot10^3 \quad B_{i,2} := y_i\cdot10^3 \quad B_{i,3} := Ip_i\cdot10^3$

Tabulating the results

Figure 3.22 Calculating the amplitude of the current in each elemental conductor

between elemental conductors 12 and 13 is set at 1 V and all other voltages set at zero. This simulates a voltage of 1 V applied to composite conductor 1 with respect to composite conductor 2.

The function *lsolve(Zloop, Vloop)* calculates the value of the current *Iloop* in every loop between adjacent conductors. When these values are known, it is a simple matter to assign a value to the current *Ip* in every elemental conductor. The

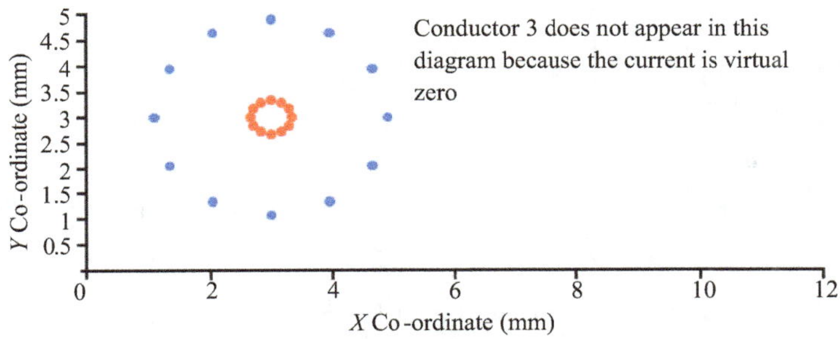

Figure 3.23 Bubble plot of current in each elemental conductor

final computation illustrated in Figure 3.22 creates a table of values of the co-ordinates of each elemental conductor and the current it carries. This table was copied to Microsoft Excel software, which was used to create a bubble plot of the results.

This bubble plot is shown in Figure 3.23. The size of each bubble is proportional to the amplitude of the current in that particular elemental conductor. Co-ordinates for conductor 3 do not appear on this diagram for the simple reason that the total current flow is infinitesimal.

This figure illustrates the fact that current flows forward along the core of the co-axial cable and back via the screen. Since no current flows on conductor 3, it would appear that the screen acts as an impenetrable shield.

It could be reasoned that this is because current only flows along the inner surface of the shield. But this does not apply to braided shields where each strand is exposed to the outer environment as well as to the central core. Nor does it apply to a solid shield, where the field inside the conductor reverberates between inner and outer surfaces. With a solid shield, it would be expected that high frequency current flows on both these surfaces.

The current which flows along the screen creates an electromagnetic field along the outer surface of the coaxial cable. Since this field radiates out into the environment, it is certain to create a voltage along the length of composite conductor 3. But the total current along this conductor is effectively zero. This can be explained by calculating all 12 partial voltages in the composite conductors of the assembly.

3.3.4 Partial voltages

The worksheet illustrated in Figure 3.24 performs the relevant computations.

The vectors ***Start*** and ***End*** at the top of the page define the first and last elemental conductors of those which simulate composite conductors 1, 2, and 3. These are used to calculate the total current flow along each composite. The results are stored in the vector ***Iq***. A current of 9.527 mA flows down the core conductor

$$\textbf{Start} := \begin{pmatrix} 1 \\ n_1 + 1 \\ n_1 + n_2 + 1 \end{pmatrix} = \begin{pmatrix} 1 \\ 13 \\ 25 \end{pmatrix}$$

Defining the first elemental conductor in each composite conductor

$$\textbf{End} := \begin{pmatrix} n_1 \\ n_1 + n_2 \\ N \end{pmatrix} = \begin{pmatrix} 12 \\ 24 \\ 36 \end{pmatrix}$$

Defining the last elemental conductor in each composite conductor

```
Iq :=  | for h ∈ 1..3
       |    | S ← 0
       |    | for i ∈ Start_h..End_h
       |    |    S ← S + Ip_i
       |    | I_h ← S
       | I
```

Calculating the total current flowing along each composite conductor

$$\textbf{Iq} \cdot 10^3 = \begin{pmatrix} 9.527 \\ -9.527 \\ 2.96 \times 10^{-6} \end{pmatrix}$$

$$(\textbf{Iq}_1 + \textbf{Iq}_2) \cdot 10^3 = -2.96 \times 10^{-6}$$

```
vq :=  | for h ∈ 1..3
       |   for k ∈ 1..3
       |      | S ← 0
       |      | for i ∈ Start_h..End_h
       |      |    for j ∈ Start_k..End_k
       |      |       S ← S + Zp_i,j · Ip_j
       |      | V_h,k ← S / n_h
       | V
```

Calculating the contribution made to the overall voltage along each conductor by current in each conductor

$$\textbf{vq} = \begin{pmatrix} 4.579 & -3.579 & 1.136 \times 10^{-6} \\ 3.579 & -3.579 & 1.137 \times 10^{-6} \\ 2.923 & -2.923 & 1.112 \times 10^{-6} \end{pmatrix}$$

```
Vq :=  | for h ∈ 1..3
       |    | S ← 0
       |    | for k ∈ 1..3
       |    |    S ← S + vq_h,k
       |    | V_h ← S
       | V
```

Calculating the voltage developed along every composite conductor

$$\textbf{Vq} = \begin{pmatrix} 1 \\ 2.437 \times 10^{-8} \\ 2.437 \times 10^{-8} \end{pmatrix}$$

Figure 3.24 Computing the partial voltages developed along each conductor

and returns via the screen. The current in the third composite conductor 3 is calculated to be 2.96 nA. In practical terms, the total voltage applied to this conductor by the electromagnetic field is zero.

The next Mathcad function calculates the voltage contributions due to mutual coupling and self-coupling and stores the results in the vector $\boldsymbol{vq}$.

The partial voltages developed along conductor 3 are as follows:

Due to self-coupling $\boldsymbol{vq}_{3,3} = 1\ \mu\text{V}$
Due to current in conductor 1 $\boldsymbol{vq}_{3,1} = 2.932$ V
Due to current in conductor 2 $\boldsymbol{vq}_{3,2} = -2.932$ V

The partial voltages due to mutual coupling are generated by currents flowing in opposite directions in every elemental conductor in composite conductor 3. Since the forward current flow is identical to the return current flow, the net current flow is infinitesimal and so is the total voltage. The vector $\boldsymbol{Vq}$ demonstrates this. The total voltage induced in conductor 3 is calculated to be $\boldsymbol{Vq}_3 = 24$ nV.

Since the current is flowing in both directions in the shield, two opposing electromagnetic waves emanate from the outer surface. These induce two voltages in conductor 3. Since these are equal in magnitude but opposite in direction, they have minimal effect on that conductor.

3.3.5 Voltages along elemental conductors

The effect can be illustrated in detail by calculating the value of each partial voltage in each elemental conductor of composite 3 and tabulating the results. This is recorded by Table 3.1.

This table shows that the current in the coaxial core of the screened cable creates a positive voltage in series with every elemental conductor of composite conductor 3, whilst the current on the cable screen creates a negative voltage in series with every elemental conductor. The net voltage induced in every elemental conductor is negligible. Since the net current flow along every elemental conductor of composite 3 is virtually zero, so is the partial voltage due to that current.

Table 3.1 Partial voltages in the elemental conductors of composite conductor 3

Conductor 3	**Voltage due to current in:**		
Element	**Core**	**Screen**	**Conductor 3**
25	2.78	−2.78	9.37E−07
26	2.823	−2.823	9.47E−07
27	2.895	−2.895	9.58E−07
28	2.992	−2.992	8.55E−07
29	3.092	−3.092	−3.76E−07
30	3.14	−3.14	5.77E−06
31	3.092	−3.092	−3.76E−07
32	2.992	−2.992	8.55E−07
33	2.895	−2.895	9.58E−07
34	2.823	−2.823	9.47E−07
35	2.78	−2.78	9.37E−07
36	2.765	−2.765	9.33E−07

The three partial voltages recorded on the bottom line of the array ***vq*** in Figure 3.24 are the average values of the partial voltages along the elemental conductors of composite 3.

3.3.6 Component values for coaxial cable

Having calculated the values of the currents flowing in each of the composite conductors and the voltage developed along each conductor, it is possible to assign values to the circuit impedances. The computations involved are illustrated in Figure 3.25, which is a copy of the final page of the worksheet.

There are two loops involved. In loop 1, current flows along conductor 1 and flows back along conductor 2. In loop 2, current flows along conductor 2 and returns via conductor 3. The driving force is a voltage of 1 V, applied between the terminations of conductors 1 and 2 at the near-end of the cable assembly. The voltage applied to loop 2 is zero.

The function at the top of Figure 3.25 calculates to values of the loop currents and stores them in the vector ***Iloop***. The total voltage developed in each loop is calculated by adding the voltages developed along send and return conductors. The

$$\mathbf{Iloop} := \left| \begin{array}{l} I_1 \leftarrow \mathbf{Iq}_1 \\ I_2 \leftarrow -\mathbf{Iq}_3 \\ I \end{array} \right. \qquad \mathbf{Iloop} = \begin{pmatrix} 9.527 \times 10^{-3} \\ -2.96 \times 10^{-9} \end{pmatrix} \quad \text{Loop voltages}$$

$$\mathbf{Vloop} := \left| \begin{array}{l} V_1 \leftarrow \mathbf{Vq}_1 - \mathbf{Vq}_2 \\ V_2 \leftarrow \mathbf{Vq}_2 - \mathbf{Vq}_3 \\ V \end{array} \right. \qquad \mathbf{Vloop} = \begin{pmatrix} 1 \\ -2.187 \times 10^{-15} \end{pmatrix} \quad \text{Loop currents}$$

$$T := \sqrt{\mu \cdot \varepsilon} = 3.336 \times 10^{-9} \qquad \text{Time for step to propagate along one metre}$$

$$Zc1 := \frac{\mathbf{Vloop}_1}{\mathbf{Iloop}_1} = 104.959 \qquad \text{Impedance of loop 1, per metre}$$

$$Lc1 := T \cdot Zc1 = 3.501 \times 10^{-7} \qquad \text{Inductance of loop 1, per metre}$$

$$\frac{Lc1}{2} = 1.751 \times 10^{-7} \qquad \text{Component of circuit model}$$

$$Cc1 := \frac{T}{Zc1} = 3.178 \times 10^{-11} \qquad \text{Component of circuit model}$$

$$Ctheory := \frac{2 \cdot \pi \cdot \varepsilon}{\ln\left(\frac{R2}{R1}\right)} = 3.178 \times 10^{-11} \qquad \text{Capacitance derived from textbook theory}$$

Figure 3.25 Calculating the value of the capacitance between composite conductors 1 and 2

minus sign in the function ***Vc*** is because current flowing back via the return conductor is defined as negative.

Since the voltage developed in loop 2 is $\boldsymbol{Vc}_2 = 2 \times 10^{-15}$, there is no point in calculating the impedance of that loop. But there is no problem in assigning a value of 105 Ω to the impedance of loop 1, and this leads to the derivation of a value of 350 nH for the inductance of loop 1 and 32 pF for the capacitance between conductors 1 and 2. The fact that the value of the capacitance is exactly the same as that derived from textbook theory provides a high degree of confidence in the modelling technique.

3.3.7 Susceptibility

External interference can be simulated by removing the voltage source from loop 1 and inserting it in loop 2. The simulation involves copying the worksheet to another file and then modifying the copy. The only change to the main computation is the definition of the function ***Vloop***. This modification is illustrated in Figure 3.26. The voltage on elemental conductor 24 is defined as −1 volt with respect to elemental conductor 25. So, the voltage on conductor 25 is positive with respect to conductor 24. The resultant current distribution is shown by the bubble chart of Figure 3.27. The red circles indicate that the current is flowing forward. Blue circles indicate that the current is flowing from the far end to the near end of the assembly.

```
Vloop :=  | for i ∈ 1..N − 1
          |     | V_i ← 0                      Locating the voltage source
          |     | V_i ← −1  if i = n1 + n2     in loop 2
          | V
```

Figure 3.26 Locating the voltage source between composite conductors 2 and 3

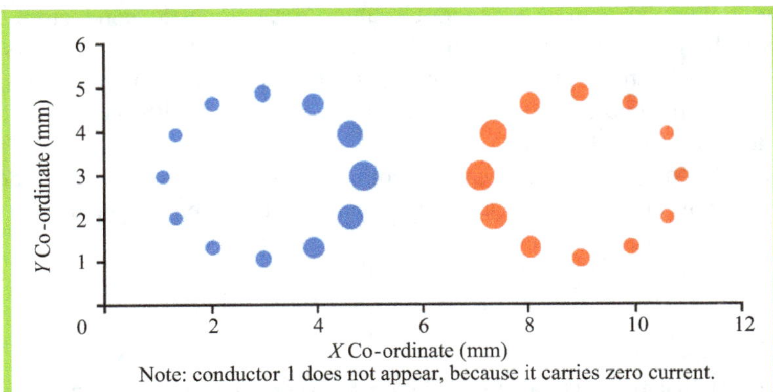

Figure 3.27 Current distribution when 1 V is applied between conductors 3 and 2

```
vq := | for h ∈ 1..3
      |   for k ∈ 1..3
      |     | S ← 0
      |     | for i ∈ Start_h..End_h
      |     |   for j ∈ Start_k..End_k
      |     |     S ← S + Zp_i,j · Ip_j
      |     | V_h,k ← S / n_h
      | V
```

Calculating the contribution made to the overall voltage along each conductor by current in each conductor

$$vq = \begin{pmatrix} 1.423\times10^{-6} & -3.041 & 2.541 \\ 1.112\times10^{-6} & -3.041 & 2.541 \\ -1.496\times10^{-6} & -2.541 & 3.041 \end{pmatrix}$$

```
Vq := | for h ∈ 1..3
      |   | S ← 0
      |   | for k ∈ 1..3
      |   |   S ← S + vq_h,k
      |   | V_h ← S
      | V
```

Calculating the voltage developed along every composite conductor

$$Vq = \begin{pmatrix} -0.5 \\ -0.5 \\ 0.5 \end{pmatrix}$$

Figure 3.28 Computing the partial voltages when voltage source is in loop 2

The partial voltages along each conductor are computed on the third page of the revised worksheet, illustrated in Figure 3.28. This shows that the voltages induced in conductor 1 are identical to those in conductor 2. That is, external electromagnetic interference penetrates the shield and induces a voltage along the core conductor. However, the current in the screen creates a voltage which matches that voltage. From Figure 3.28, $Vq_1 = Vq_2 = -0.5$ V.

3.3.8 External components

Calculating the value of the L and C values for composite conductors 2 and 3 involves the same process as that used to calculate that between conductors 1 and 2. Figure 3.29 illustrates the computations. In this case, the value of $\boldsymbol{Vloop}_1$ is far too small to calculate reliable values for those associated with the loop between composite conductors 1 and 2.

The value derived for *Cc3* can be compared to that derived in textbooks of electromagnetic theory for the capacitance between two tubular conductors:

$$Ctheory = \frac{\pi \cdot \varepsilon}{\ln\left(\frac{b+\sqrt{b^2-r^2}}{r}\right)} = 2.701 \times 10^{-11}\ \text{F/m} \tag{3.13}$$

where $b = \frac{sep}{2}$ and $r = \mathbf{R2} = 1.9 \times 10^{-11}$.

The fact that there is close correlation between *Cc3* and *Ctheory* provides a high level of confidence in the method of composite conductors.

3.3.9 Circuit model

Having computed values for the circuit components of both the internal and external loops, there is now enough information to construct a circuit model of the assembly under review as depicted in Figure 3.30.

The values for *Lc1* and *Cc1* are derived from Figure 3.25, whilst *Lc3* and *Cc3* are derived from Figure 3.29.

It would seem that the only component coupling energy through the shield is the resistance *Rc2* of the screen. But that is an illusion. The voltage induced in the shield by the interfering source is cancelled out by the self-induced voltage.

$$\mathbf{Iloop} := \begin{vmatrix} I_1 \leftarrow \mathbf{Iq}_1 \\ I_2 \leftarrow -\mathbf{Iq}_3 \\ I \end{vmatrix} \qquad \mathbf{Iloop} = \begin{pmatrix} 2.96 \times 10^{-9} \\ -8.094 \times 10^{-3} \end{pmatrix}$$

Loop currents

$$\mathbf{Vloop} := \begin{vmatrix} V_1 \leftarrow \mathbf{Vq}_1 - \mathbf{Vq}_2 \\ V_2 \leftarrow \mathbf{Vq}_2 - \mathbf{Vq}_3 \\ V \end{vmatrix} \qquad \mathbf{Vloop} = \begin{pmatrix} 2.665 \times 10^{-15} \\ -1 \end{pmatrix}$$

Loop voltages

$$T := \sqrt{\mu \cdot \varepsilon} = 3.336 \times 10^{-9}$$

Time for step to propagate along one metre

$$Zc3 := \frac{\mathbf{Vloop}_2}{\mathbf{Iloop}_2} = 123.543$$

Impedance of loop 1, per metre

$$Lc3 := T \cdot Zc3 = 4.121 \times 10^{-7}$$

Inductance of loop 1, per metre

$$\frac{Lc3}{2} = 2.06 \times 10^{-7}$$

Component of circuit model

$$Cc3 := \frac{T}{Zc3} = 2.7 \times 10^{-11}$$

Component of circuit model

Figure 3.29 Computing values for inductance and capacitance of external loop

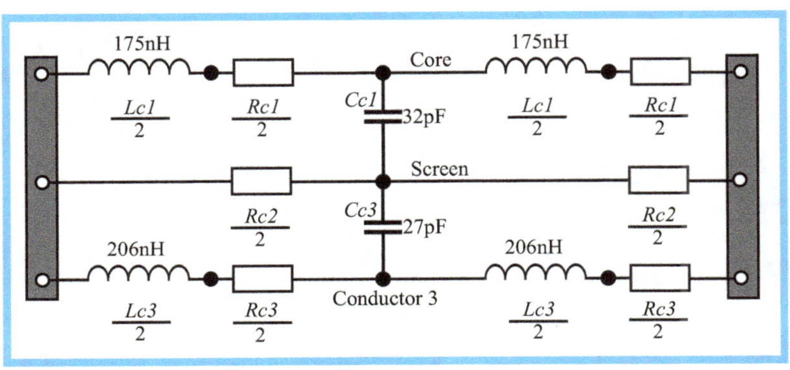

Figure 3.30 Circuit model of 1 metre length of cable assembly under review

3.3.10 Conclusion

It has been shown that the action of the screen is to neutralise the interference, whether the source of that interference is internal or external to the signal carried by the coaxial cable.

Since the analysis has been derived from electromagnetic theory and since it predicts the actual behaviour of the system under review, it can reasonably be concluded that the mechanisms involved in the propagation of EMI are as described in the introduction.

A circuit model has been derived to simulate the coupling between the internal and external loops.

The technique can be developed to simulate the cross coupling between the differential mode and common mode loops of any cable assembly.

Chapter 4
Antenna models

4.1 The radiation resistance

4.1.1 Introduction

At its half-wave frequency, a dipole antenna provides a load on the RF transmitter which is purely resistive; the radiation resistance. At other frequencies, the antenna behaves as a series LCR circuit.

If a voltage source is located at the centre of a long conductor, this assembly will behave in essentially the same way. The current flowing in the conductor is a measure of the electromagnetic energy delivered to the environment. A circuit model is developed which accurately simulates the frequency response of such an assembly. It is entirely possible to extend the use of the technique of circuit modelling to analyse radiated interference.

4.1.2 The dipole antenna

The basic construction of a dipole antenna is shown in Figure 4.1. Each monopole acts in the same way as a transmission line with an open-circuit at the far end. If the frequency of the output of the transmitter is the same as the quarter-wave resonance of each monopole, the amplitude of the current will peak at the junction between antenna and feeder cable but will be zero at the tip of the monopole. The distribution of current along each section will be that of a cosine wave.

It is then possible to define the magnetic potential at any point P in the surrounding space by summing the effects of elements dz along the length of each conductor.

Invoking the mathematics of electromagnetic field propagation leads to the derivation of formulae for the value of the longitudinal component of the magnetic field and the latitudinal component of the electric field on the surface of a sphere which encloses the antenna. Multiplying these two components gives a value for the power density S at every point on the sphere. The power density of a spherical surface is shown in Figure 4.2. Summing the values of the power density over the entire surface gives a value for the total power P delivered to the environment. Since power is equal to a resistance multiplied by the square of the current, a value can be calculated for that resistance.

$$Rrad = 73\ \Omega,\ \text{approximately} \tag{4.1}$$

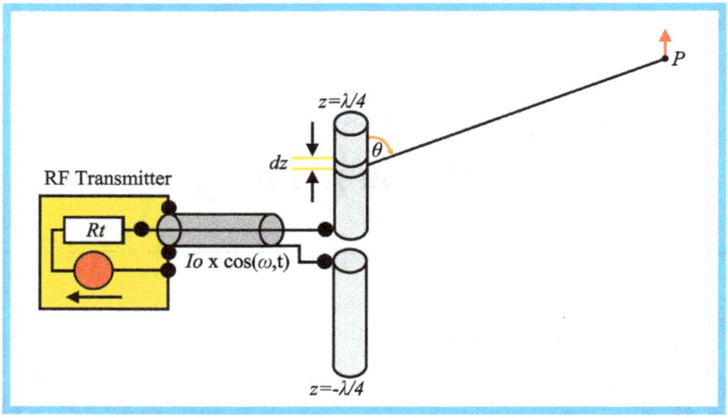

Figure 4.1 Analysis of the half-wave dipole

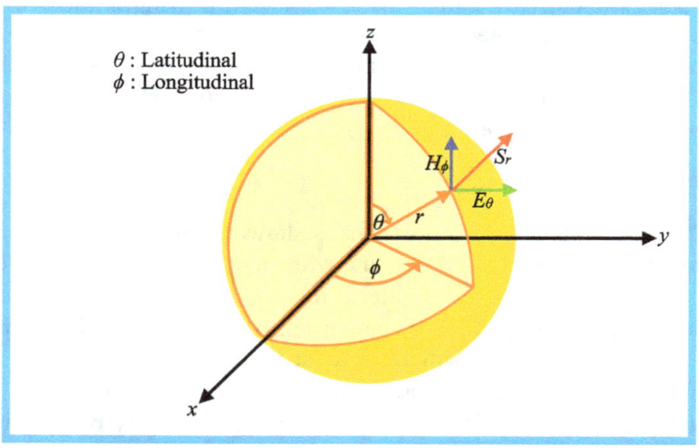

Figure 4.2 Power density on a spherical surface

4.1.3 Isolated conductor

If the configuration is changed to that of a single conductor with a voltage source *Vgen* located at the centre and each monopole is represented as a T-network, this leads to the circuit model of Figure 4.3. At resonance, a series LCR circuit behaves as a pure resistance.

To cater for the fact that the current varies along the length of the conductor, it is necessary to invoke the use of distributed parameters. The model of Figure 4.3 can be transformed into that of Figure 4.4.

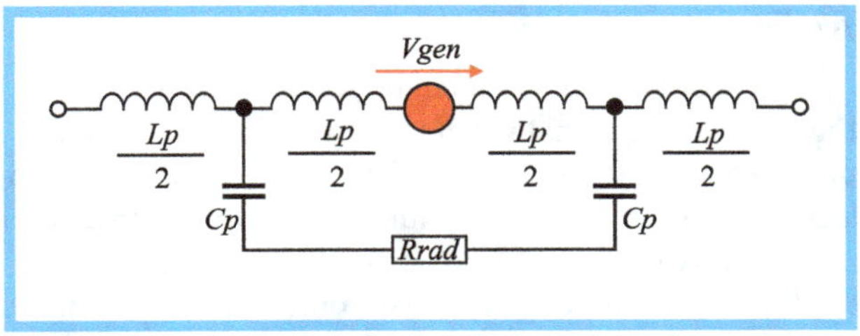

Figure 4.3 *Circuit model of the coupling between a single conductor and the environment*

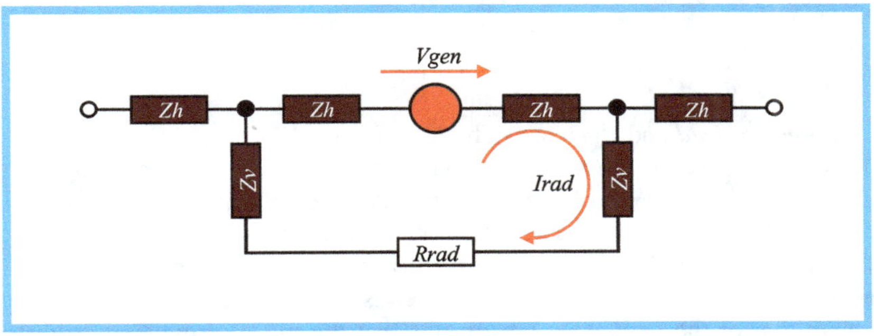

Figure 4.4 *Distributed parameter model of Figure 4.3*

The equation relating current *Irad* to voltage *Vgen* can be derived from Figure 4.4:

$$Vgen = (2 \cdot Zh + 2 \cdot Zv + Rrad) \cdot Irad \tag{4.2}$$

where

$$\begin{aligned} Zh &= Zo \cdot \tanh\left(\frac{\theta}{2}\right) \\ Zv &= Zo \cdot \operatorname{cosech}(\theta) \end{aligned} \tag{4.3}$$

And

$$\begin{aligned} Zo &= \sqrt{\frac{Rp + j \cdot \omega \cdot Lp}{Gp + j \cdot \omega \cdot Cp}} \\ \theta &= \sqrt{(Rp + j \cdot \omega \cdot Lp) \cdot (Gp + j \cdot \omega \cdot Cp)} \end{aligned} \tag{4.4}$$

The derivation of the above equations is provided in Section 2.2. In this particular application, it is assumed that the series resistance of the conductors is minimal and that the conductance of the capacitors is zero.

4.1.4 Measurement

The response of the configuration under review can be measured by the setup illustrated in Figure 4.5. The signal generator provides a sinusoidal waveform over a range of frequencies and the voltage transformer acts as a voltage source in series with the conductor under test. A monitor turn on this transformer allows the peak-to-peak amplitude of this voltage to be measured by channel 1 of the oscilloscope.

The current flowing in the conductor is monitored by the current transformer which provides a signal input to channel 2 of the oscilloscope. The voltages monitored by channel 1 and 2 are recorded at each frequency.

This data is recorded on a Mathcad worksheet which calculates the ratio *Yt* of the current *Irad* in the conductor to the applied voltage *Vgen* at each frequency. A copy of this page is illustrated in Figure 4.6.

$$Yt = \frac{Irad}{Vgen} \tag{4.5}$$

The parameters *Irad* and *Vgen* are defined in Figure 4.4.

The ***data*** array on the worksheet page of Figure 4.6 is a record of the test results, and the program used to process it is defined on the right-hand side.

The *L*, *C*, and *R* variables are derived from the circuit model of the current transformer. This model had been derived from a similar set of tests of the frequency response of that transformer and the response of the model correlates closely with that of the hardware. This means that the model provides a calibration curve relating the voltage at channel 2 *Vch2* to the current *Irad* flowing in the conductor. The calibration of this device is described in Section 2.5.

The variable *s* identifies the number of the row in the ***data*** array. F_s is the spot frequency of that particular record. The program to the right of Yt_s calculates the admittance of the loop-under-test at that frequency.

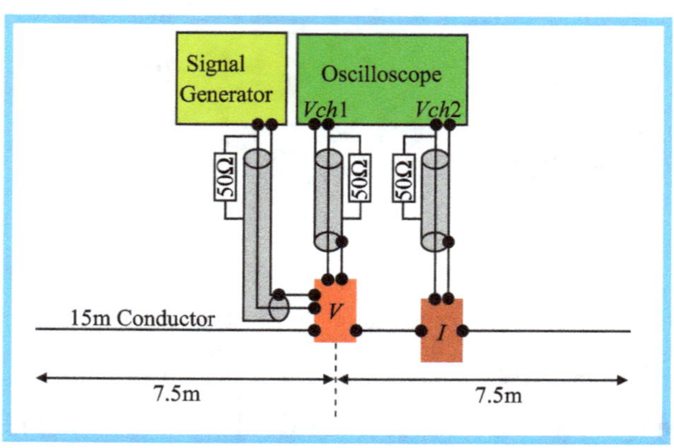

Figure 4.5 Set-up to measure radiation resistance of an isolated conductor

$$\mathbf{data} := \begin{pmatrix} 1 & 400 & 0.4 \\ 2 & 400 & 0.4 \\ 3 & 400 & 1 \\ 4 & 395 & 1.9 \\ 5 & 390 & 3.4 \\ 6 & 390 & 5.9 \\ 6.5 & 385 & 8.4 \\ 7 & 385 & 14.5 \\ 7.5 & 380 & 26 \\ 7.83 & 360 & 36 \\ 8 & 350 & 31.5 \\ 8.5 & 345 & 16.5 \\ 9 & 345 & 10.5 \\ 10 & 345 & 6.2 \\ 11 & 345 & 6 \\ 12 & 345 & 4.4 \\ 13 & 340 & 3.4 \\ 14 & 340 & 3 \\ 15 & 330 & 3 \\ 16 & 330 & 2.4 \\ 17 & 320 & 1.4 \\ 18 & 315 & 2.2 \\ 19 & 310 & 3.4 \\ 19.6 & 305 & 4.2 \end{pmatrix}$$

Column 1: Frequency, MHz
Column 2: Channel 1 voltage, mV
Column 3: Channel 2 voltage, mV

From Figure 2.29 of Section 2.5
'The Current Monitor Transformer':

$R1 := 300\,\Omega \qquad R2 := 51\Omega$

$R3 := 50\Omega \qquad R4 := 850\Omega$

$L1 := 200 \cdot 10^{-6} \qquad C1 := 60 \cdot 10^{-12}$

$Turns := 10$

$s := 1 .. rows(\mathbf{data}) \qquad F_s := \mathbf{data}_{s,1} \cdot 10^6$

$$Yt_s := \left| \begin{array}{l} \omega \leftarrow 2 \cdot \pi \cdot F_s \\ Z1 \leftarrow R4 + \dfrac{1}{j \cdot \omega \cdot C1} \\ Y2 \leftarrow \dfrac{1}{R1} + \dfrac{1}{R2} + \dfrac{1}{R3} + \dfrac{1}{j \cdot \omega \cdot L1} + \dfrac{1}{Z1} \\ Vch2 \leftarrow \mathbf{data}_{s,3} \cdot 10^{-3} \\ Isec \leftarrow |Y2| \cdot Vch2 \\ Irad \leftarrow Isec \cdot Turns \\ Vch1 \leftarrow \mathbf{data}_{s,2} \cdot 10^{-3} \\ Vgen \leftarrow \dfrac{50 + 51}{50} \cdot Vch1 \\ \dfrac{Irad}{Vgen} \end{array} \right.$$

Figure 4.6 Worksheet used to calculate the response of the transfer admittance Yt

The first three steps in the program calculate the value of the admittance *Y2* of the circuit model of the current transformer at that particular spot frequency. If the amplitude of the output voltage of the secondary of that transformer was 1 Volt, this would be a measure of the current in the secondary winding. Since this voltage is actually *Vch2*, the current in the secondary *Isec* can be calculated using

$$Isec = |Y2| \cdot Vch2 \tag{4.6}$$

Since the loop-under-test constitutes a single turn and the number of turns on the secondary is *Turns*, the current *Irad* in the primary winding is easily calculated:

$$Irad = Isec \cdot Turns \tag{4.7}$$

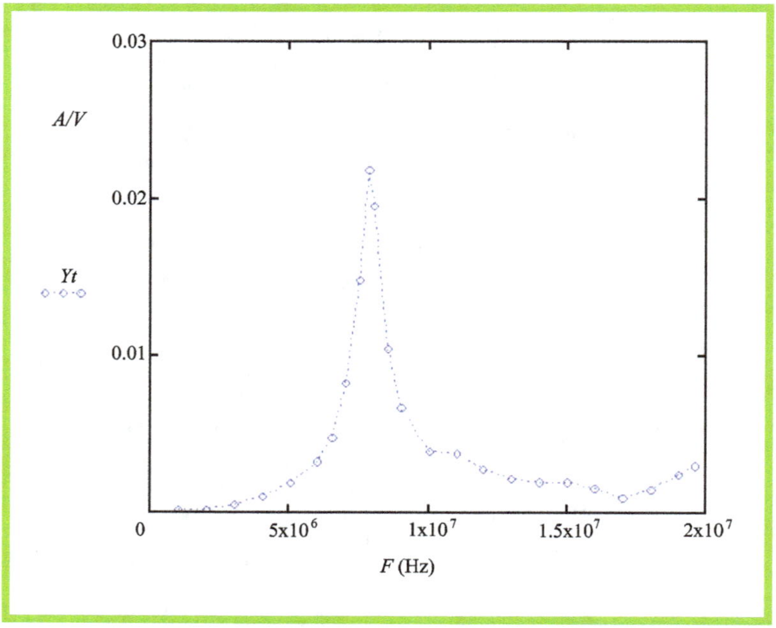

Figure 4.7 Radiated emission characteristic of a 15 m conductor

The design of the circuitry of the monitor turn of the voltage transformer includes a 51 Ω series resistance. This ensures that the output impedance matches the input impedance of the oscilloscope. Hence, the relationship between the voltage *Vgen* induced in the loop-under-test and the voltage *Vch1* monitored by channel 1 of the oscilloscope is as follows:

$$Vgen = \frac{50 + 51}{50} \cdot Vch1 \tag{4.8}$$

The transfer admittance *Yt* of the loop-under-test at the spot frequency *F* is as defined by (4.5). The parameter *Yt* can be defined as the transfer admittance derived from test results and is the ratio of current output to voltage input. It is a measure of the current delivered to the environment. That is, it defines the radiation emission characteristics of the conductor-under-test over the range of frequencies where it is most likely to cause EMC problems.

Figure 4.7 is a graph which illustrates the frequency response. The blue circles are the admittance values calculated at each spot frequency.

4.1.5 Modelling

The next task is to assign values to the circuit model of Figure 4.3, and this is done using the Mathcad worksheet of Figure 4.8. It is assumed that the series resistance

$$Rp := 0.2 \qquad Gp := 0 \qquad Fq := 7.83 \cdot 10^{6}$$

$$Rrad := 45 \qquad Lp := 12 \cdot 10^{-6} \qquad Cp := \frac{1}{16 \cdot Fq^{2} \cdot Lp} = 8.495 \times 10^{-11}$$

$$i := 1 .. 200 \qquad f_i := i \cdot 10^{5}$$

$$Ym_i := \left| \begin{array}{l} \omega \leftarrow 2 \cdot \pi \cdot f_i \\ \theta \leftarrow \sqrt{(Rp + j \cdot \omega \cdot Lp) \cdot (Gp + j \cdot \omega \cdot Cp)} \\ Zo \leftarrow \sqrt{\dfrac{Rp + j \cdot \omega \cdot Lp}{Gp + j \cdot \omega \cdot Cp}} \\ Zh \leftarrow Zo \cdot tanh\left(\dfrac{\theta}{2}\right) \\ Zv \leftarrow Zo \cdot csch(\theta) \\ Z3 \leftarrow 2 \cdot Zh + 2 \cdot Zv + Rrad \\ \dfrac{1}{|Z3|} \end{array} \right.$$

Figure 4.8 Second page of worksheet. Assigning values to the circuit model.

Rp of the conductor is negligible. It is assumed that the conductance *Gp* is also zero. (Conductance is due to a resistance in parallel with each capacitance.)

There are two unknown parameters; the radiation resistance *Rrad* and the primitive inductance *Lp*. The capacitance *Cp* can be described as a dependent parameter, since it is a function of the quarter-wave resonant frequency *Fq* and the inductance.

The relationship is derived from the following equation:

$$\sqrt{Lp \cdot Cp} = T = \frac{1}{4 \cdot Fq} \tag{4.9}$$

where *T* is the time taken for a pulse to propagate along the length of a monopole, and *Fq* is the quarter-wave frequency. Since *Fq* is defined by the peak frequency in the curve of Figure 4.7, it is possible to calculate the value of *Cp* by re-arranging (4.9).

The third row of this page of the worksheet defines the range of frequencies to be simulated; from 1 MHz to 20 MHz in steps of 100 kHz. The program to the right of Ym_i uses (4.4), (4.3), and (4.2) to calculate a value for the impedance *Z3* at each frequency. The amplitude of the inverse of *Z3* gives a value for the magnitude of the admittance *Ym* derived from the model.

By assigning guess values to the parameters *Lp* and *Rrad*, running this program to calculate the frequency response curve for *Ym*, plotting the results on the same graph as *Yt*, it is possible to create a curve which is similar in shape to that derived from test results.

Altering the value of *Rrad* changes the amplitude of the peak of *Ym*. It is then possible to use the technique of successive approximations to bring the two peaks close together. Applying the same process to *Lp* allows the shapes of the two curves

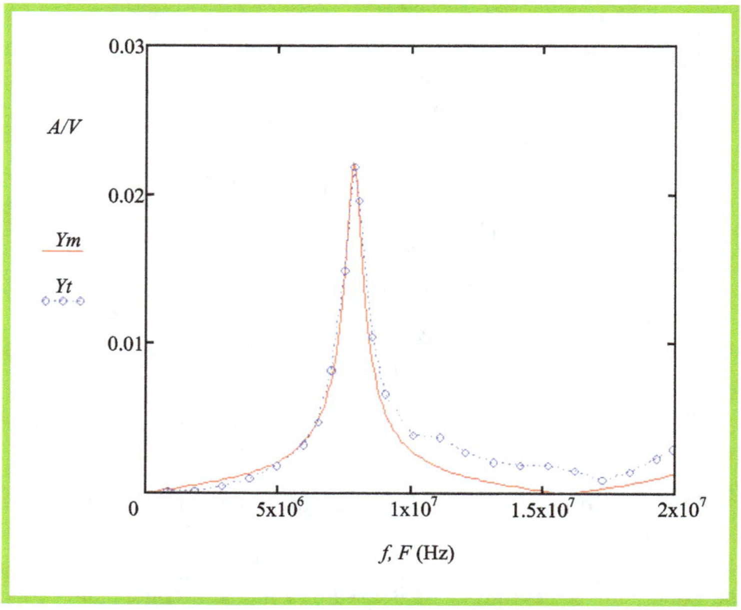

Figure 4.9 Correlating the theoretical results with the actual results

to correlate quite closely. The technique is similar to that used to solve non-linear equations.

The end result of the process is shown in Figure 4.9. The blue circles are a record of the test results, *Yt*. The red curve illustrates the response of the circuit model.

4.1.6 Definitive model

Good correlation has been achieved between the simulation and the actual results up to and beyond the frequency of resonance.

The actual results indicate that the radiated current is higher than theory would predict at frequencies higher than 10 MHz. This is because the conductor is a length of conductor routed through the house and out of the windows on either side. There are many curves and kinks in the path followed by the current as it flows forward and back. These discontinuities create reflections which cause minor resonances in the response.

Since a good correlation has been achieved, the values of *Lp*, *Cp*, and *Rrad* can be copied from the definitions at the top of the second page of the worksheet (Figure 4.8). Applying these values to the components of Figure 4.3 leads to Figure 4.10. This can be described as a definitive circuit model of the assembly-under-review. It effectively defines the EMI characteristics of this particular cable.

The value of the radiation resistance is significantly less than the 73 Ω predicted by textbook theory. This is because the current arriving back at the generator

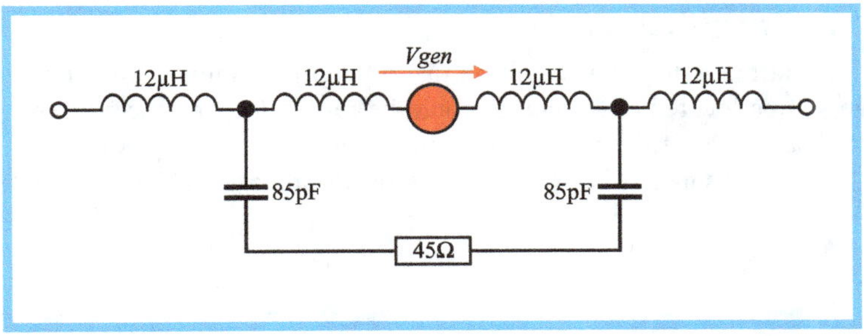

Figure 4.10 Definitive circuit model of conductor under review

from one monopole continues to flow into the other monopole. So energy is stored in the reactive components. The only place for this energy to go is out into the environment. With a dipole designed as a transmitting antenna, the system configuration is as shown in Figure 4.1. The current arriving back at the centre of the antenna is absorbed by the source resistance *Rt* of the RF Transmitter.

4.1.7 Conclusion

This section has shown that a circuit model can be created to simulate the radiated emission of an isolated length of cable over a wide range of frequencies; a range which encompasses worst-case conditions where interference reaches a maximum level.

Full details of the test method, test results, data processing, and the creation of the definitive model have been provided. It is possible for any electronic equipment designer to repeat this demonstration.

4.2 The virtual conductor

4.2.1 Introduction

A circuit model is developed to simulate the emission from a twin-conductor cable. Antenna-mode current flows out into a virtual conductor.

Three of the building blocks of all circuit models are the primitive inductance, the primitive capacitance, and the series resistance. The primitive inductance relates the energy level of the magnetic field caused by current in a conductor to the voltage developed along that conductor. The primitive capacitance relates the voltage on the surface of a conductor to the electric field.

The series resistor relates the current and voltage along a conductor to the power loss due to thermal effects.

The fourth building block is the radiation resistance. This simulates the power loss due to current flowing out into the environment.

This section brings all four parameters together in the form of a simple network.

4.2.2 The single conductor

Section 4.1 started with the assumption that a length of single-core conductor with a voltage source located at the centre could be represented by the circuit model of Figure 4.11(a). Then it described a test which enabled the values of all the components to be measured. The focus here is on the theoretical derivation for such a model.

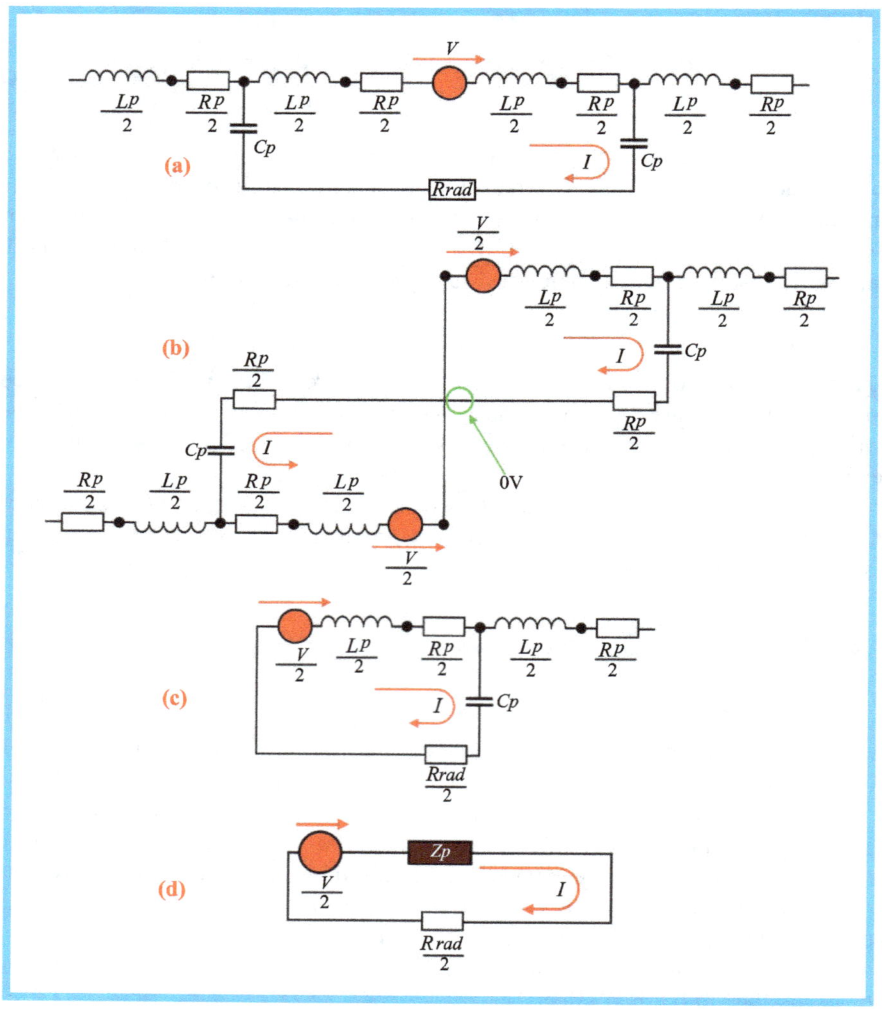

Figure 4.11 Simplifying the model of a single conductor: (a) the basic model, (b) separating the two sections, (c) focussing on the right-hand section, and (d) simulating the reactive components as an impedance

Section 2.3 derives formulae for the reactive components of Figure 4.11a:

$$Lp = \frac{\mu \cdot l}{2 \cdot \pi} \cdot \ln\left(\frac{l}{r}\right) \tag{4.10}$$

$$Cp = \frac{2 \cdot \pi \cdot \varepsilon \cdot l}{\ln\left(\frac{l}{r}\right)} \tag{4.11}$$

where r is the radius of the conductor, μ is the permeability of the conductor, ε is the permittivity of the insulation, and l is the length of the conductor-under-review. The parameter *Rrad* represents the effective resistance of the environment. Voltage developed across *Rrad* is a measure of the energy loss due to radiation. The parameter *Rp* simulates the series resistance of each section of the conductor.

If the voltage along each section is assumed to be the same, the voltage at the centre can be defined as zero volts. Figure 4.11(b) illustrates this.

It is then possible to focus on the behaviour of the right hand section, as illustrated by the equivalent circuit of Figure 4.11(c).

If it is assumed that the model is lossless, it can be simplified further, to that shown in Figure 4.11(d), where

$$Zp = \sqrt{\frac{Lp}{Cp}} \tag{4.12}$$

The parameter *Zp* correlates with the characteristic impedance *Zo* of a lossless transmission line.

4.2.3 *Relating L and C to Z*

Section 2.2 describes the derivation of (4.12). It also relates the characteristic impedance of a lossless line to the time T taken by the front edge of a signal to propagate along the length l.

$$T = \sqrt{Lp \cdot Cp} \tag{4.13}$$

From (4.12)

$$Cp = \frac{Lp}{Zp^2} \tag{4.14}$$

Substituting for *Cp* in (4.13)

$$T = \frac{Lp}{Zp}$$

Hence

$$Lp = T \cdot Zp$$

Substituting for Lp in (4.14)

$$Cp = \frac{T}{Zp}$$

Hence

$$Zp = \frac{T}{Cp} \tag{4.15}$$

And

$$Zp = \frac{Lp}{T} \tag{4.16}$$

4.2.4 Twin conductor assembly

Taking the model of Figure 4.11(d) as a template for each conductor and ignoring the radiation resistance, the simplified model of a twin-conductor assembly becomes as Figure 4.12(a). This is not a circuit model. Rather, it is a depiction of the relationships of electromagnetic theory:

$$\begin{aligned} Vp_1 &= Zp_{1,1} \cdot Ip_1 + Zp_{1,2} \cdot Ip_2 \\ Vp_2 &= Zp_{2,1} \cdot Ip_1 + Zp_{2,2} \cdot Ip_2 \end{aligned} \tag{4.17}$$

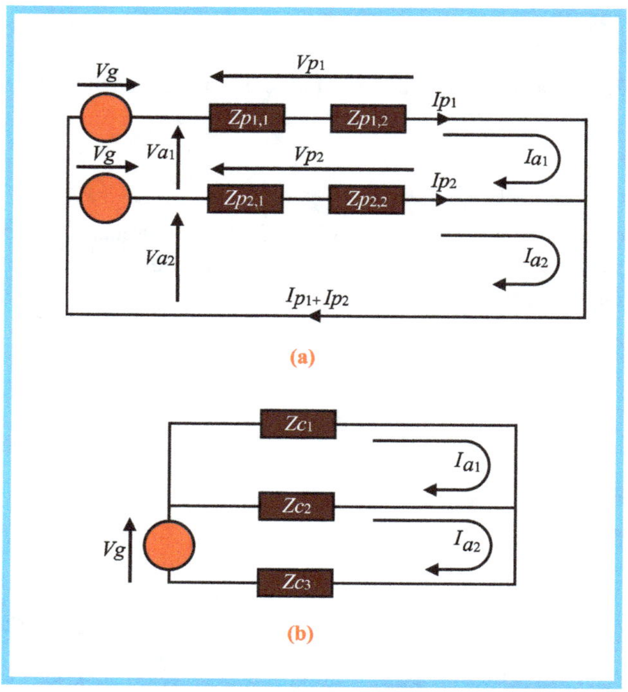

Figure 4.12 Creating a circuit model of a twin-conductor assembly: (a) diagrammatic representation of the primitive equations and (b) circuit model derived from the primitive equations

Since the impedance parameters in (4.17) are the simplest possible, they are defined as 'primitives'. They relate the voltage along each conductor to the energy level of the electromagnetic field created by the currents in both conductors.

In electrical systems, the currents and voltages monitored by the test equipment are the loop currents and loop voltages, and the relationship between primitive parameters and loop parameters is illustrated also in Figure 4.12(a).

The relationship between the currents is defined as

$$\begin{aligned} Ip_1 &= Ia_1 \\ Ip_2 &= Ia_2 - Ia_1 \end{aligned} \tag{4.18}$$

The relationship between the voltages is as follows:

$$\begin{aligned} Va_1 &= Vp_1 - Vp_2 \\ Va_2 &= Vp_2 \end{aligned} \tag{4.19}$$

Using (4.18) to substitute for Ip_1 and Ip_2 in (4.17)

$$\begin{aligned} Vp_1 &= Zp_{1,1} \cdot Ia_1 + Zp_{1,2} \cdot (Ia_2 - Ia_1) \\ Vp_2 &= Zp_{2,1} \cdot Ia_1 + Zp_{2,2} \cdot (Ia_2 - Ia_1) \end{aligned}$$

Rearranging:

$$\begin{aligned} Vp_1 &= \left(Zp_{1,1} - Zp_{1,2}\right) \cdot Ia_1 + Zp_{1,2} \cdot Ia_2 \\ Vp_2 &= \left(Zp_{2,1} - Zp_{2,2}\right) \cdot Ia_1 + Zp_{2,2} \cdot Ia_2 \end{aligned}$$

Using (4.19) to determine the loop voltages

$$\begin{aligned} Va_1 &= \left(Zp_{1,1} - Zp_{1,2} - Zp_{2,1} + Zp_{2,2}\right) \cdot Ia_1 + \left(Zp_{1,2} - Zp_{2,2}\right) \cdot Ia_2 \\ Va_2 &= \left(Zp_{2,1} - Zp_{2,2}\right) \cdot Ia_1 + Zp_{2,2} \cdot Ia_2 \end{aligned} \tag{4.20}$$

Defining the loop impedances as

$$\begin{aligned} Za_{1,1} &= Zp_{1,1} - Zp_{1,2} - Zp_{2,1} + Zp_{2,2} \\ Za_{1,2} &= Zp_{1,2} - Zp_{2,2} \\ Za_{2,1} &= Zp_{2,1} - Zp_{2,2} \\ Za_{2,2} &= Zp_{2,2} \end{aligned} \tag{4.21}$$

leads to the loop equations:

$$\begin{aligned} Va_1 &= Za_{1,1} \cdot Ia_1 + Za_{1,2} \cdot Ia_2 \\ Va_2 &= Za_{2,1} \cdot Ia_1 + Za_{2,2} \cdot Ia_2 \end{aligned} \tag{4.22}$$

For a two-conductor cable,

$$Zp_{1,2} = Zp_{2,1}$$

So, from (4.21)

$$Za_{1,2} = Za_{2,1} \tag{4.23}$$

Equation (4.22) is similar, but not identical, to the equations of circuit theory. At this point, it was necessary to apply lateral thinking and postulate the existence of a circuit model which can be used to create a similar pair of equations. This task does not take long. The mesh equations for Figure 4.12(b) are as follows:

$$\begin{aligned} Va_1 &= (Zc_1 + Zc_2)\cdot Ia_1 - Zc_2 \cdot Ia_2 \\ Va_2 &= -Zc_2 \cdot Ia_1 + (Zc_2 + Zc_3)\cdot Ia_2 \end{aligned} \tag{4.24}$$

Correlating (4.22) and (4.24) enables loop impedances to be related to the circuit impedances:

$$\begin{aligned} Za_{1,1} &= Zc_1 + Zc_2 \\ Za_{1,2} &= -Zc_2 \\ Za_{2,2} &= Zc_2 + Zc_3 \end{aligned}$$

Defining circuit impedances in terms of loop impedances

$$\begin{aligned} Zc_1 &= Za_{1,1} + Za_{1,2} \\ Zc_2 &= -Za_{1,2} \\ Zc_3 &= Za_{2,2} + Za_{1,2} \end{aligned} \tag{4.25}$$

Using (4.21) to substitute primitive impedances for loop impedances:

$$\begin{aligned} Zc_1 &= \left(Zp_{1,1} - Zp_{1,2} - Zp_{2,1} + Zp_{2,2}\right) + \left(Zp_{1,2} - Zp_{2,2}\right) \\ Zc_2 &= Zp_{2,2} - Zp_{1,2} \\ Zc_3 &= Zp_{2,2} + \left(Zp_{1,2} - Zp_{2,2}\right) \end{aligned}$$

This leads to

$$\begin{aligned} Zc_1 &= Zp_{1,1} - Zp_{1,2} \\ Zc_2 &= Zp_{2,2} - Zp_{1,2} \\ Zc_3 &= Zp_{1,2} \end{aligned} \tag{4.26}$$

4.2.5 Circuit parameters

Using (4.16) to substitute for Zp and Zc in (4.26)

$$\begin{aligned} Lc_1 &= Lp_{1,1} - Lp_{1,2} \\ Lc_2 &= Lp_{2,2} - Lp_{1,2} \\ Lc_3 &= Lp_{1,2} \end{aligned} \tag{4.27}$$

For two or more conductors,

$$Lp_{i,j} = \frac{\mu \cdot l}{2 \cdot \pi} \cdot \ln\left(\frac{l}{r_{i,j}}\right) \tag{4.28}$$

And

$$Cp_{i,j} = \frac{2 \cdot \pi \cdot \varepsilon \cdot l}{\ln\left(\frac{l}{r_{i,j}}\right)} \tag{4.29}$$

where $r_{i,j}$ is the separation between the centres of conductors i and j, and $r_{i,i}$ is the radius of conductor i. Using (4.28) to substitute for the primitive parameters in (4.27)

$$\begin{aligned} Lc_1 &= \frac{\mu \cdot l}{2 \cdot \pi} \cdot \ln\left(\frac{r_{1,2}}{r_{1,1}}\right) \\ Lc_2 &= \frac{\mu \cdot l}{2 \cdot \pi} \cdot \ln\left(\frac{r_{1,2}}{r_{2,2}}\right) \\ Lc_3 &= \frac{\mu \cdot l}{2 \cdot \pi} \cdot \ln\left(\frac{l}{r_{1,2}}\right) \end{aligned} \tag{4.30}$$

Using (4.15) to substitute for the primitive parameters in (4.27) and rearranging

$$\begin{aligned} \frac{1}{Cc_1} &= \frac{1}{Cp_{1,1}} - \frac{1}{Cp_{1,2}} \\ \frac{1}{Cc_2} &= \frac{1}{Cp_{2,2}} - \frac{1}{Cp_{1,2}} \\ \frac{1}{Cc_3} &= \frac{1}{Cp_{1,2}} \end{aligned} \tag{4.31}$$

Using (4.29) to substitute for the primitive parameters in (4.31)

$$\begin{aligned} Cc_1 &= \frac{2 \cdot \pi \cdot \varepsilon \cdot l}{\ln\left(\frac{r_{1,2}}{r_{1,1}}\right)} \\ Cc_2 &= \frac{2 \cdot \pi \cdot \varepsilon \cdot l}{\ln\left(\frac{r_{1,2}}{r_{2,2}}\right)} \\ Cc_1 &= \frac{2 \cdot \pi \cdot \varepsilon \cdot l}{\ln\left(\frac{l}{r_{1,2}}\right)} \end{aligned} \tag{4.32}$$

4.2.6 Conclusion

Replacing each *Z*-parameter of Figure 4.12(b) with the T-network of *Lc* and *Cc* parameters, adding the resistors *Rc* to cater for the fact that the conductors possess series resistance, and incorporating the left hand section into the model lead to the complete circuit model: Figure 4.13. The voltage sources can be created by a transformer coupled round both conductors.

The striking feature of this model is that, for each monopole, it includes a T-network of L, C, and R components: arrayed in exactly the same way as those representing an actual conductor. A value can be assigned to every one of them. Since this conductor does not exist in the real world, the most appropriate name must be the 'Virtual Conductor'.

There are two current loops involved; the differential mode loop which carries signal (or power) current along the send conductor and back via the return

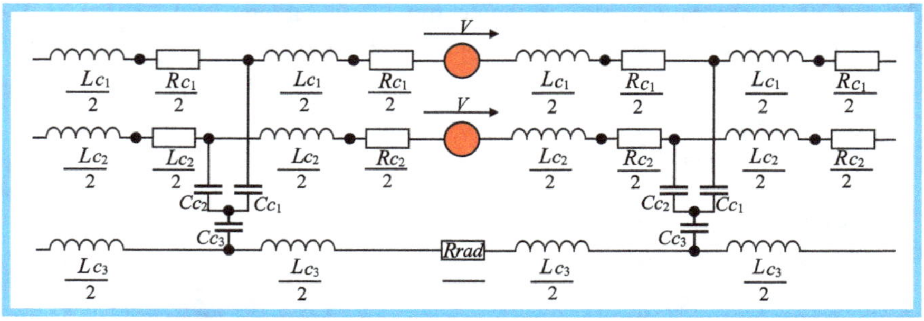

Figure 4.13 The complete circuit model of a twin-conductor assembly

conductor; and the antenna-mode loop which carries current along both conductors and out into the environment.

A significant feature of this model is that the parameters assigned to each conductor are exactly the same as those derived for a twin-conductor assembly by traditional analytical techniques, where it is assumed that antenna-mode current does not exist. The reason for this is that partial currents flow in both directions along every conductor. If there is any imbalance in the forward flowing currents (or the backward-flowing currents), the only place for the unbalanced current to flow is out into the environment.

The existence of partial currents can be catered for in circuit models, since the Principle of Superposition allows for just such an eventuality.

It is useful to note that the effective radius of the virtual conductor is the separation between the two actual conductors; $r_{1,2}$.

4.3 Cable characterisation

4.3.1 Introduction

Section 4.2 derived a theoretical basis for a circuit model which simulated the coupling between a cable and the environment. This section shows how values can be assigned to the components of that model.

A set of tests is described which measure the characteristics of a 15 m length of twin-conductor cable. The data is then used to create a model which replicates the response of the hardware. It shows how both antenna-mode and differential-mode radiation can be identified and measured. Full details of the process are provided.

4.3.2 Test method

Figure 4.14 illustrates the setup for measuring the frequency response of the radiated emission. The voltage source is a ten-turn transformer clamped on to one of the conductors, halfway along the cable. Apart from the transformer coupling between cable and test equipment, both conductors of the cable are completely isolated.

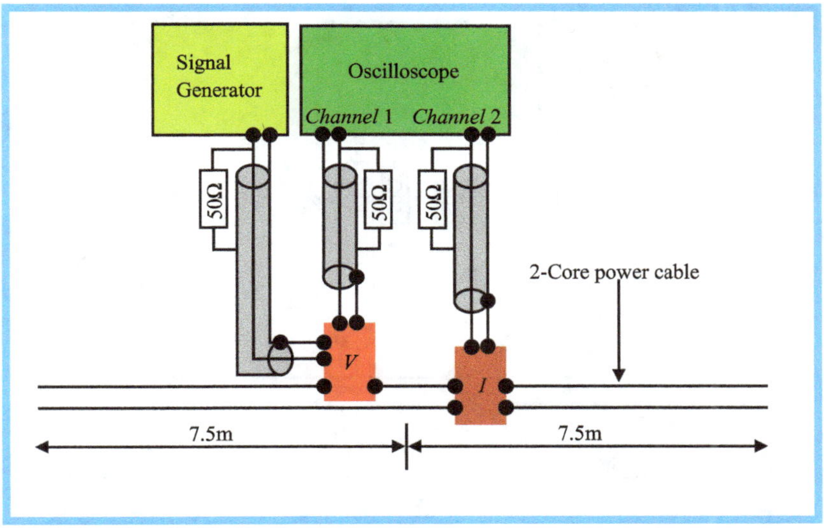

Figure 4.14 Test to measure emitted radiation characteristics of cable

A monitor turn on the voltage transformer is connected to channel 1 of the oscilloscope to enable the peak to peak voltage of the source to be measured. Section 2.4 provides details of the construction and characterisation of this unit.

A current transformer clamped round the cable is used to monitor the unidirectional flow of current in both conductors. Since there is no ground return conductor, the only place for this current to go is out into the environment. It can be defined as the antenna-mode current.

The output of the current transformer is connected to channel 2 of the oscilloscope. The peak-to peak amplitude of this signal provides a measure of the antenna-mode current. Section 2.5 provides details of the construction and calibration of this unit.

The ratio of output current to input voltage has the dimensions of admittance; Y. Since the current being measured and the voltage source are located in different circuit loops, it can be designated 'Transfer Admittance' YT. Since this value has been derived from test results, it is logical to identify this parameter as YTt. That is

From Figure 4.14:

$$YTt = \frac{Iout}{Vin} \tag{4.33}$$

The parameter YTt is a measure of the radiated emission of the cable. Tests were carried out, over a set of spot frequencies, of the input voltage and output current and the transfer admittance calculated for each frequency.

The current transformer was then unclamped from the cable and clamped round the second conductor as shown in Figure 4.15. Then a further set of tests

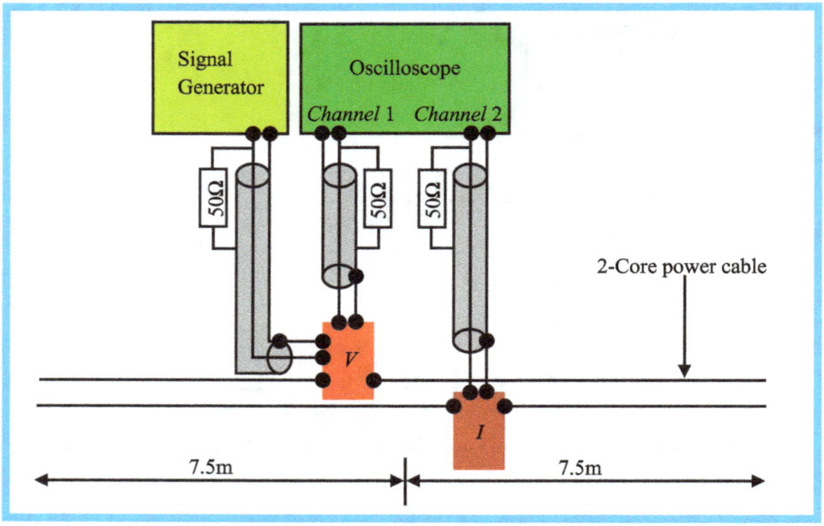

Figure 4.15 Test set-up to measure the cable conductance

provided a record of the admittance Yt of the cable over the same range of frequencies.

From Figure 4.15

$$Yt = \frac{Iout}{Vin} \tag{4.34}$$

4.3.3 Circuit modelling

Figure 4.16 is essentially a copy of the model derived in Section 4.2. This model can be transformed into the distributed parameter model (Figure 4.17) by invoking the relationships derived in Section 2.2.

That is

$$\begin{aligned} Zh_i &= Zo_i \cdot \tanh\left(\frac{\theta_i}{2}\right) \\ Zv_i &= Zo_i \cdot \operatorname{cosech}(\theta_i) \end{aligned} \tag{4.35}$$

where

$$\begin{aligned} \theta_i &= \sqrt{(Rc_i + j \cdot \omega \cdot L_i) \cdot (Gc_i + j \cdot \omega \cdot C_i)} \\ Zo_i &= \sqrt{\frac{Rc_i + j \cdot \omega \cdot L_i}{Gc_i + j \cdot \omega \cdot C_i}} \end{aligned} \tag{4.36}$$

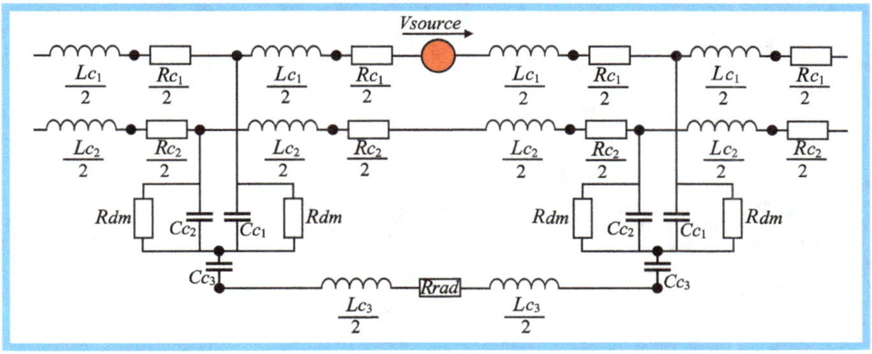

Figure 4.16 Lumped parameter model

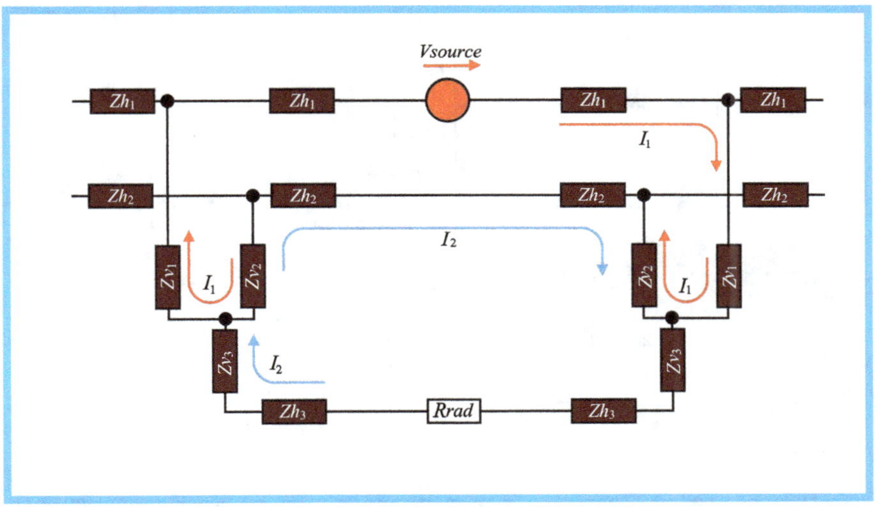

Figure 4.17 Distributed parameter model of Figure 4.16

4.3.4 Computations

The computations for this analysis were carried out using Mathcad. This software has the advantages that the equations are presented in the same way as a textbook on mathematics and that all the calculations can be carried out on a single worksheet.

Figure 4.18 is a copy of the first page of the worksheet. Data from the tests is provided in the form of an array. The parameters *R1, R2, R3, R4, L1*, and *C1* on the right-hand side are definitions of the components of the circuit model used to calibrate the frequency response of the current transformer. Given a value for the voltage *Vch2* at the input of the oscilloscope, it is possible to calculate the amplitude of the current in the secondary winding of this transformer. The current in the primary winding is the secondary current times *Turns*, the number of turns on the secondary.

$$data1 := \begin{pmatrix} 1 & 420 & 0.4 \\ 2 & 420 & 0.4 \\ 3 & 420 & 0.8 \\ 4 & 420 & 1.4 \\ 5 & 435 & 2.4 \\ 6 & 350 & 3.8 \\ 6.5 & 365 & 5.8 \\ 7 & 370 & 11 \\ 7.38 & 370 & 15 \\ 7.5 & 370 & 14 \\ 7.7 & 365 & 11 \\ 8 & 365 & 7.6 \\ 9 & 365 & 3.6 \\ 10 & 360 & 2 \\ 11 & 360 & 1.8 \\ 12 & 360 & 1.8 \\ 13 & 360 & 1.8 \\ 14 & 355 & 1.6 \\ 15 & 360 & 1 \\ 16 & 375 & 2.6 \\ 17 & 320 & 2 \\ 18 & 270 & 1 \\ 19 & 280 & 1 \\ 19.8 & 285 & 1 \end{pmatrix}$$

Data from test on setup of Figure 4.14

Column 1: Frequency, MHz
Column 2: Peak-to-peak voltage at channel 1, mV
Column 3: Peak-to-peak voltage at channel 2, mV

See Figure 2.26 of Section 2.5 'The Current Monitor Transformer'

$R1 := 300\,\Omega \quad R2 := 51\,\Omega \quad R3 := 50\,\Omega \quad Turns := 10$

$R4 := 850\,\Omega \quad L1 := 200\cdot 10^{-6}\mathrm{H} \quad C1 := 60\cdot 10^{-12}\mathrm{F}$

$s := 1 .. rows(data1) \qquad f1_s := data1_{s,1}\cdot 10^6$

$$YTt_s := \left| \begin{array}{l} Vch1 \leftarrow data1_{s,2}\cdot 10^{-3} \\ Vin \leftarrow \dfrac{51+50}{50}\cdot Vch1 \\ Vch2 \leftarrow data1_{s,3}\cdot 10^{-3} \\ \omega \leftarrow 2\cdot\pi\cdot f1_s \\ Z1 \leftarrow R4 + \dfrac{1}{j\cdot\omega\cdot C1} \\ Y2 \leftarrow \dfrac{1}{R1} + \dfrac{1}{R2} + \dfrac{1}{R3} + \dfrac{1}{j\cdot\omega\cdot L1} + \dfrac{1}{Z1} \\ Isec \leftarrow |Y2|\cdot Vch2 \\ Iprim \leftarrow Isec\cdot Turns \\ Yt \leftarrow \dfrac{Iprim}{Vin} \end{array} \right.$$

Figure 4.18 Creating a vector of the transfer admittance values; YTt

The voltage delivered to the cable under test is also the voltage delivered to a monitor turn on the voltage transformer. Since this monitor turn is in series with a 51 Ω resistor and the load at the input terminals of channel 2 of the oscilloscope is loaded by a 50 Ω resistor, it is easy to calculate the value *Vin*. The output of the function which processes the data is the vector *YTt*.

Figure 4.19 is a copy of the second page of the worksheet. It records the data taken from the set-up illustrated in Figure 4.15 and processes that data in the same way as described above to create the vector *Yt*.

The third page of the worksheet is illustrated in Figure 4.20. It first defines the parameters for the lumped-parameter model; *r11* and *r22* are the radii of the cable conductors; *r12* is the separation between the centres of those conductors; *Rc1, Rc2, Rc3, Rdm,* and *Rrad* are the resistance values assigned to the components of

Data from tests carried out on setup of Figure 4.3.2

$$data := \begin{pmatrix} 1 & 420 & 2.8 \\ 2 & 420 & 6.4 \\ 3 & 420 & 11 \\ 4 & 430 & 20.5 \\ 5 & 460 & 60 \\ 5.2 & 480 & 88 \\ 5.5 & 420 & 135 \\ 5.7 & 325 & 110 \\ 6 & 330 & 63.5 \\ 7 & 370 & 25 \\ 8 & 380 & 18.5 \\ 9 & 375 & 5.2 \\ 10 & 375 & 3.6 \\ 11 & 380 & 1.8 \\ 12 & 375 & 0.6 \\ 13 & 380 & 3 \\ 14 & 380 & 6.6 \\ 15 & 390 & 11.5 \\ 15.5 & 405 & 17 \\ 16 & 415 & 26 \\ 16.5 & 395 & 43 \\ 16.8 & 325 & 47 \\ 17 & 280 & 44 \\ 18 & 270 & 20 \\ 19 & 290 & 11.5 \\ 19.8 & 295 & 8.2 \end{pmatrix}$$

$$f_s := data_{s,1} \cdot 10^6$$

$$s := 1 .. rows(data)$$

$$Yt_s := \left| \begin{aligned} & Vch1 \leftarrow data_{s,2} \cdot 10^{-3} \\ & Vin \leftarrow \frac{51+50}{50} \cdot Vch1 \\ & Vch2 \leftarrow data_{s,3} \cdot 10^{-3} \\ & \omega \leftarrow 2 \cdot \pi \cdot f_s \\ & Z1 \leftarrow R4 + \frac{1}{j \cdot \omega \cdot C1} \\ & Y2 \leftarrow \frac{1}{R1} + \frac{1}{R2} + \frac{1}{R3} + \frac{1}{j \cdot \omega \cdot L1} + \frac{1}{Z1} \\ & Isec \leftarrow |Y2| \cdot Vch2 \\ & Iprim \leftarrow Isec \cdot Turns \\ & Yt \leftarrow \frac{Iprim}{Vin} \end{aligned} \right.$$

Figure 4.19 Creating a vector of the admittance values Yt

Figure 4.16. The parameters *Lc1, Lc2*, and *Lc3* are values calculated for the inductors using the equations derived in Section 4.2.

The values of *Vch2* are recorded on the right hand columns of the data arrays in Figures 4.18 and 4.19 and *fqa* and *fqb* are the frequencies at which this voltage reaches a peak value. They are the quarter-wave frequencies where resonance occurs. Since

$$\sqrt{L \cdot C} = T = \frac{4}{fq} \tag{4.37}$$

$$\mu := 4 \cdot \pi \cdot 10^{-7} \quad r11 := 0.48 \cdot 10^{-3} \quad r22 := r11 \quad r12 := 1.95 \cdot 10^{-3} \quad len := 7.5$$

$$Rc_1 := 0.02 \quad Rc_2 := 0.02 \quad Rc_3 := 0 \quad Rdm := 350 \quad Rrad := 52$$

$$G_1 := \frac{1}{Rdm} \quad G_2 := G_1 \quad G_3 := 0$$

$$Lc_1 := \frac{\mu \cdot len}{2 \cdot \pi} \cdot ln\left(\frac{r12}{r11}\right) \quad Lc_2 := Lc_1 \quad Lc_3 := \frac{\mu \cdot len}{2 \cdot \pi} \cdot ln\left(\frac{len}{r12}\right)$$

$$fqa := 5.66 \cdot 10^6 \quad fqb := 7.55 \cdot 10^6$$

$$Cc_1 := \frac{1}{16 \cdot fqa^2 \cdot Lc_1} \quad Cc_2 := Cc_1 \quad Cc_3 := \frac{1}{16 \cdot fqb^2 \cdot Lc_3}$$

$$Zbranch(f) := \left| \begin{array}{l} \omega \leftarrow 2 \cdot \pi \cdot f \\ \text{for } k \in 1..3 \\ \quad \left| \begin{array}{l} Gc_k \leftarrow G_k \cdot \dfrac{f}{fqa} \\ \theta \leftarrow \sqrt{(Rc_k + j \cdot \omega \cdot Lc_k) \cdot (Gc_k + j \cdot \omega \cdot Cc_k)} \\ Zo \leftarrow \sqrt{\dfrac{Rc_k + j \cdot \omega \cdot Lc_k}{Gc_k + j \cdot \omega \cdot Cc_k}} \\ Zh_k \leftarrow Zo \cdot tanh\left(\dfrac{\theta}{2}\right) \\ Zv_k \leftarrow Zo \cdot csch(\theta) \end{array} \right. \\ (Zh \quad Zv) \end{array} \right.$$

$$Zloop(f) := \left| \begin{array}{l} (Zh \quad Zv) \leftarrow Zbranch(f) \\ Z11 \leftarrow 2 \cdot (Zh_1 + Zv_1 + Zh_2 + Zv_2) \\ Z12 \leftarrow -2 \cdot (Zh_2 + Zv_2) \\ Z22 \leftarrow 2 \cdot (Zh_2 + Zv_2 + Zh_3 + Zv_3) + Rrad \\ \begin{pmatrix} Z11 & Z12 \\ Z12 & Z22 \end{pmatrix} \end{array} \right.$$

$$V := \begin{pmatrix} 1 \\ 0 \end{pmatrix}$$

Figure 4.20 Defining the input parameters and the functions Zbranch and Zloop

Then

$$C = \frac{1}{16 \cdot fq^2 \cdot L} \tag{4.38}$$

The function Zbranch(*f*) of Figure 4.20 uses (4.35) and (4.36) to calculate an impedance value for each branch of Figure 4.17. The function Zloop(f) uses these to create an array of loop impedances. *Z11* is the sum of the impedances in the loop carrying the current I_1. *Z22* is the sum of the impedances carrying the current I_2.

Z12 is the sum of the impedances carrying both currents. It is negative because they are flowing in opposite directions. The vector V assigns a voltage of 1 Volt in loop 1 and 0 Volts in loop 2.

This process has reduced the complexities of the interactions to a pair of simultaneous equations:

$$\begin{aligned} V_1 &= Z11 \cdot I_1 + Z12 \cdot I_2 \\ V_2 &= Z12 \cdot I_1 + Z22 \cdot I_2 \end{aligned} \tag{4.39}$$

Solving them can be achieved by invoking the built-in function lsolve(Z, V). Since this needs to be done over a range of frequencies, the first action at the top of the final page of the worksheet (Figure 4.21) is to specify F_i as a set of frequencies between 100 kHz and 20 MHz. The function Ym_i calculates the values of I_1 and I_2 at each frequency, then creates a vector of values of the magnitude of I_1, the differential-mode current. The function YTm_i carries out the same process to create a set of values for I_2, the antenna-mode current.

There are two plots on each of the graphs in Figure 4.21. On the upper graph, the first plot is a set of data points defining the transfer admittance *YTt*, derived from the test results. The second plot is the response of the transfer admittance *YTm*, derived from the response of the model of Figure 4.17. On the lower graph, the actual admittance *Yt* is compared with the simulation *Ym*.

That is, the upper graph of Figure 4.21 compares the response of the setup of Figure 4.14 with that of the model. The lower graph compares the response of the setup of Figure 4.15 with that of the model. The same model simulates the response of the cable under review, no matter how it is tested.

This correlation did not happen when the program was run for the first time. Although the shapes of the actual responses were similar to those of the model, there was a significant discrepancy in the amplitude of the peaks.

By making successive alterations to the value of *Rrad* between program runs, it was possible to cause the response of the model to coincide with the peak value derived from the results of tests on the configuration of Figure 4.14. Making successive alterations to the value of *Rdm* enabled coincidence to occur between the simulation of the model and the results recorded from tests on the configuration of Figure 4.15.

The results of this process are recorded at the bottom of Figure 4.21. These include values for the inductors Lc_1, Lc_2, and Lc_3. These values were derived directly from measurements of the radius and separation of the conductors, and from a measurement of the length of the cable. The values calculated for the capacitors *Cc1, Cc2*, and *Cc3* were derived from measurements of the frequencies of resonance. The process was similar to that used in successive approximations in solving non-linear equations.

To cater for the fact that the second peak in the response of *Yt* was much smaller than that of the first peak, it was necessary to assume that the resistors *Rdm* reduced as the frequency increased. This makes sense if it is also assumed that these resistors simulate the radiated current which disappears into the environment.

$i := 1 \,..\, 200$

$F_i := i \cdot 10^5$

$$YTm_i := \left| \begin{array}{l} f \leftarrow F_i \\ Z \leftarrow Zloop(f) \\ I \leftarrow lsolve(Z, V) \\ |I_2| \end{array} \right. \qquad Ym_i := \left| \begin{array}{l} f \leftarrow F_i \\ Z \leftarrow Zloop(f) \\ I \leftarrow lsolve(Z, V) \\ |I_1| \end{array} \right.$$

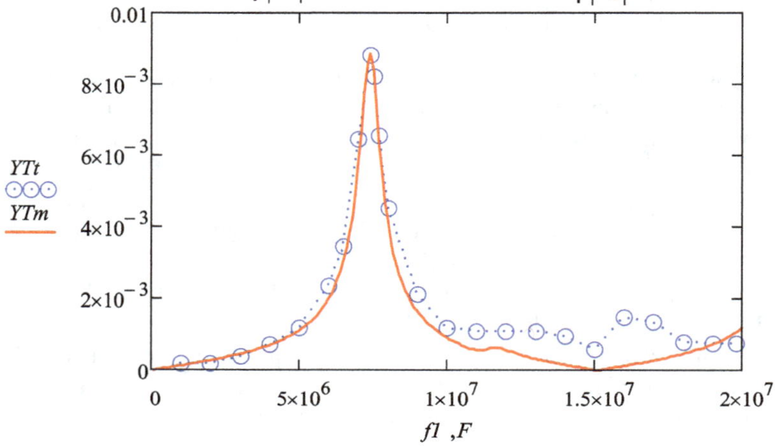

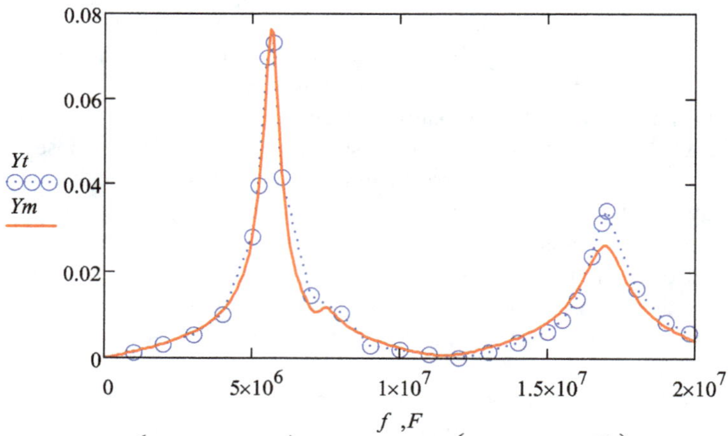

$Rrad = 52$

$Rdm = 350$

$$\frac{Lc}{2} = \begin{pmatrix} 1.051 \times 10^{-6} \\ 1.051 \times 10^{-6} \\ 6.191 \times 10^{-6} \end{pmatrix} \qquad Cc = \begin{pmatrix} 9.278 \times 10^{-10} \\ 9.278 \times 10^{-10} \\ 8.855 \times 10^{-11} \end{pmatrix}$$

Figure 4.21 Correlating the circuit model with the test data

That is, differential-mode current departs from both conductors but some of it does not arrive at the other. As the frequency increases, more and more energy is lost. As the frequency reduces, the resistance *Rdm* increases. Under steady state conditions, the resistance of the insulating material of cables is extremely large.

4.3.5 Assessment

The resistance *Rrad* is a measure of the current departing from the cable when it is acting like a dipole antenna. That is, when the partial current in each conductor is flowing in the same direction, the resistors *Rdm* are a measure of the differential-mode losses.

The frequency of resonance of the differential-mode current is lower than that of antenna-mode resonance because the dielectric constant of the cable insulation is higher than that of air. So the propagation velocity of the electromagnetic field is reduced.

At frequencies above 10 MHz, resonances occur in each section of the cable due to the discontinuity created by the test equipment at the centre. So the emission is higher than that of the simulation.

It is useful to note that constant values are assigned to the inductors and capacitors in this model, and that the values assigned to resistors vary with frequency.

Since the inductance, capacitance, resistance, and admittance of the cable are proportional to length, it is a simple matter to calculate the values which would be assigned to different cable lengths.

4.3.6 Conclusion

A circuit model has been developed which can simulate the coupling between a twin-conductor cable and the environment.

4.4 Transfer admittance

4.4.1 Introduction

The previous Section 4.3 described a model which simulated the frequency response of the emission of a twin-conductor cable. This section shows that this response is identical to the susceptibility characteristic.

In any analysis of EMI, the most critical parameter is the transfer admittance – the ratio of the current delivered to the victim loop to the voltage applied to the culprit loop.

In the case of the configuration under review, the culprit loop is that carrying the differential-mode signal and the victim loop is that between the cable and the virtual conductor simulating the environment. Here, the transfer admittance defines the emission characteristic.

If the voltage source was in the antenna-mode loop, the transfer admittance would be the ratio of the current induced in the differential-mode loop to the amplitude of that voltage source. This is a measure of the susceptibility of the configuration.

It is shown that the transfer admittance is the same, whether the measurement is of radiation susceptibility or radiated emission. This is also true for conducted susceptibility and conducted emission.

Although transfer admittance is an essential parameter in any analysis of EMC, it cannot be measured by any formal EMC test. The only engineers with the freedom to carry out such analysis are those who design the equipment under test.

4.4.2 Analysis

The loop equations for the circuit shown in Figure 4.22 are as follows:

$$V_1 = (Z1 + Z2) \cdot I_1 - Z2 \cdot I_2 \tag{4.40}$$

$$0 = -Z2 \cdot I_1 + (Z2 + Z3) \cdot I_2 \tag{4.41}$$

Re-arranging (4.41)

$$I_1 = \frac{Z2 + Z3}{Z2} \cdot I_2 \tag{4.42}$$

Substituting for I_1 in (4.40)

$$V_1 = \frac{(Z1 + Z2) \cdot (Z2 + Z3)}{Z2} \cdot I_2 - Z2 \cdot I_2 \tag{4.43}$$

This expands to

$$V_1 = \frac{Z1 \cdot Z2 + Z1 \cdot Z3 + Z2 \cdot Z2 + Z2 \cdot Z3 - Z2 \cdot Z2}{Z2} \cdot I_2 \tag{4.44}$$

Giving

$$\frac{I_2}{V_1} = \frac{Z2}{Z1 \cdot Z2 + Z1 \cdot Z3 + Z2 \cdot Z3} \tag{4.45}$$

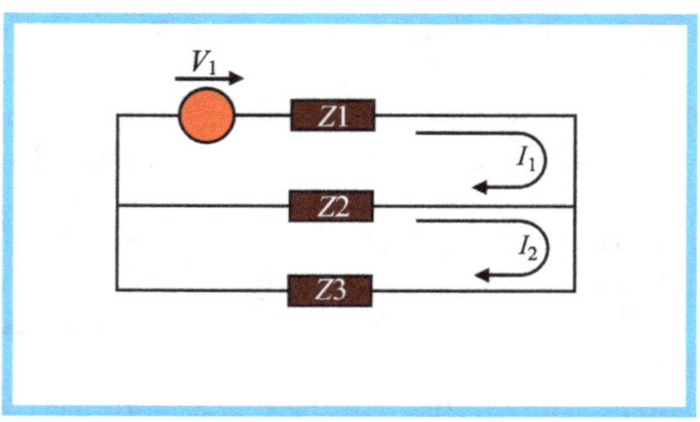

Figure 4.22 Calculating transfer admittance

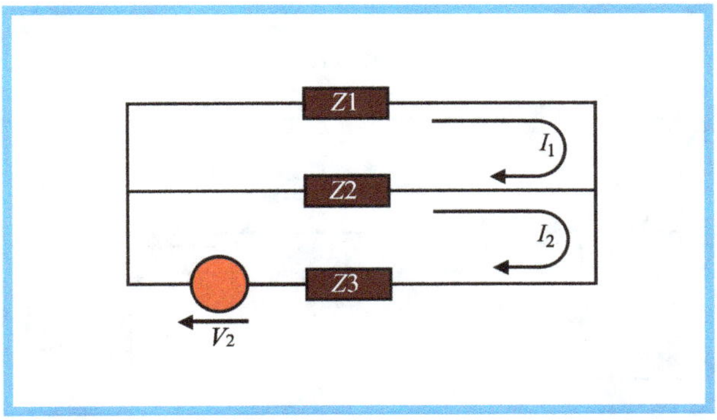

Figure 4.23 Voltage source in loop 2

If the voltage source is located in the second loop, as shown in Figure 4.23, the same process leads to

$$\frac{I_1}{V_2} = \frac{Z2}{Z1 \cdot Z2 + Z1 \cdot Z3 + Z2 \cdot Z3} \tag{4.46}$$

This means that the transfer admittance YT is the same, whether the voltage source is located in loop 1 or 2:

$$YT = \frac{I_1}{V_2} = \frac{I_2}{V_1} \tag{4.47}$$

Since the coupling between the culprit and victim loops can be simulated by a circuit model of the configuration under review and since the transfer admittance is the same if the roles are interchanged, it follows that the response to a measurement of the emission characteristic can be used to define the susceptibility.

4.4.3 Principle of reciprocity

Reviewing (4.45) and (4.46) reveals the fact that the numerator on the right-hand side is the only impedance common to both circuit loops and that the denominator is the sum of the products of every pair of impedances in the two loops. This property will apply to every pair of loops which share a common impedance.

For a network with many loops, this interrelation continues to be valid. It follows that the ratio of the current in loop i to the voltage in loop j is identical to the ratio of the current in loop j to a voltage source in loop i:

$$\frac{I_i}{V_j} = \frac{I_j}{V_i} \tag{4.48}$$

4.4.4 Radiated emission and radiation susceptibility

Figure 4.24 is a copy of the circuit model used to measure the radiated emission characteristic of an isolated length of twin-conductor cable. The red curve on Figure 4.25

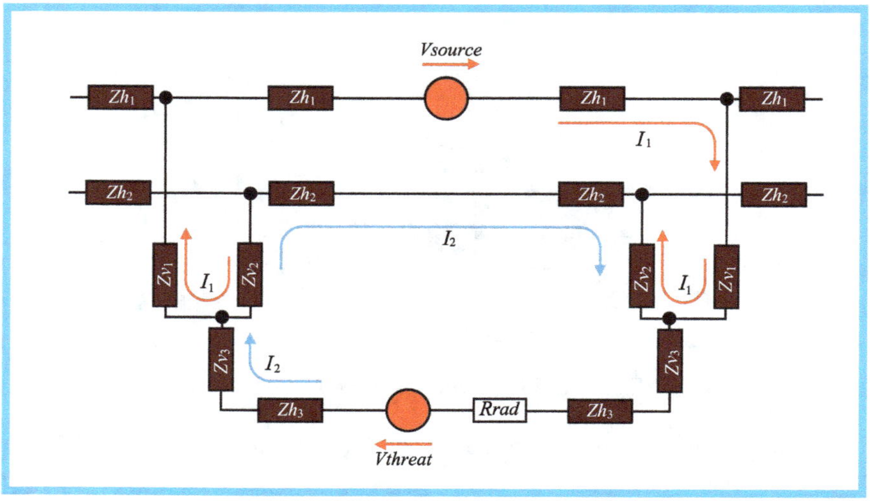

Figure 4.24 Circuit model used to simulate radiated emission characteristic of twin-core cable

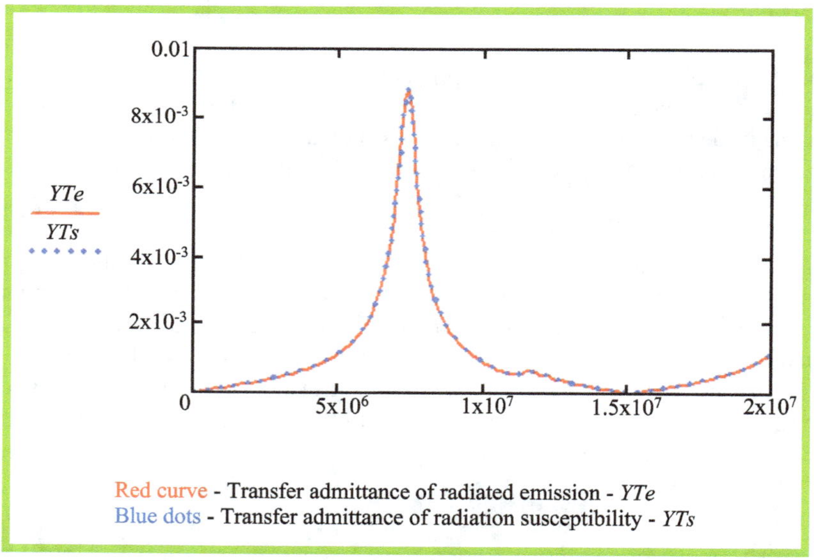

Figure 4.25 Response of circuit model of cable characterisation test red curve – transfer admittance of radiated emission – Yte blue dots – transfer admittance of radiation susceptibility – YTs

shows the frequency response of the conducted emission when a voltage source *Vsignal* is applied and the voltage source *Vthreat* is zero.

In Figure 4.24, the radiated emission is defined as

$$YTe = \frac{I_2}{Vsignal} \tag{4.49}$$

And the radiation susceptibility is defined as

$$YTs = \frac{I_1}{Vthreat} \tag{4.50}$$

The fact that the red curve *YTe* simulates the actual response of the hardware is demonstrated by the Section 4.3. The fact that the blue dots of *YTs* lie precisely on the red curve demonstrates that the analysis of the simple model of Figures 4.22 and 4.23 applies to the more complex model of Figure 4.24.

The curve of *YTe* versus frequency can be described as the radiated emission characteristic of the cable-under-test. The curve of *YTs* can be described as the radiation susceptibility of that same cable, the differential-mode current induced in the cable by a voltage of 1 Volt created by an external electromagnetic field.

That is, a test of the radiated emission can be used to predict the radiation susceptibility of the configuration under review.

4.4.5 Conducted emission and conducted susceptibility

Figure 4.26 is the circuit model developed to simulate the cross-coupling between the differential-mode signal and the induced current in the common-mode loop. Details of the actual test and the creation of this model are described in Section 2.6.

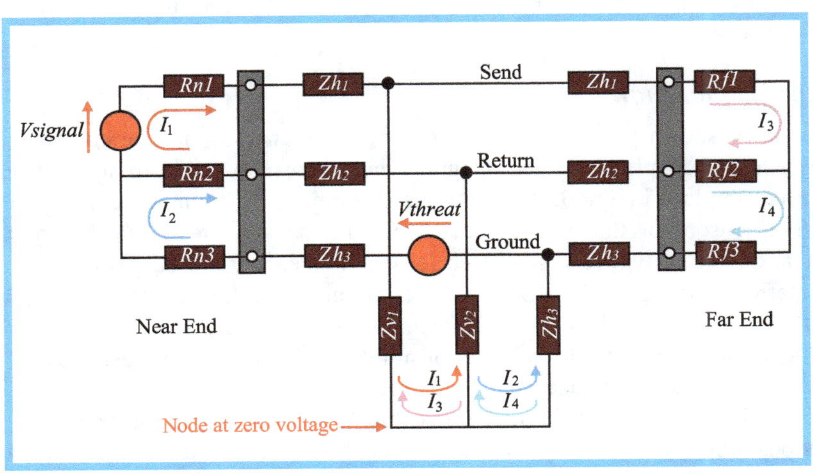

Figure 4.26 Circuit model of signal link

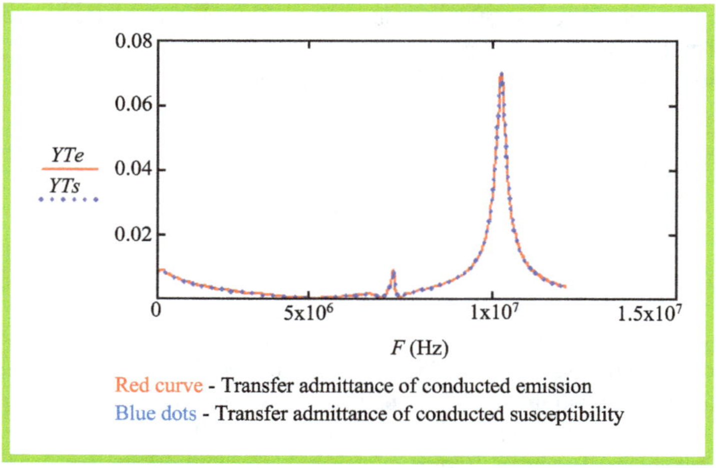

Figure 4.27 Response of circuit model of signal link

The red curve of Figure 4.27 is a copy of the transfer admittance *YTe* of that particular setup – the curve relating the current I_2 induced in the common-mode loop by a voltage *Vsignal* in the differential-mode loop. In the parlance of EMC requirements, this can be described as a conducted emission test.

The blue dotted curve *YTs* of Figure 4.27 is the ratio of the current I_1 induced in the differential-mode loop by a voltage *Vthreat* in the common-mode loop, that is, the predicted response of a test of the conducted susceptibility of the signal link.

This means that a test of conducted emission can be used to predict the results of a conducted susceptibility of any configuration under review – and vice versa.

4.4.6 Conclusion

It has been shown that the transfer admittance is an essential parameter in any analysis of the coupling of electromagnetic interference to and from the EUT. Even so, it has no place in any of the multitude of formal EMC requirements.

The reason for this is the demarcation in the way these requirements are written. Of necessity, there is no need for the engineers at the Test House to know the internal workings of the EUT. The test equipment and the use of that equipment are the sole responsibility of technicians at the test facility. The EUT and its functional behaviour are the sole responsibility of the equipment designer.

For susceptibility requirements, a severe electromagnetic environment is created by the EMC test equipment. Then the EUT is tested to check that it still functions properly. It some cases, the functional behaviour of the EUT is monitored by the equipment designer during the testing to confirm that no abnormal behaviour was detected.

For emission requirements, the EUT is cycled through a set of functions, designed to create maximum electromagnetic interference and the test equipment is monitored to ensure that none of the specified test limits are exceeded.

The problem with this demarcation is the need for the regulatory requirements to be framed in such a way that all eventualities are covered. This means that the formal documents are replete with legalistic jargon and require significant expertise in interpreting their precise meaning. Working groups update the regulations at regular intervals.

So there are three groups of engineers: those who formulate the EMC requirements, those who carry out the tests, and those who design the EUT.

Equipment designers soon came to the conclusion that EMC requirements are so complex that there is no point in trying to understand the phenomena. This led naturally to the creation of a fourth group; EMC experts who offer their expertise in the form of multiple and varied guidelines as to how to design equipment to meet those requirements. But the experts who offer this advice are long gone when the manufactured equipment is submitted to the EMC Test House.

4.5 Modelling radiated interference

4.5.1 Introduction

A circuit model is derived which can simulate the differential-mode response of a signal link when it is exposed to electromagnetic interference (EMI).

The concept of the radiation resistance is developed to relate the current in an isolated conductor to that of the current in a dipole transmitting antenna. By invoking worst-case conditions, a simple equation can be used to define the maximum *H*-field of the transmitted wave in terms of the current in the conductor. The *E*-field can then be determined by multiplying the *H*-field by the intrinsic impedance of free space. When the *E*-field interacts with a conductor in the far distance, a voltage is induced in that conductor. A simple equation is derived which relates that voltage to the magnitude of the *E*-field. The power received is then calculated by dividing the square of the voltage by the radiation resistor.

Representing a twin-conductor cable as two transmission lines in series allows a circuit model to be created which simulates the differential-mode current induced in the signal link.

4.5.2 Emission

Figure 4.28 illustrates the model of an isolated conductor when a voltage source *Vgen* is located half-way along the length. This model treats the conductor as two monopoles. At low frequencies, the current *I* will be negligible; but as the frequency increases towards resonance, this current will increase rapidly. At resonance, the current is limited only by the resistance *Rrad*. Section 4.1 provides an example of the response of a 15 m conductor. The parameters *Lp* and *Cp* represent the properties of a single monopole of length 7.5 m, and the parameter *Rrad*

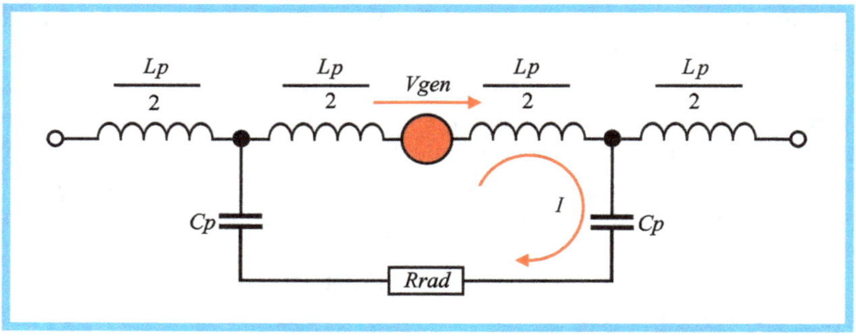

Figure 4.28 Simulating radiated emission

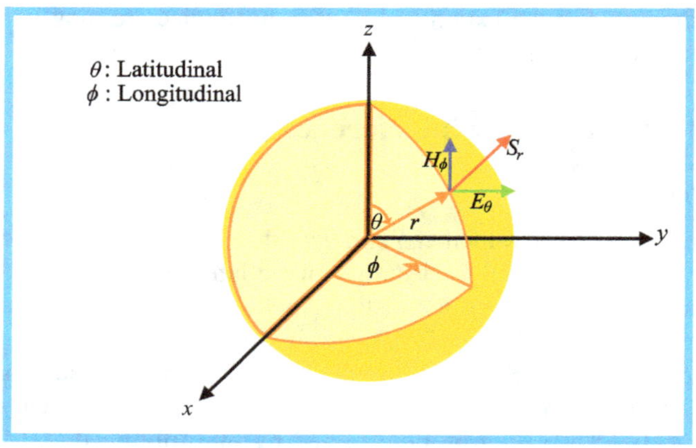

Figure 4.29 Power density on a spherical surface

represents the effect of current radiating out into the environment in the form of an electromagnetic field.

4.5.3 Propagation

Figure 4.29 shows the relationship between the electric field E and magnetic field H at a point on the surface of a sphere of radius r. The power density S is the product:

$$S = E \cdot H \text{ Watts per square metre} \tag{4.51}$$

This diagram also shows that these three vectors are perpendicular to each other and that the energy propagates outwards along the radius r. These are the only components of the electromagnetic field which radiate out into the far distance.

The power density varies with latitude θ and is a maximum when $\theta = 90°$. With any analysis of electromagnetic interference, it is maximum values which are most critical. EMC tests are usually defined in terms of pass/fail criteria. If the measured parameter is less than a specified limit, the system under review will pass that particular test. Hence, it is only necessary to analyse worst-case conditions.

In this case, it can be assumed that the maximum magnetic field H is in the x–y plane, and the maximum electric field E is parallel to the z axis. So

$$H = \frac{I}{2 \cdot \pi \cdot r} \text{ A/m} \tag{4.52}$$

The characteristic impedance of free space is as follows:

$$Zo = \frac{E}{H} = \sqrt{\frac{\mu_o}{\varepsilon_o}} = 377 \ \Omega \tag{4.53}$$

Hence, the maximum value of the electric field is as follows:

$$E = Zo \cdot H \text{ V/m} \tag{4.54}$$

Substituting for H

$$E = Zo \cdot \frac{I}{2 \cdot \pi \cdot r} \text{ V/m} \tag{4.55}$$

If the amplitude of the current at the source and the separation distance from that source are known, the maximum intensity of the electric field can be calculated using (4.55).

4.5.4 The threat voltage

If it is assumed that the same conductor now behaves in essentially the same way as a dipole receiver when it is exposed to an external electromagnetic field E, this effect can be represented as the sum of a number of discrete voltage sources. An incremental voltage dV is induced in series with each element dz of the conductor. Figure 4.30 illustrates this.

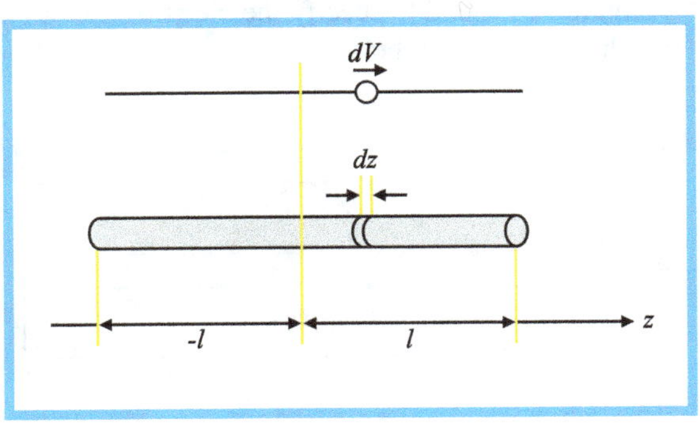

Figure 4.30 Conductor represented as two dipoles

The incremental voltage dV is as follows:

$$dV = E \cdot dz \tag{4.56}$$

If it is assumed that the waveform is sinusoidal, the distribution in space will also be sinusoidal. Figure 4.31 illustrates this. For such a waveform, the relationship can be defined as

$$E = Emax \cdot \cos\left(2 \cdot \pi \cdot \frac{z}{\lambda}\right) \tag{4.57}$$

where λ is the wavelength. Over the length of the cable the total voltage is as follows:

$$Vthreat = \int_{-l}^{l} Emax \cdot \cos\left(2 \cdot \pi \cdot \frac{z}{\lambda}\right) \cdot dz \tag{4.58}$$

That is

$$Vthreat = Emax \cdot \frac{\lambda}{\pi} \cdot \sin\left(2 \cdot \pi \cdot \frac{l}{\lambda}\right) \tag{4.59}$$

If $l > \frac{\lambda}{4}$, the waveform will include negative values, as illustrated in Figure 4.31. The curve relating *Vthreat* to wavelength will have a series of peaks and troughs. The envelope curve which touches the peaks can be obtained by setting:

$$\sin\left(2 \cdot \pi \cdot \frac{l}{\lambda}\right) = 1 \tag{4.60}$$

Then

$$Vthreat = Emax \cdot \frac{\lambda}{\pi} \tag{4.61}$$

Equation (4.61) defines worst-case conditions.

Figure 4.30 shows the effect of an external electromagnetic field on a conductor; in the figure EMC requirements tend to specify the electric field strength and a range of frequencies to which the EUT must be subjected and require that the EUT operates

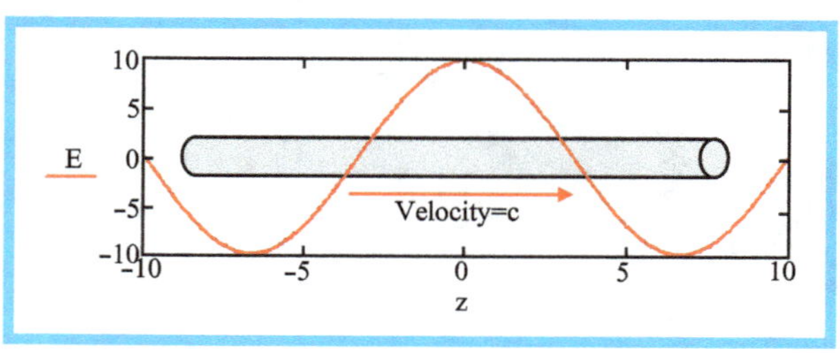

Figure 4.31 Distribution of electric field along conductor

normally under these conditions. The threat voltage can be calculated using (4.61) and by invoking the relationship: Where c is the velocity of light in a vacuum:

$$\lambda = \frac{c}{f} \tag{4.62}$$

Giving the electric field distribution shown in Figure 4.31.

4.5.5 Power received

With a dipole receiving antenna as illustrated in Figure 4.32, the power delivered to the radio frequency receiver can be modelled as shown in Figure 4.33.

For this configuration, the power received $Prec$ at the amplifier input when the dipole is operating at its half-wave frequency is as follows:

$$Prec = \frac{1}{Rrec} \cdot \left(\frac{Vthreat}{2}\right)^2 \tag{4.63}$$

If the wiring of the system-under-review is not configured as a dipole receiving antenna, the power received will be:

$$Prec = \frac{Vthreat^2}{Rrec} \tag{4.64}$$

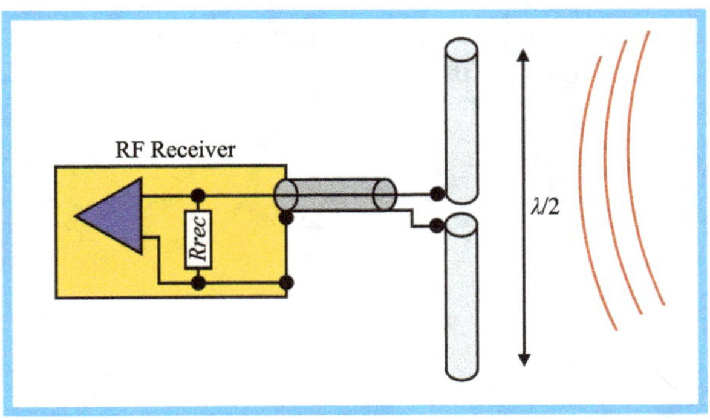

Figure 4.32 Illustration of a receiver assembly

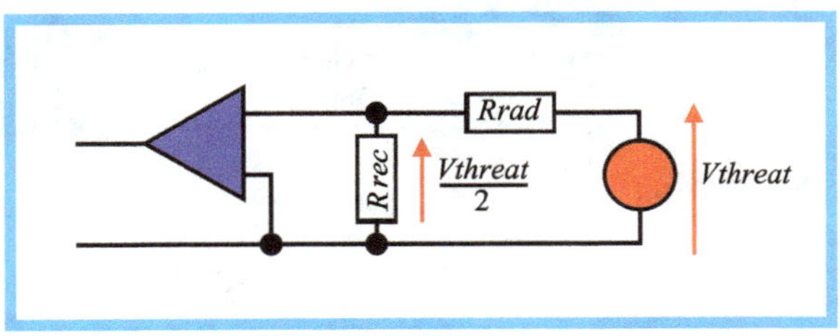

Figure 4.33 Circuit model of receiver assembly

This is because the energy is stored in the conductor in the form of oscillating current, rather than delivered to the input of the RF amplifier.

4.5.6 The signal link

In any practical configuration, the cable is connected to a signal source which is mounted on a conducting framework, as illustrated in Figure 4.34. This shows a set-up where a processing unit is transmitting a signal along a twin-conductor cable to a remote transducer. It is assumed that the length l of the cable is 15 m and that the amplitude E of the interference is 10 V/m. Under worst-case conditions, the length of the structure would be the same as that of the cable. So, maximum voltage would exist along the cable when the frequency of the external radiation was creating half-wave resonance along the assembly, that is, when it was creating quarter-wave resonance along the cable.

Since resonance is due to charges propagating back and forth between the extremities of the assembly, it takes several cycles of the incoming wave before the energy stored in the cable builds up and the current reaches a peak level. This does not take very long, since the charges are moving at near-light velocity. At frequencies lower than that of the quarter-wave resonance of the assembly, the impulses supplied by the threat field cannot sustain resonance. So the induced current is reduced.

It is fair to assume that the threat voltage is zero under steady-state conditions and increases as the frequency of the interference increases, up to that of quarter-wave resonance fq, and decreases as the frequency increases further.

At quarter-wave resonance, the wavelength λq is as follows:

$$\lambda q = 4 \cdot l \tag{4.65}$$

The frequency is as follows:

$$fq = \frac{c}{\lambda q} \tag{4.66}$$

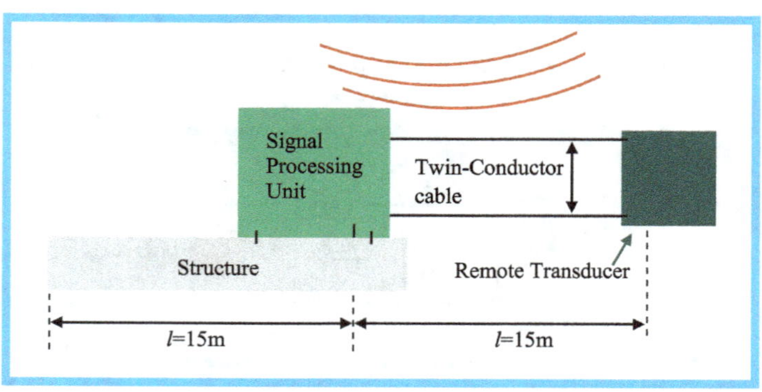

Figure 4.34 Cable exposed to external radiation

$c := 300 \cdot 10^6$ $\quad E := 10$ $\quad l := 15$

$\lambda q := 4 \cdot l = 60$ $\quad fq := \frac{c}{\lambda q}$ $\quad Vq := \frac{\lambda q}{\pi} \cdot E = 190.986$

$i := 1 .. 950$ $\quad F_i := i \cdot 10^5$

$$Vthreat_i := \left| \begin{array}{l} f \leftarrow F_i \\ V \leftarrow Vq \cdot \frac{f}{fq} \\ V \leftarrow Vq \cdot \frac{fq}{f} \quad \text{if } f > fq \end{array} \right.$$

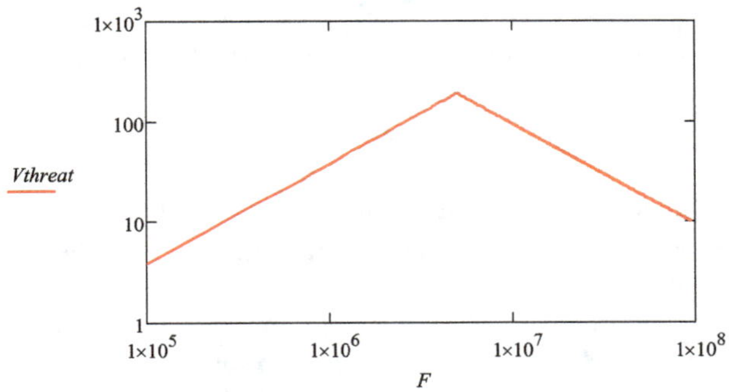

Figure 4.35 Graph of the variation of threat voltage with frequency

And the threat voltage is as follows:

$$Vq = \frac{\lambda q}{\pi} \cdot E \tag{4.67}$$

At frequencies less than fq

$$Vthreat = Vq \cdot \frac{f}{fq} \tag{4.68}$$

And at frequencies higher than fq

$$Vthreat = Vq \cdot \frac{fq}{f} \tag{4.69}$$

The graph at the bottom of the Mathcad worksheet of Figure 4.35 illustrates how the amplitude of *Vthreat* varies with frequency.

4.5.7 Circuit model

The model of Figure 4.33 indicates that the action of an external electromagnetic field on a dipole receiver can be represented as a voltage source $\frac{Vthreat}{2}$ across the input resistor. With a single conductor, the action of the field can be represented by a voltage source *Vthreat* in series with *Rrad*; both in series with the conductor.

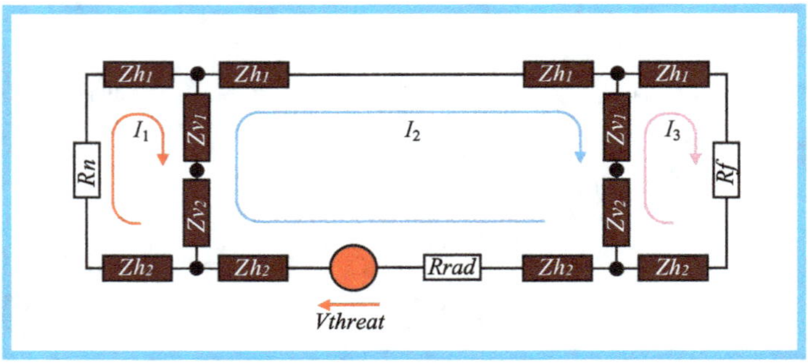

Figure 4.36 Simulating the response of a twin-conductor line to external interference

For the two-conductor signal link depicted in Figure 4.34, the response of the system to the presence of an external electromagnetic field can be analysed using the circuit model of Figure 4.36.

This is effectively a model of two transmission lines connected in series, with the voltage source located at the interface between them and connected in such a way that it creates a current along the entire length of one of the conductors. Coupling between the conductors creates a differential-mode current I_2. In turn, this creates the currents I_1 and I_3 which flow in the resistors *Rn* and *Rf*. The relationship between the impedance parameters and the L, C, and R components is defined in Section 4.1.

If the amplitude of *Vthreat* and the value of the resistor *Rrad* are set at zero, this model will behave as a transmission line. Replacing the components *Rn* and *Rf* with circuitry at the interfaces at the near and far ends will allow the functional response of this signal link to be simulated.

It then becomes possible to relate the response of the signal link under normal conditions to its response in the presence of interference.

4.5.8 Conclusion

A step-by-step method has been described which introduces all the key parameters involved in propagating an electromagnetic field from the source of interference to the conductors which experience that interference. The relationship between the current in the source and the magnetic field has been defined in terms of a circuit model, as has the relationship between the voltage induced in the victim circuit and the electric field.

Since worst-case analysis has been employed, the relationships are the simplest possible. It also means that the simulated response can be compared directly with the required EMC performance of the system-under-review. This opens up the way for radiated emission and radiation susceptibility to be subjected to the same design process as any other system requirement.

4.6 Absorbing EMI

4.6.1 Introduction

Circuit modelling can be used to create designs which ensure the safety of the system under review when exposed to EMI. The process involved in designing a firing circuit which can withstand high levels of EMI is described in detail.

An electro-explosive device (EED) can be triggered by connecting it to a voltage source via length of twin-conductor cable. A control unit containing a battery and a switch provides the necessary power.

The susceptibility of such a system to EMI is analysed using a circuit model. If the interconnecting cable is 15 m long and exposed to a field of 10 V/m, that cable will pick up more than enough energy to ignite the device even when the switch is open.

By terminating each end of the cable with a resistor equal in value to its characteristic resistance, the power delivered to the cable will be absorbed. By inserting a small value inductor in series with the EED, it can be guaranteed that there is a significant margin of safety between the no-fire-threshold (NFT) and the power delivered to the device.

4.6.2 EED characteristics

Figure 4.37 illustrates a simplified model of the firing circuit for an EED. There are several types, namely bridge wire, film bridge, conducting composition, and semiconductor initiator.

A bridge wire device is activated by passing a current through a resistive bridge between two supply conductors. The bridge is in close thermal contact with an explosive charge rather like a conductor inside a match-head. Power dissipation in the bridge creates a temperature rise in the explosive material, causing a rapid release of chemical energy. The release of energy creates a rapid release of heat and

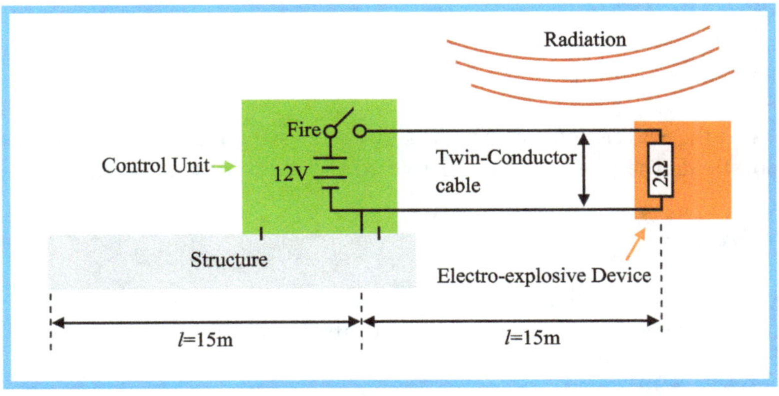

Figure 4.37 Firing system

light, and this creates a shock wave in the surrounding material. If the shock wave is less than the velocity of sound, the device is classed as an igniter. If the shock wave exceeds the velocity of sound, the output is defined as detonative.

Every type of device is subjected to a range of tests and analyses and defined in terms of its resistance, NFT, and temperature co-efficient. As far as immunity to alternating signals is concerned, the critical parameter is the NFT Power: '*The power required to produce a 0.1% probability of fire when applied to the EED for a time which is long compared with the thermal time constant of the device*'. Such a definition applies to its use for non-hazardous failures. A safety factor is included in any assessment of hazardous failures.

This analysis assumes that the resistance of the device is 2 Ω, the no-fire power is 100 mW, and that the safety margin required is 10 dB.

4.6.3 Firing system

In most systems, the procedural requirements for system safety are met by having Safe, Arm, and Fire switches of various types in the control unit: mechanical, relay and semiconductor.

In the illustration of Figure 4.37, only the firing switch is shown. The control unit is mounted on a conducting structure, and the return conductor is connected to that structure. The link between this unit and the EED is a 15 m length of twin-conductor cable.

This system acts as a dipole antenna when exposed to an external electromagnetic field. Current will flow forwards and backwards along the 30 m length. Under worst-case conditions, it will resonate, with maximum voltage appearing between the extremities of the system and maximum current existing at the interface between structure and cable. When the exposed end of the structure experiences negative voltage with respect to the control unit, the voltage at the EED will experience a positive voltage; and vice versa. If it is assumed that the voltage at the control unit reference terminal is zero, the waveform of the voltage between the EED and this reference terminal will be sinusoidal.

Under worst-case conditions, the amplitude of this threat voltage *Vthreat* is as follows:

$$Vthreat = \frac{\lambda}{\pi} \cdot Emax \ \ \mathrm{V/m} \tag{4.70}$$

where λ is the wavelength in metres, and *Emax* is the magnitude of the electromagnetic field. The power *Pthreat* delivered to the cable is as follows:

$$Pthreat = \frac{Vthreat^2}{Rrad} \ \ \mathrm{W} \tag{4.71}$$

where *Rrad* is the radiation resistance of the cable.

At quarter-wave frequency, the wavelength λq is as follows:

$$\lambda q = 4 \cdot l \ \ \mathrm{m} \tag{4.72}$$

And the frequency is as follows:

$$fq = \frac{c}{\lambda q} \text{ Hz} \tag{4.73}$$

At frequencies above *fq*, *Vthreat* reduces as the frequency *f* increases. At frequencies below *fq*, the amplitude of *Vthreat* increases as the frequency increases.

A derivation of (4.70)–(4.73) can be found in Section 4.5.

4.6.4 Lumped parameter model

The response of the system can be modelled by the circuit of Figure 4.38. This represents the 15 metre cable as two transmission lines connected in series. So the length of each section is $len = \frac{l}{2} = 7.5$ metres. Each conductor of each section is represented as a T-network with the inductive and resistive components (*Lc* and *Rc*) arrayed horizontally and the capacitive components *Cc* arrayed vertically. The effect of the external radiation on the return conductor is simulated by the voltage source *Vthreat* in series with the resistor *Rrad*, both located at the centre of that conductor. Since the near end of the cable is open-circuit, the resistance between the terminals *Rn* can be assumed to be 10MΩ. At the far end, the resistance *Rf* represents the load presented by the EED; 2 Ω.

4.6.5 Distributed parameter model

Since the frequency response of the lumped parameter model of Figure 4.38 is very limited, it is necessary to transform this into the model illustrated in Figure 4.39. This is described as the distributed parameter model since it caters for the fact that each component is distributed along the length of each conductor. It also caters for the fact that current is flowing in both directions along each conductor. (Since there are reflections at each of the cable terminations, there must be incident and reflected currents, described in transmission line theory as 'partial currents').

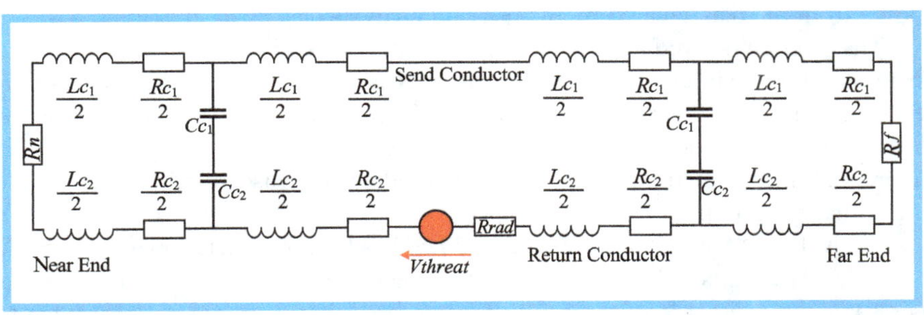

Figure 4.38 Lumped parameter model of firing system

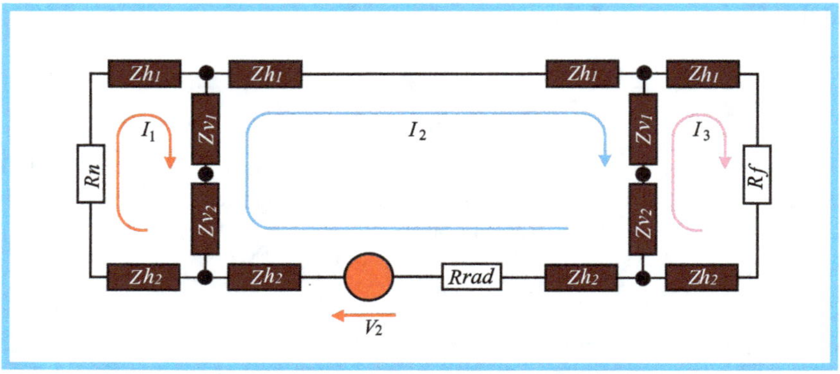

Figure 4.39 Distributed parameter model of Figure 4.38

The relationship between each branch of Figure 4.39 and the components of Figure 4.38 is as follows:

$$Zh_i = Zo_i \cdot \tanh\left(\frac{\theta_i}{2}\right)$$
$$Zv_i = Zo_i \cdot \text{cosech } \theta_i \tag{4.74}$$

where

$$\theta_i = \sqrt{(Rc_i + j \cdot \omega \cdot Lc_i) \cdot (Gc_i + j \cdot \omega \cdot Cc_i)}$$
$$Zo_i = \sqrt{\frac{Rc_i + j \cdot \omega \cdot Lc_i}{Gc_i + j \cdot \omega \cdot Cc_i}} \tag{4.75}$$

And where *i* identifies the conductor.

4.6.6 Mathcad worksheet: page 1

Figure 4.40 is a copy of the first page of the Mathcad worksheet used to analyse the response of the model of Figure 4.39. The values for *Lc*, *Rc,* and *Cc* have been copied from the data created by the test and analysis described in Section 4.3.

That section derived a model which simulated the emission characteristics of a 15 m length of twin-conductor cable. The fact that the actual measurements correlated closely with the simulation of the model provides a fair level of confidence in the validity of the technique. This same model is the basis for the model of Figure 4.39 in this section; a model which simulates the susceptibility of the cable to external radiation.

The value of the radiation resistance *Rrad* has been rounded down, from 52 Ω to 50 Ω. The value of *Rn* simulates an open-circuit termination. Resistor *Rf* represents the resistance of the EED. It is assumed that there are no losses due to radiated emission. So the value of each conductance in *Gc* is set at zero.

$$Rn := 10^7 \qquad Rf := 2 \qquad Rrad := 50$$

$$Lc := \begin{pmatrix} 2.103 \\ 2.103 \end{pmatrix} \cdot 10^{-6} \quad Cc := \begin{pmatrix} 927.8 \\ 927.8 \end{pmatrix} \cdot 10^{-12} \quad Rc := \begin{pmatrix} 0.02 \\ 0.02 \end{pmatrix} \quad Gc := \begin{pmatrix} 0 \\ 0 \end{pmatrix}$$

$$Zbranch(f) := \left| \begin{array}{l} \omega \leftarrow 2 \cdot \pi \cdot f \\ \text{for } k \in 1..2 \\ \quad \left| \begin{array}{l} \theta \leftarrow \sqrt{(Rc_k + j \cdot \omega \cdot Lc_k) \cdot (Gc_k + j \cdot \omega \cdot Cc_k)} \\ Zo \leftarrow \sqrt{\dfrac{Rc_k + j \cdot \omega \cdot Lc_k}{Gc_k + j \cdot \omega \cdot Cc_k}} \\ Zh_k \leftarrow Zo \cdot tanh\left(\dfrac{\theta}{2}\right) \\ Zv_k \leftarrow Zo \cdot csch(\theta) \end{array} \right. \\ (Zh \quad Zv) \end{array} \right.$$

$$Zloop(f) := \left| \begin{array}{l} (Zh \quad Zv) \leftarrow Zbranch(f) \\ \omega \leftarrow 2 \cdot \pi \cdot f \\ Z11 \leftarrow Zh_1 + Zv_1 + Zv_2 + Zh_2 + Rn \\ Z12 \leftarrow -(Zv_1 + Zv_2) \\ Z22 \leftarrow 2 \cdot (Zh_1 + Zv_1 + Zv_2 + Zh_2) + Rrad \\ Z23 \leftarrow -(Zv_1 + Zv_2) \\ Z33 \leftarrow Zh_1 + Zh_2 + Zv_1 + Zv_2 + Rf \\ \begin{pmatrix} Z11 & Z12 & 0 \\ Z12 & Z22 & Z23 \\ 0 & Z23 & Z33 \end{pmatrix} \end{array} \right.$$

Figure 4.40 Defining the loop impedance components at each frequency

The function *Zbranch(f)* assigns values to every distributed impedance of Figure 4.39 at each frequency, using (4.74) and (4.75). The output is two vectors, *Zh* and *Zv*.

The function *Zloop(f)* takes as input the values of *Zh* and *Zv* at each frequency and assigns values to the loop impedances. It is created by inspecting the network of Figure 4.39. For example, *Z11* is the sum of the impedances through which the current I_1 flows and *Z12* is the sum of the impedances through which currents I_1 and I_2 flow. *Z12* is negative because these currents are flowing in opposite directions. The output of the function is an array of impedance components which can be used in vector analysis.

4.6.7 Mathcad worksheet: page 2

Figure 4.41 is a copy of the second page of the worksheet. It defines the variation of threat voltage with frequency in the range 100 kHz–2 MHz. Both the voltage and frequency scales are linear.

$c := 300 \cdot 10^6$ $\quad E := 10$ $\quad l := 15$

$\lambda q := 4 \cdot l = 60$ $\quad fq := \dfrac{c}{\lambda q} = 5 \times 10^6$ $\quad Vq := \dfrac{\lambda q}{\pi} \cdot E = 190.986$

$i := 1 .. 200$ $\quad F_i := i \cdot 10^5$

$$Vthreat_i := \left| \begin{array}{l} f \leftarrow F_i \\ V \leftarrow Vq \cdot \dfrac{f}{fq} \\ V \leftarrow Vq \cdot \dfrac{fq}{f} \quad \text{if } f > fq \end{array} \right.$$

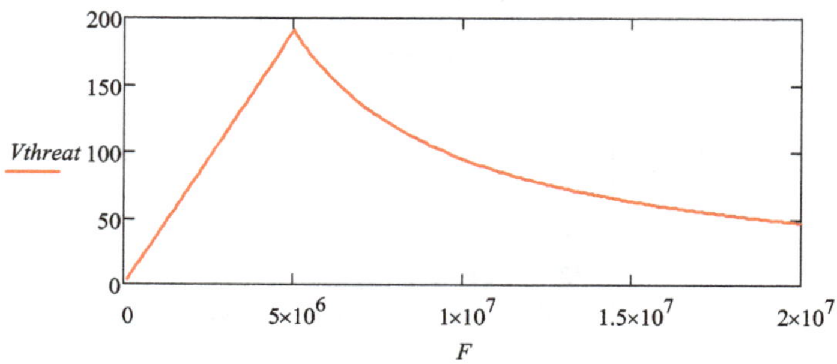

Figure 4.41 Defining the threat voltage

Given the assumption that the speed of light c is 3×10^8 m/s, the requirement that the firing system must remain safe when the amplitude of the external electromagnetic field E is 10 V/m, and the knowledge that the length l of the cable is 15 m; it can be deduced that the quarter-wavelength λq of the field under worst-case conditions is 60 m. This allows the frequency fq at which peak voltage will occur to be calculated by invoking (4.73). At this frequency, the voltage induced in the return cable is defined by replacing λ with λq in (4.70).

The magnitude of *Vthreat* will increase with frequency f from 100 kHz to fq. Then it will decrease exponentially as the frequency increases. The graph at the bottom of Figure 4.41 illustrates this.

4.6.8 Mathcad worksheet: page 3

This page invokes the functions defined in the first and second pages of the worksheet to calculate the frequency response of currents I_1 and I_2 of the network of Figure 4.39.

In the function defined on the right-hand side of $I2_i$, the first action is to look at the table of values of the vector F to select the frequency defined by the control variable i and assign it to the variable f. The second action is to select the threat voltage $Vthreat_i$ and assign that value to the variable $V2$. Since $Vthreat_i$ has been calculated at frequency F_i, *V2* is automatically assigned the value of the threat voltage at frequency *f*. This value is then assigned to the variable V_2 in the vector *V*.

The elements of the array *Z* are then defined as the set of loop impedances calculated for the frequency *f*. The command *lsolve*(*Z*,*V*) calculates the values of the three currents I_1, I_2, and I_3 at this frequency. But only the magnitude of I_2 is selected as the output of the function. This value is stored in row *i* of the vector *I2*. This vector defines all the values of the current in the middle loop of Figure 4.39 over the frequency range 100 kHz to 20 MHz. This current is that induced in the return conductor of the cable by the external electromagnetic field.

The same process is repeated to create the vector *I3*: the current flowing in the loop at the far end of the cable over the same range of frequencies, that is, the current flowing in the EED.

The graph near the bottom of Figure 4.42 shows how the currents *I1* and *I2* vary with *F*, the frequency.

$$I2_i := \left| \begin{array}{l} f \leftarrow F_i \\ V2 \leftarrow Vthreat_i \\ V \leftarrow \begin{pmatrix} 0 \\ V2 \\ 0 \end{pmatrix} \\ Z \leftarrow Zloop(f) \\ I \leftarrow lsolve(Z,V) \\ |I_2| \end{array} \right. \qquad I3_i := \left| \begin{array}{l} f \leftarrow F_i \\ V2 \leftarrow Vthreat_i \\ V \leftarrow \begin{pmatrix} 0 \\ V2 \\ 0 \end{pmatrix} \\ Z \leftarrow Zloop(f) \\ I \leftarrow lsolve(Z,V) \\ |I_3| \end{array} \right.$$

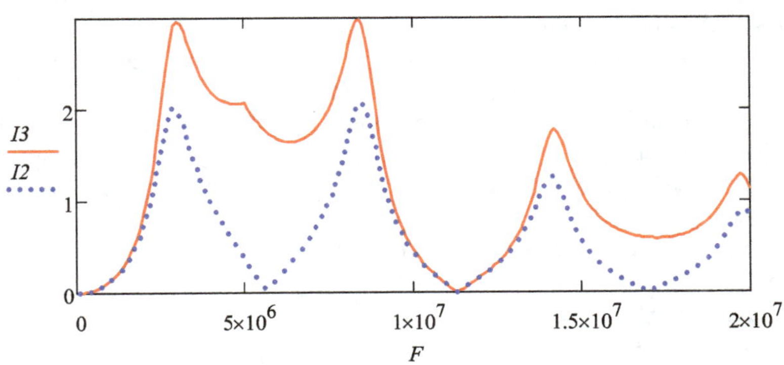

$I2max := max(I2) = 2.078$ $\qquad$ $I3max := max(I3) = 2.993$

Figure 4.42 Variation of currents with frequency

One of the built-in functions of Mathcad is *max*(*A*). This returns the largest element in vector A. So the final line in the worksheet defines the maximum value of each current.

The most interesting feature is the fact that there are two peaks of almost equal amplitude, and that neither peak coincides with the peak voltage illustrated by the curve of Figure 4.41.

4.6.9 Initial assessment

The graph at the bottom of Figure 4.42 shows the variation of each current over the selected range of frequencies. The dotted blue curve shows how the current I_2 in the middle loop of Figure 4.38 varies. This is the current induced in the return conductor by the external electromagnetic field.

The explanation of this must be that the external field induces a current which flows back and forth between the extremities of the assembly illustrated in Figure 4.37. As far as the external field is concerned, that assembly acts as a half-wave dipole. The cable itself acts as a transmission line. At every instant, the voltage at the near end doubles and the current at the far end changes direction. At a frequency of about 2.9 MHz, stored energy builds up in the cable. After several cycles of the threat voltage, the system resonates at about the quarter-wave frequency of the transmission line.

At just over 5 MHz, partial currents in each conductor of the middle loop are about the same amplitude and flowing in opposite directions. So there is a null in the response.

Since no energy is delivered to *Rn* at the near end of the cable, that energy is reflected back to the far end. So the current delivered to the EED is significantly higher than that induced by the field. The red curve on the graph of Figure 4.42 illustrates the response of this current.

The current I_3 in the EED peaks at about 3 A. So the power delivered to the device can be as high as $3^2 \times 2 = 18$ W. Since the NFT for power is 100mW, the device would certainly ignite.

This first assessment of the response of the system provides valuable information which can be used to improve the design.

Since the fundamental problem is due to the fact that the system resonates, the first objective must be to avoid resonant peaks in the response of the transmission line. This can be done by installing resistors equal to the characteristic impedance at each end.

The value of this impedance is as follows:

$$Ro = 2 \cdot \sqrt{\frac{Lc_1}{Cc_1}} = 95\ \Omega \tag{4.76}$$

An inductor can be inserted in series with the EED to reduce the current at high frequencies. A first estimate of the value to use would be 40 μH, since this is ten times the inductance of the cable.

4.6.10 Modified system

When these modifications are incorporated, the wiring diagram of the assembly would be as shown in Figure 4.43.

4.6.11 Modified model

As far as the distributed parameter model is concerned, the only change is the insertion of a fourth circuit loop appended to loop 3. Figure 4.44 illustrates this.

In this model, *L1* is depicted as a lumped parameter because it is constructed as a multi-turn component. The photons (the electromagnetic field) emitted from any particular turn will interact with every other turn, but the electrons will be constrained to flow along the surface of the conductor. Even though the components which make up loop 3 are represented as distributed parameters and those of loop 4 are lumped parameters, the interaction between these loops in the circuit model is as defined by circuit theory.

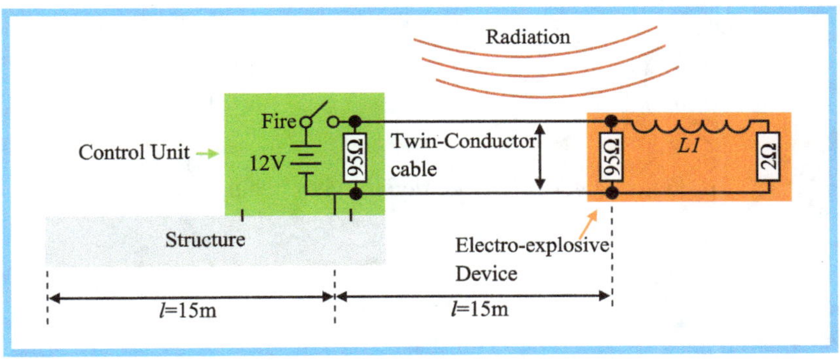

Figure 4.43 Modified firing system

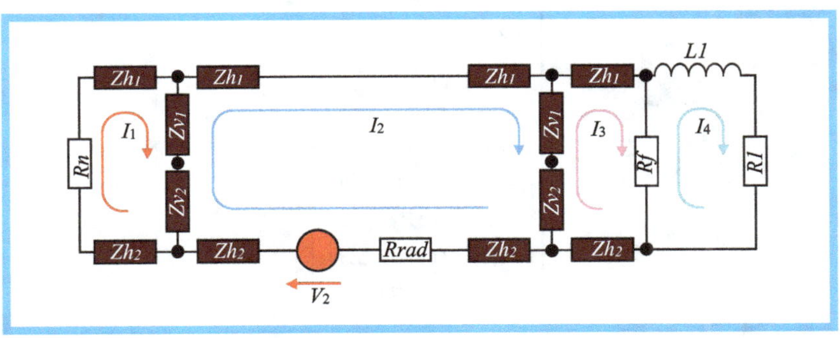

Figure 4.44 Modified circuit model

4.6.12 Modified worksheet

The addition of a resistor at each end of the cable and the addition of the inductor in series with the EED calls for a redefinition of the components at the beginning of the first page of the worksheet. Figure 4.45 is a record of this change.

Since the cable depicted in Figure 4.43 is exactly the same as that of Figure 4.37, the definition of the branch impedances, *Zbranch(f)*, remains the same.

The addition of a fourth loop in the model calls for an increase in the number of loop parameters. The function *Zloop(f)* needs now to include a definition of *Z44* and *Z34*. Figure 4.46 illustrates this. The output of this subroutine is a four-by-four array of impedance parameters.

The function *Vthreat* on the second page of the worksheet remains unchanged (Figure 4.41).

Since there are now four loops in the model, the definition of the voltage vector *V* in the functions *I2* also needs to reflect this change. The current flowing in the EED is now in loop 4, so the second function on page 3 of the worksheet is also changed (Figure 4.42) and renamed *I4*.

$$Rn := 95 \quad Rf := 95 \quad Rrad := 50 \quad L1 := 40\cdot 10^{-6} \quad R1 := 2$$

$$Lc := \begin{pmatrix} 2.103 \\ 2.103 \end{pmatrix}\cdot 10^{-6} \quad Cc := \begin{pmatrix} 927.8 \\ 927.8 \end{pmatrix}\cdot 10^{-12} \quad Rc := \begin{pmatrix} 0.02 \\ 0.02 \end{pmatrix} \quad Gc := \begin{pmatrix} 0 \\ 0 \end{pmatrix}$$

Figure 4.45 Modified definition of components

$$Zloop(f) := \left| \begin{array}{l} (Zh \;\; Zv) \leftarrow Zbranch(f) \\ \omega \leftarrow 2\cdot\pi\cdot f \\ Z11 \leftarrow Zh_1 + Zv_1 + Zv_2 + Zh_2 + Rn \\ Z12 \leftarrow -(Zv_1 + Zv_2) \\ Z22 \leftarrow 2\cdot(Zh_1 + Zv_1 + Zv_2 + Zh_2) + Rrad \\ Z23 \leftarrow -(Zv_1 + Zv_2) \\ Z33 \leftarrow Zh_1 + Zh_2 + Zv_1 + Zv_2 + Rf \\ Z34 \leftarrow -Rf \\ Z44 \leftarrow Rf + j\cdot\omega\cdot L1 + R1 \\ \begin{pmatrix} Z11 & Z12 & 0 & 0 \\ Z12 & Z22 & Z23 & 0 \\ 0 & Z23 & Z33 & Z34 \\ 0 & 0 & Z34 & Z44 \end{pmatrix} \end{array} \right.$$

Figure 4.46 Modified definition of the function Zloop(f)

$$I2_i := \left| \begin{array}{l} f \leftarrow F_i \\ V2 \leftarrow Vthreat_i \\ V \leftarrow \begin{pmatrix} 0 \\ V2 \\ 0 \\ 0 \end{pmatrix} \\ Z \leftarrow Zloop(f) \\ I \leftarrow lsolve(Z, V) \\ |I_2| \end{array} \right.$$

$$I4_i := \left| \begin{array}{l} f \leftarrow F_i \\ V2 \leftarrow Vthreat_i \\ V \leftarrow \begin{pmatrix} 0 \\ V2 \\ 0 \\ 0 \end{pmatrix} \\ Z \leftarrow Zloop(f) \\ I \leftarrow lsolve(Z, V) \\ |I_4| \end{array} \right.$$

$I2max := max(I2) = 0.784$

$I4max := max(I4) = 0.061$

$P2 := I2max^2 \cdot Rf = 58.395$

$P4 := I4max^2 \cdot R1 = 7.406 \times 10^{-3}$

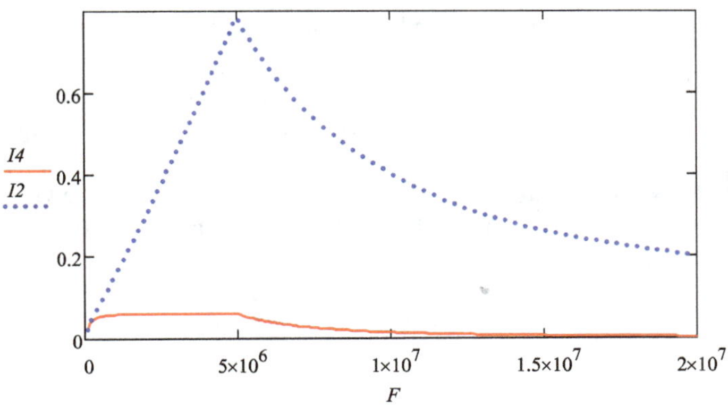

Figure 4.47 Response of modified model

It was found that, for each current, there was only one peak in the frequency response. This is illustrated on Figure 4.47.

4.6.13 Final assessment

There is only one resonant peak in the response of the modified system: that due to the cable and structure acting in the same way as a dipole antenna. The dotted curve for *I2*, showing the current flowing along the return conductor and back along the send conductor, demonstrates this. The peak current and the peak voltage occur at the same frequency: 5 MHz. This is because the resistors *Rn* and *Rf* absorb all the energy delivered by the electromagnetic field.

Also, the amplitude of the maximum value of *I2* has reduced from 2.1 A in the original design to 0.8 A in the modified system. The reduction in the peak value of the current in the EED has reduced from 3A in the original design to 61 mA in the final version.

The NFT power for the device has been defined as 100 mW. The power delivered to the EED in the final version is 7.4 mW. That is, there is a fair margin of safety in the final design.

There is still a problem. The modified design requires the resistor at each end to be able to absorb 60 W. The next section shows how this problem can be avoided.

4.6.14 Conclusion

A step-by-step method has been described which enables a circuit model to predict the response of a firing control system when that system is exposed to a defined level of electromagnetic interference over a defined frequency range. It has been shown that the basic system is unsafe.

Assessment of the mechanisms involved in the transfer of power from the threat field to the EED identifies a simple modification which could improve matters.

Analysis of the modified system indicates that it will ensure that it will remain safe under the defined hazard conditions.

The supreme benefit to be derived from such an approach is that the analysis can be carried out before any hardware is constructed. Measurements of the response of a simple test rig can be used to validate that analysis. The results of such a set of tests can then be subject to a design review before drawings of the final assembly are started. That is, EMC requirements can be included in the normal design process.

Chapter 5
Transient analysis

5.1 Time step analysis

5.1.1 Introduction

The classical approach to time-domain analysis of EMI is to define the spectrum of the *E*-field, analyse the response of the system in the frequency domain, and then transform the results back into the time domain. There is a simpler way; stay in the time domain.

Even so, the approach taken by academia would be to invoke the use of Laplace transforms or Heaviside transforms. Again, this would lead to some heavy-going mathematics.

Such complexity can be bypassed by using time-step analysis. This is similar to the use of iterative techniques used to solve simultaneous equations. Simulation Programs with integrated circuit emphasis (SPICE) have used this method for decades.

But there is still a fundamental problem; SPICE software is based on nodal analysis which invokes the concept of the equipotential ground, that is a conductor at which all points are at zero voltage. This can be represented as a node to which all other voltages in the system are referred. Such a concept contradicts all the lessons learned from electromagnetic theory.

This problem can be overcome by using mesh analysis. This relates the strength of the *E*-field to a voltage source in series with the victim conductor, calculates the current in each circuit loop, and defines the response of the interface circuitry at each end of the signal link.

This section provides an introduction to the technique.

5.1.2 Basic equations

Time step analysis is similar to the use of the iterative method used in mathematics to solve non-linear equations, where each step in the process yields a more accurate value for the unknown variables. The essential difference is that each calculation defines the condition of the circuitry at a later time. By defining the size of each time step to be sufficiently small, it can be assumed that the rate of change of each current and each voltage will be linear.

The basic equations are the simplest possible:
For an inductor

$$V = L \cdot \frac{dI}{dt} \tag{5.1}$$

For a resistor

$$V = R \cdot I \tag{5.2}$$

For a capacitor

$$V = \frac{Q}{C} \tag{5.3}$$

Where

$$Q = \sum_{0}^{t} I \cdot dt \tag{5.4}$$

It follows from (5.4) that

$$I = \frac{dQ}{dt} \tag{5.5}$$

Given that t is the elapsed time, dt is a finite increment of time, dI is a finite increment of current, and dQ is a finite increment of charge.

5.1.3 Series LCR circuit

For the circuit illustrated in Figure 5.1, it is assumed that the input Vin is a step voltage of 1 V, and the objective is to determine the waveform of the current I. The first step in the process is to set out the voltage equation:

$$Vin = L \cdot \frac{dI}{dt} + R \cdot I + \frac{Q}{C} \tag{5.6}$$

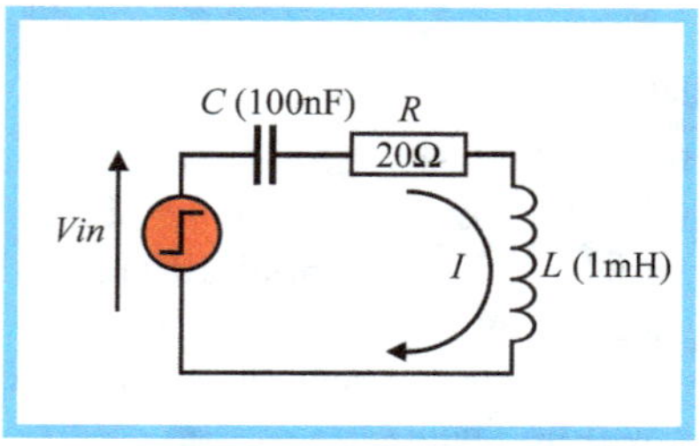

Figure 5.1 Series LCR circuit

The next step is to rearrange this equation to put the variable dI on the left-hand side:

$$dI = \frac{dt}{L} \cdot \left(Vin - R \cdot I - \frac{Q}{C} \right) \tag{5.7}$$

If the initial values of I and Q are zero, apart from dt, the values of all the parameters on the right-hand side of the equation are known. A guess-value for this parameter can be obtained by determining the time constant $\sqrt{L \cdot C} = 1 \times 10^{-5}$.

Dividing this time constant by 10 gives the guess value: $dt = 10^{-6}$.

The task then is to compile a computer program which calculates the value of dI during the time interval dt and use it to update the values of I and Q. Then these new values can be assigned to the right-hand side of (5.7) and the process repeated. The value of the current I can be recorded after each time step. Figure 5.2 illustrates the response.

5.1.4 Low pass filter

A slightly more complex example is that of Figure 5.3, which simulates a low-pass filter. To enable the model to simulate steady-state conditions, it is necessary to place a high value of resistance in parallel with the capacitance.

The loop equations for this filter are as follows:

$$\begin{aligned} Vin &= \mathbf{R1} \cdot Ia + L1 \cdot \frac{dIa}{dt} + \frac{Ia \cdot dt}{C1} - \frac{Ib \cdot dt}{C1} \\ 0 &= -\frac{Ia \cdot dt}{C1} + \frac{Ib \cdot dt}{C1} + R2 \cdot Ib - \mathbf{R2} \cdot Ic \\ 0 &= -\mathbf{R2} \cdot Ib + \mathbf{R2} \cdot Ic + L2 \cdot \frac{dIc}{dt} + R3 \cdot Ic \end{aligned} \tag{5.8}$$

Let

$$Q = (Ia - Ib) \cdot dt \tag{5.9}$$

Re-arranging (5.8) and invoking (5.9)

$$\begin{aligned} dIa &= \frac{dt}{\mathbf{L1}} \cdot \left[Vin - \mathbf{R1} \cdot Ia - \frac{Q}{C1} \right] \\ Ib &= \frac{1}{R2} \cdot \left[\frac{Q}{C1} + R2 \cdot Ic \right] \\ dIc &= \frac{dt}{L2} \cdot [\mathbf{R2} \cdot Ib - \mathbf{R2} \cdot Ic - R3 \cdot Ic] \end{aligned} \tag{5.10}$$

The input variables of the Mathcad worksheet are derived from the circuit model of Figure 5.3. The function *next(Ia,Ib,Ic,Q)* incorporates the set of equations defined by (5.10). The main program calculates a set of values for the output voltage *Vout* across 100 Ω load *R3*.

The graph at the bottom of the worksheet (Figure 5.4) illustrates the response of this particular low pass filter to a 1 V step input.

$L := 1 \cdot 10^{-3}$ $\quad C := 100 \cdot 10^{-9}$ $\quad R := 20$ $\quad \sqrt{L \cdot C} = 1 \times 10^{-5}$

$Vin := 1$ $\quad dt := 10^{-6}$ $\quad N := 100$

$I := 0$ $\quad Q := 0$

$$next(I, Q) := \left| \begin{array}{l} dI \leftarrow \frac{dt}{L} \cdot \left(Vin - R \cdot I - \frac{Q}{C} \right) \\ I \leftarrow I + dI \\ Q \leftarrow Q + I \cdot dt \\ (I \quad Q) \end{array} \right.$$

$$Iout := \left| \begin{array}{l} \text{for } i \in 1..N \\ \quad \left| \begin{array}{l} Imon_i \leftarrow I \\ (I \quad Q) \leftarrow next(I, Q) \end{array} \right. \\ Imon \end{array} \right.$$

$i := 1..N$ $\quad t_i := (i-1) \cdot dt$

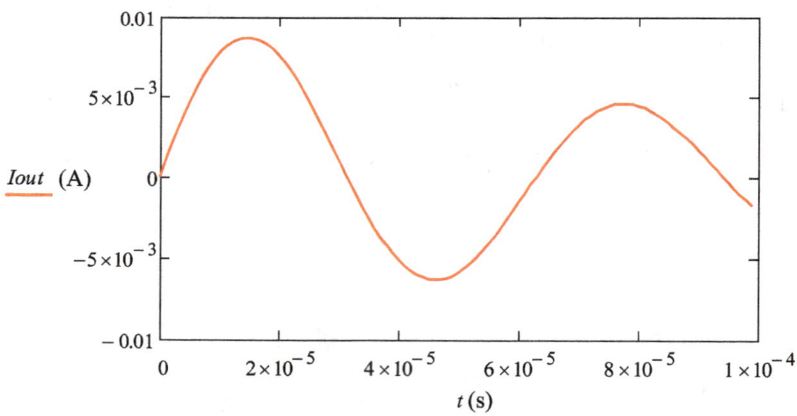

Figure 5.2 Transient response of series LCR circuit

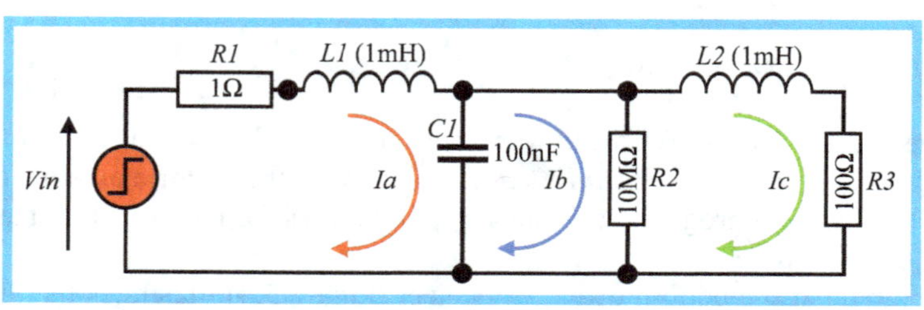

Figure 5.3 Low pass filter

$L1 := 1 \cdot 10^{-3}$ $\quad L2 := 1 \cdot 10^{-3}$ $\quad C1 := 100 \cdot 10^{-9}$

$R1 := 1$ $\quad R2 := 10^{7}$ $\quad R3 := 100$

$Vin := 1$ $\quad dt := 10^{-6}$ $\quad N := 200$

$Ia := 0$ $\quad Ib := 0$ $\quad Ic := 0$ $\quad Q := 0$

$$next(Ia, Ib, Ic, Q) := \left| \begin{array}{l} dIa \leftarrow \dfrac{dt}{L1} \cdot \left(Vin - R1 \cdot Ia - \dfrac{Q}{C1} \right) \\ Ib \leftarrow \dfrac{Q}{R2 \cdot C1} + Ic \\ dIc \leftarrow \dfrac{dt}{L2} \cdot [R2 \cdot Ib - (R2 + R3) \cdot Ic] \\ Ia \leftarrow Ia + dIa \\ Ic \leftarrow Ic + dIc \\ Q \leftarrow Q + (Ia - Ib) \cdot dt \\ (Ia \quad Ib \quad Ic \quad Q) \end{array} \right.$$

$$Vout := \left| \begin{array}{l} for \ i \in 1..N \\ \quad \left| \begin{array}{l} Vout_i \leftarrow R3 \cdot Ic \\ (Ia \quad Ib \quad Ic \quad Q) \leftarrow next(Ia, Ib, Ic, Q) \end{array} \right. \\ Vout \end{array} \right.$$

$i := 1..N$ $\qquad t_i := (i - 1) \cdot dt$

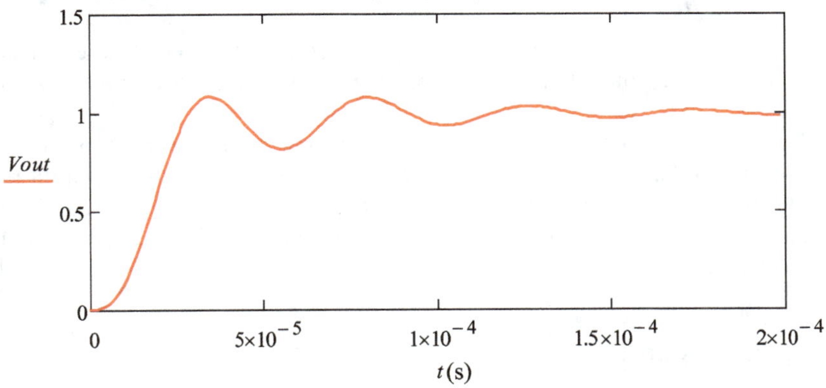

Figure 5.4 Transient response of low-pass filter

5.1.5 Assessment

A review of the calculations reveals that the parameters are time, voltage, charge, current, resistance, inductance, and capacitance. The concepts of frequency, impedance, admittance, and phase angle have no existence in any analysis in the time domain.

The process used to create and run the program can be summarised:

Define the values of the components of the network.

Define the loop equations, the value of the time step, and the waveform of the voltage source.

To calculate the values of the parameters after each time-step:

Define the new value of the voltage source.
Define the step change of current in each inductor.
Define the new current in each resistor.
Define the new charge in each capacitor.
Provide the new values of the currents and charges as the output of the function.

To create the output waveform:

Define the timescale.
Define the current or voltage in the selected component.
Provide the value of the selected parameter as the output after each time-step.

When the program is run, check that the output is plausible. If necessary, reduce the value of the time-step. If the process takes too long, increase the time-step.

It usually takes less than a second for the software to carry out the computations. This means that the designer can alter the value of any component and note how this affects the output waveform.

5.1.6 Conclusion

A method of using mesh analysis to compile programs to simulate the transient behaviour of circuit models has been described and illustrated using two simple examples.

The calculations used in the above two examples can be carried out by any SPICE program. With such software, the engineer must rely on its infallibility.

The advantages of mesh analysis are that it can be developed to simulate all the mechanisms involved in the propagation of transient interference, and that the designer remains in complete control of the process.

5.2 Charges and photons

5.2.1 Introduction

By invoking the concept of the shift register, a program is developed to simulate the propagation of charges back and forth along a cable. By adding the equations which define the incident, reflected, and absorbed currents, a program is developed which can simulate the response of any transmission line to a transient waveform.

The response of an open-circuit line to a 40 ns pulse is simulated, and a snapshot is provided of the distribution of current and voltage along the line after an elapsed time of 60 ns. A similar snapshot illustrates the status of a short-circuited line.

The program itself provides an insight into the mechanism involved. It reveals that voltage and current at any point at any instant are each a function of charge. Further reasoning leads to the conclusion that charge itself is a manifestation of the interaction between photons and conducting material.

5.2.2 Synchronisation

Figure 5.5 illustrates a schematic diagram of a twin-conductor transmission line. The voltage source *Vgen* at the near end of the line creates a current *In*. This current propagates along the line at a velocity similar to that of light. The current *If* arriving at the far end generates a voltage *Vf* across that load. Analysis in the frequency domain leads to the derivation of formulae for the characteristic impedance *Zo* and the phase constant λ. A significant relationship which emerges is that:

$$\frac{\delta I}{\delta x} = \frac{1}{Zo} \cdot \frac{\delta V}{dx} \tag{5.11}$$

Where V is the voltage between the send and return conductors at the point x and I is the loop current at the same location. This relationship is derived in any book on electromagnetic theory; or in Section 2.2.

The more familiar relationship is as follows:

$$I = \frac{1}{Zo} \cdot V \tag{5.12}$$

Equation (5.11) means that the rate of change of current is proportional to the rate of change of voltage. Equation (5.12) means that the current is also proportional to the voltage. The only way for these two equations to be valid is for the phase difference between current and voltage to be zero; at least during forward propagation of the signal. Figure 5.6 illustrates this.

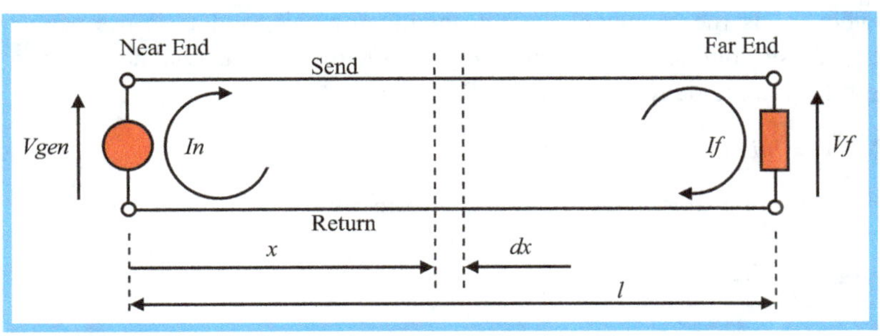

Figure 5.5 Schematic diagram of a twin conductor transmission line

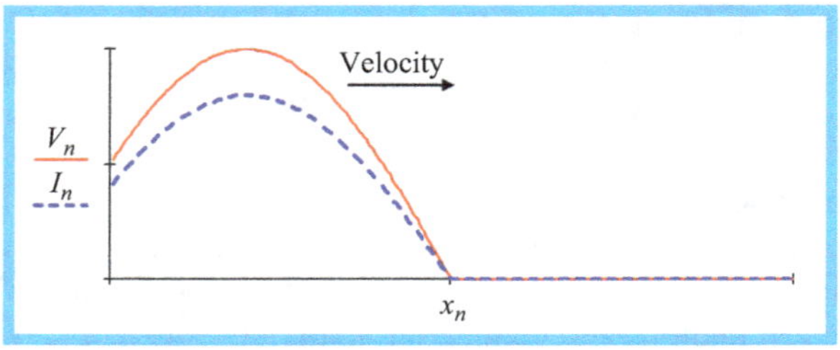

Figure 5.6 Relationship between voltage and current waveforms

5.2.3 Delay line model

Time-step analysis is based on the assumption that the current is constant for a finite time *dt*. If it is assumed that there are no losses in the line, the assembly of Figure 5.5 can be modelled as a set of *n* segments, each of length:

$$dx = \frac{l}{n} \tag{5.13}$$

If *T* is the time taken for a pulse to travel from one end of the line to the other, the time taken for it to propagate along one segment is as follows:

$$dt = \frac{T}{n} \tag{5.14}$$

If a voltage *Vn* is applied to the near end of the line, the current *In* flowing into the line would be

$$In = \frac{Vn}{Ro} \tag{5.15}$$

Where

$$Ro = \sqrt{\frac{L}{C}} \tag{5.16}$$

And where *L* is the inductance of the conductor pair, and *C* is the capacitance between those conductors. During the time *dt*, the charge *Qni* delivered to one segment by the incident current *Ini* is as follows:

$$Qni = Ini \cdot dt \tag{5.17}$$

This charge propagates along the line at a velocity *v*

$$v = \frac{dx}{dt} = \frac{l}{T} \tag{5.18}$$

At the far end, some of this charge is delivered to the interface circuitry at the receiver. What is not absorbed is reflected. The reflected charge *Qf* travels back to

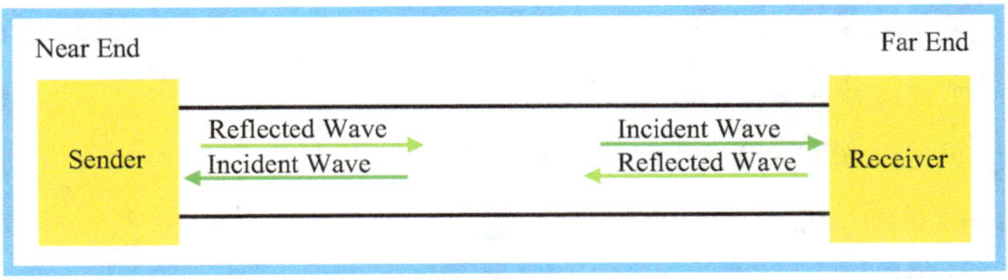

Figure 5.7 Block diagram illustrating the propagation of electromagnetic energy

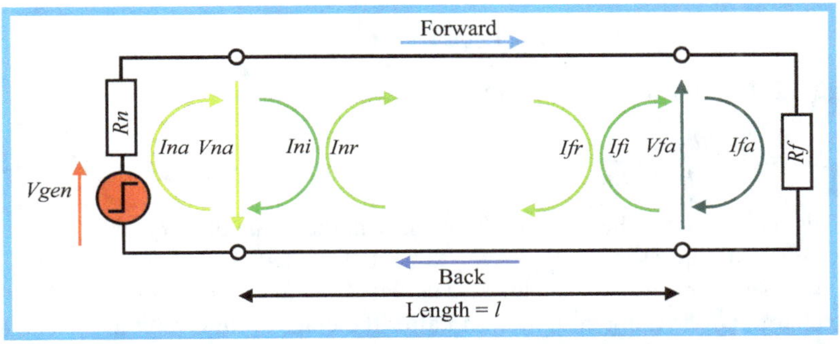

Figure 5.8 Delay line model

the near end and appears as an incident voltage *Vni*. In terms of electromagnetic theory, this energy is carried by an electromagnetic wave, as illustrated by the block diagram of Figure 5.7.

5.2.4 *Reflections*

There are now two voltage sources appearing at the near end: The generator *Vgen* and the voltage *Vni* delivered by the incident wave. To determine the response at this end, it is necessary to define the parameters and their relationships.

In Figure 5.8, *Ini* is the incident current arriving at the near end, and *Inr* is the current reflected back into the line. The current absorbed at the near end is *Ina*. Similarly, *Ifi*, *Ifr*, and *Ifa* are incident, reflected, and absorbed currents at the far end. These currents are depicted as loop currents flowing along the send conductor and back via the return conductor. All current loops are defined as clockwise. Rn is the source resistance at the near end, and Rf is the load resistance at the far end. The relationships between currents and voltages are as follows:

$$Vni + Vnr = Vna \tag{5.19}$$

$$Inr + Ini = Ina \tag{5.20}$$

$$Vni = Ro \cdot Ini \tag{5.21}$$

$$Vnr = -Ro \cdot Inr \tag{5.22}$$

$$Vna + Vgen = Rn \cdot Ina \tag{5.23}$$

Adding (5.21) and (5.22) and invoking (5.20)

$$Vni - Vnr = Ro \cdot Ina \tag{5.24}$$

Using (5.23) to substitute for *Vna* in (5.19)

$$Vni + Vnr = Rn \cdot Ina - Vgen \tag{5.25}$$

Adding (5.24) and (5.25)

$$2 \cdot Vni = (Ro + Rn) \cdot Ina - Vgen \tag{5.26}$$

Using (5.21) to substitute for *Vni* in (5.26) and rearranging

$$Ina = \frac{2 \cdot Ro \cdot Ini + Vgen}{Ro + Rn} \tag{5.27}$$

This means that if the value of *Ini* is known, the value of *Ina* can be calculated. Invoking (5.20) allows the value of *Inr* to be determined. The charge *Qnr* delivered to the first segment of the line can be calculated using (5.17). This gives the value of the charge *Qfi* arriving at the far end after the elapsed time *T* (at any instant, the values of *Qnr* and *Qfi* are usually quite different).

The current *Ifi* arriving at the far end is as follows:

$$Ifi = \frac{Qfi}{dt} \tag{5.28}$$

The current at the far end can be calculated using the relationship

$$Ifa = \frac{2 \cdot Ro \cdot Ifi}{Ro + Rf} \tag{5.29}$$

5.2.5 Computation

Having defined all the relationships involved in simulating the response of the line, the next step is to create a Mathcad worksheet to simulate the mechanism involved.

Figure 5.9 is a copy of the first page. As with any program, the first action is to define the variables used on the page; the voltage *Vg* of the generator, the characteristic resistance *Ro*, and the values of the resistors *Rn* and *Rf* at the near and far ends. It is assumed that the length of the line *l* is 15 m and that the velocity of propagation *v* is 300 m/μs. This gives a value for the time *T* it takes for a pulse to travel from one end to the other. It is assumed that the number of segments *n* of the line is 100. This gives a value of *dt*, the time it takes a charge to move from one segment to the next.

The function *near(Qni , Vg)* invokes (5.17) and (5.27) to calculate the value of *Ina*, then (5.20) and (5.17) to evaluate *Qnr*. The output of this function is a vector containing the variables *Qnr* and *Ina*.

$$Vg := 100 \qquad Ro := 100 \qquad Rn := 10 \qquad Rf := 10^7 \qquad l := 15$$

$$n := 100 \qquad v := 3 \cdot 10^8 \qquad T := \frac{l}{v} = 5 \times 10^{-8} \qquad dt := \frac{T}{n} = 5 \times 10^{-10}$$

```
near(Qni, Vg) :=  | Ini ← Qni/dt
                  | Ina ← (2·Ro·Ini + Vg)/(Ro + Rn)
                  | Qnr ← (Ina − Ini)·dt
                  | (Qnr  Ina)

forward(F, Qn) := | F_1 ← Qn
                  | for x ∈ n + 1..1
                  |   F_(x+1) ← F_x
                  | Qf ← F_(n+2)
                  | (F  Qf)

far(Qfi) :=       | Ifi ← Qfi/dt
                  | Ifa ← 2·Ro·Ifi/(Ro + Rf)
                  | Qfr ← (Ifa − Ifi)·dt
                  | (Qfr  Ifa)

back(B, Qf) :=    | B_(n+1) ← Qf
                  | for x ∈ 2..n + 1
                  |   B_(x−1) ← B_x
                  | Qn ← B_1
                  | (B  Qn)
```

Figure 5.9 First page of Mathcad worksheet

The parameter $\boldsymbol{F}$ in the input to the function *forward*($\boldsymbol{F}$,*Qn*) is a vector of n lines and one column. It holds the value of the charge in every segment of the line. The first action of the function is to place the value *Qn* in the first segment. Then it moves the value of the charge in every segment to the next segment. This leaves the first segment free to accept the next input. The value of the parameter *Qf* in the final segment is provided as one output of the function. The other output is the new set of values in the vector $\boldsymbol{F}$.

The function *far*(*Qfi*) takes as input the value of the charge delivered to the end of the line. Then it calculates the value of the current *Ifa* and that of the charge *Qfr* reflected back into the line. These values are provided as the output of this subroutine.

The function *back*($\boldsymbol{B}$,*Qf*) processes the value of *Qf* in much the same was as *forward*($\boldsymbol{F}$,*Qn*) shifts *Qn* forward one segment. But this time, the process is reversed. *Qfr* is shifted back one segment. This means that the output *Qni* arriving at the near end of the line is that which was delivered to the far end at time $(t - T)$.

The functions *forward*(***F***,*Qn*) and *back*(***B***,*Qf*) behave in essentially the same was as shift registers.

The main program is defined in Figure 5.10, the second page of the Mathcad worksheet. This simulates the behaviour of the transmission line when a single pulse of 100 V is applied to the line via a 10 Ω resistor (*Rn*). The duration of this pulse, *Tmark* is defined as 40 ns. Since the time taken for a charge to move from segment to segment is 0.5 ns, the number of segments, *width*, to which this charge is delivered, is 80.

The simulation shows that the leading edge of the pulse propagates forward to the far end, is reflected, and then propagates back towards the near end. The objective of this particular simulation is to determine the distribution of charges along the line at the instant *Tfreeze* = 60 ns. This is essentially the same as freezing

$$Tmark := 40\cdot 10^{-9} \qquad width := \mathrm{floor}\left(\frac{Tmark}{dt}\right) = 80$$

$$Tfreeze := 60\cdot 10^{-9} \qquad Nfreeze := floor\left(\frac{Tfreeze}{dt}\right) = 120$$

$$N := 2\cdot n - 1 \qquad i := 1..N \qquad t_i := i\cdot dt \qquad x_i := i\cdot\frac{l}{n}$$

```
(Qfd  Qbk) :=  | F[n+2] ← 0                      Main Program
               | B[n+1] ← 0
               | for i ∈ 1..N − 1
               |     | Vgen ← Vg  if i > 1
               |     | Vgen ← 0  if i > width
               |     | (Qnr  Ina) ← near(Qni, Vgen)
               |     | (F  Qfi) ← forward(F, Qnr)
               |     | (Qfr  Ifa) ← far(Qfi)
               |     | (B  Qni) ← back(B, Qfr)
               |     | if i = Nfreeze
               |     |     | Qfd ← F
               |     |     | Qbk ← B
               | (Qfd  Qbk)

(Ix  Vx) :=  | for x ∈ 1..n + 1
             |     | I[x] ← (Qfd[x] + Qbk[x]) / dt
             |     | V[x] ← ((Qfd[x] − Qbk[x]) / dt)·Ro
             | (I  V)
```

Figure 5.10 Second page of Mathcad worksheet

a video picture. Hence, the number of time steps which will have occurred up to this instant would be *Nfreeze*. This is equivalent to 120 line segments. (The Mathcad function 'floor' ensures that the value returned is a whole number.)

The number N of time steps of the simulation ensures that pulse at $t = 1$ will arrive back at the near end. The time taken is $2 \cdot T$. The vector t provides a value for the horizontal axis of any graph of the waveform of any voltage or current. The vector x allows the distribution of current or voltage along the line to be depicted.

The program first defines the vector ***F*** which holds the value of the forward moving charge at every segment of the line at any instant. The vector ***B*** holds the values of the charges moving back along the line.

The repetitive part of the program first defines the waveform of the generator voltage, *Vgen*. Then it calculates the values of the current at each termination and the charges located at every segment at every instant. The output is two vectors; ***Qfd*** holds the values of the forward-moving charge at each line segment at the time *Tfreeze*, and ***Qbk*** holds the values of the charges moving back towards the near end.

The final function defined in Figure 5.9 calculates the sum and difference between the forward and backward moving charges and computes the value of the current ***Ix*** and the voltage ***Vx*** at each location of the line at the instant *Tfreeze*.

Since the charge Qfd_x is propagating forward, whilst Qbk_x is flowing backward, the sum of these parameters gives a value for the amount of charge flowing along that particular segment. (*Qbk* is a negative quantity since it is flowing backwards). Equally, the difference between these parameters gives a value for the charge accumulated at that location.

5.2.6 Results

Figure 5.11 depicts the graphs of the variation of *Vx* and *Ix* with distance x along the length of the line. *Vx* is the voltage between the conductors, and *Ix* is the value of the current flowing forward along the send conductor and back via the return conductor. The leading edge is propagating back towards the near end of the line, whilst the trailing edge is still moving forward. The voltage at the far end doubles, whist the current drops to zero. The arrows illustrate the direction of propagation of each edge of the waveform.

Figure 5.12 depicts these same two parameters when the resistance at the far end of the line is zero. That is, when the terminals at the far end are shorted together. This time, the voltage at the far end drops to zero whilst the amplitude of the current doubles.

5.2.7 Assessment

The program defined in Figures 5.9 and 5.10 calculates the amplitude of the current *Ina* at the near end of the line. This current flows in the resistor *Rn*. It is possible to calculate the value of the voltage across this resistor at any instant. The same is true of the current and voltage at the far end. Any of these parameters can be selected as

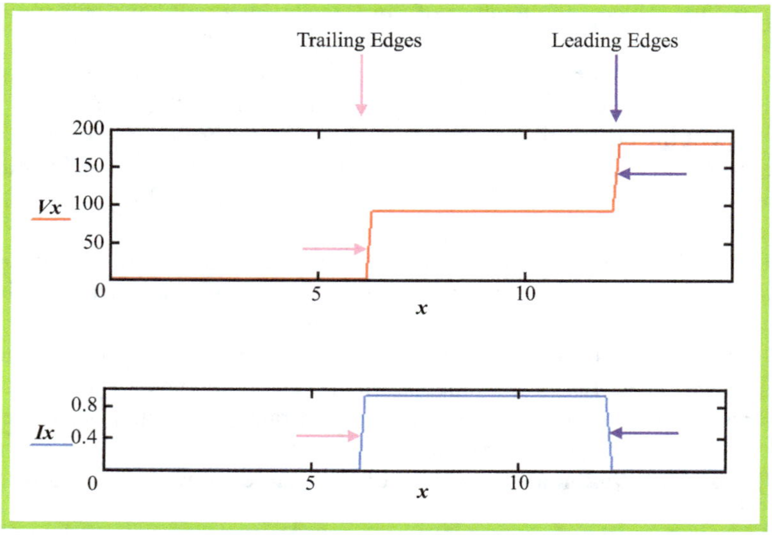

Figure 5.11 Distribution of voltage and current at t = 60 ns: far end open-circuit

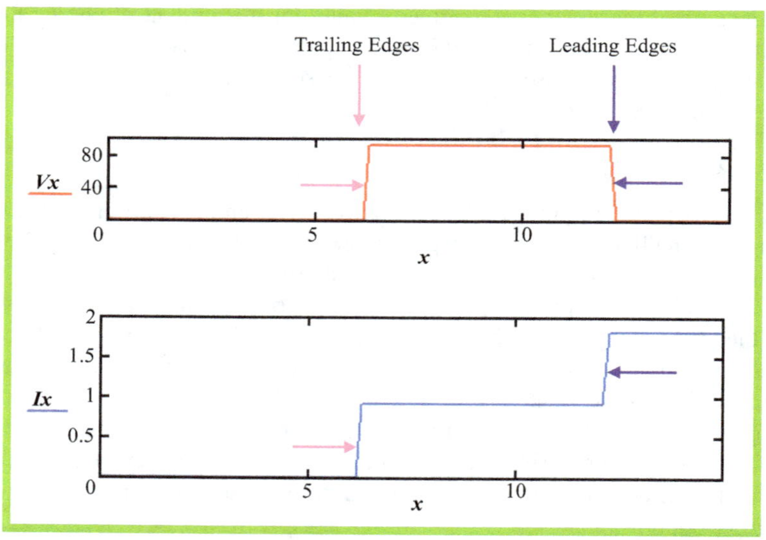

Figure 5.12 Distribution of voltage and current at t = 60 ns: far end short-circuit

the output of the main program. The waveform of the input voltage *Vgen* can be defined as a voltage which varies with time. So it is possible to simulate the response of the signal link between the sender and receiver to any transient waveform.

5.2.8 Underlying mechanism

The results illustrated in Figures 5.11 and 5.12 are entirely in accordance with the conclusions of transmission line theory. Voltage doubles at an open-circuit; current doubles at a short circuit.

The significance of this illustration is that it identifies the fact that voltage and current are manifestations of the action of charges. Voltage is a function of the number of charges accumulated at a particular segment. Current is a function of the flow of charges through that segment.

But charged particles do not move at the velocity of light along conductors. Nor do they move freely through the insulating material enclosing conductors. The entities with that ability are photons, and photons do not carry charge. But they can apply force. The logical conclusion must be that charge itself is a function of the interaction of photons with atoms. When they interact, a force is applied to the conducting material and hence, between the conductors.

However, it is unnecessary to understand the intricacies of quantum physics to design electronic circuitry. Nor is it necessary to be able to handle the complexities of full-field analysis, as developed by Electromagnetic Theory.

The concept of the shift register is well understood by circuit designers. Invoking this concept allows the mechanisms involved in the transmission of energy along and between transmission lines to be analysed using the equations of Circuit Theory.

5.2.9 Conclusion

A program has been developed which allows the transient response of any transmission line to be simulated.

The analysis also identifies the fact that currents and voltages are manifestations of the action of charges. Charges themselves are due to the interaction of photons with conductors.

5.3 Power line transients

5.3.1 Introduction

Transient spikes due to load switching are a prolific source of interference in any electronic system. Circuit models can be created to simulate the interaction between equipment and cable.

Power line filters are an essential part of the design of any unit of electronic equipment because they form a barrier between the Electromagnetic Interference (EMI) which has been picked up by the supply line and the equipment to which the power is being delivered. But they do this by reflecting that EMI back down the line. This doubles the level of unwanted power carried by the cable and doubles the EMI created by that line. Moreover, when the equipment unit is switched off, the transient power stored in the filter has only one place to go; out into the environment.

This section analyses the effect of an inductive load and of a capacitive load on the cable. It concludes that a filter with reactive components will interact with the supply cable to create high-frequency resonance, and this resonance will be a rich source of EMI.

5.3.2 Capacitive load

If a switch at the input end of a power line is closed, as illustrated by Figure 5.13, a step current will start to flow along a path defined by the routing of the cable. This step will propagate at a velocity similar to that of the speed of light in a vacuum. If it is assumed that the line is lossless, the amplitude of the step arriving at the far end of the line will be the same as that delivered to the near end.

The current *In* delivered to the near end is as follows:

$$In = \frac{Vn}{Ro} \tag{5.30}$$

where *Vn* is the voltage at the near end, and *Ro* is the characteristic resistance of the line. If *l* is the length of the line and *v* is the velocity of propagation, the time *T* taken for the step to traverse from the near end to the far end will be

$$T = \frac{l}{v} \tag{5.31}$$

So the current arriving at the far end *If* will be the same as that which existed at the near end *T* seconds previously. By treating the line as a number *n* of segments of equal length, this propagation can be simulated by a shift register. The time taken for the step to move from one segment to the next is as follows:

$$dt = \frac{T}{n} \tag{5.32}$$

where *dt* is a finite period of time.

It is assumed that the generator voltage *Vg* is constant at 100 V and that the switch closes at time 0+*dt* and switches off at time *Toff*. The number of time steps

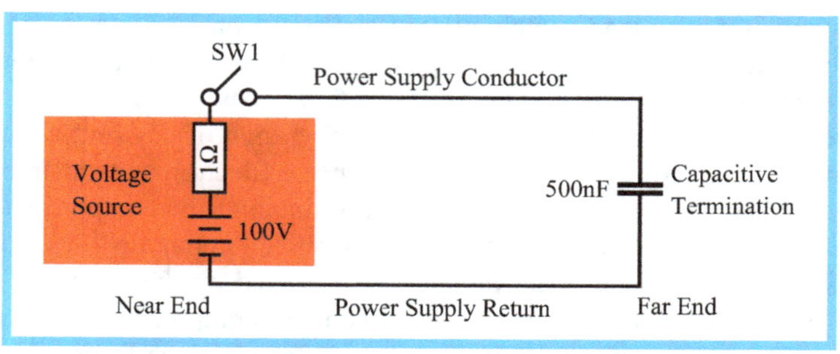

Figure 5.13 Power line wiring diagram, capacitive termination

Noff at that instant is as follows:

$$Noff = \frac{Toff}{dt} \tag{5.33}$$

The parameters involved in simulating the propagation of energy back and forth are defined in Figure 5.14. The far end equations can be derived from the circuit of Figure 5.15.

From Figure 5.15

$$2 \cdot Ro \cdot Ifi = \frac{Qfa}{C} + (Ro + Rf) \cdot Ifa \tag{5.34}$$

This leads to the definition of the Mathcad function *far*(*Ifi*,*Qfa*), shown on Figure 5.16. This takes the parameters *Ifi* and *Qfa* as input, calculates what the values of *Ifr*, *Ifa,* and *Qfa* would be after time *dt*, and provides these as the output.

Having defined this function, the task of computing the response of the system can be carried out by a Mathcad worksheet similar to that used in Section 5.2. So it was just a matter of copying that program onto a blank worksheet and altering a few details. The result is the worksheet shown in Figures 5.17 and 5.18.

The basic parameters are defined at the top of Figure 5.17. It is assumed that the source voltage is 100 V, the characteristic resistance of the line is 100 Ω, the

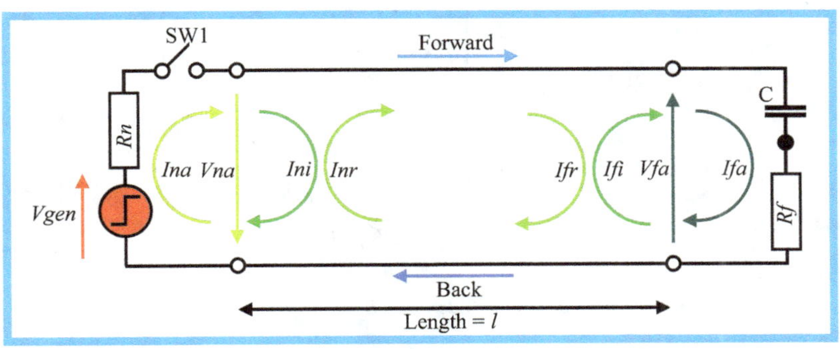

Figure 5.14 Circuit model of transmission line

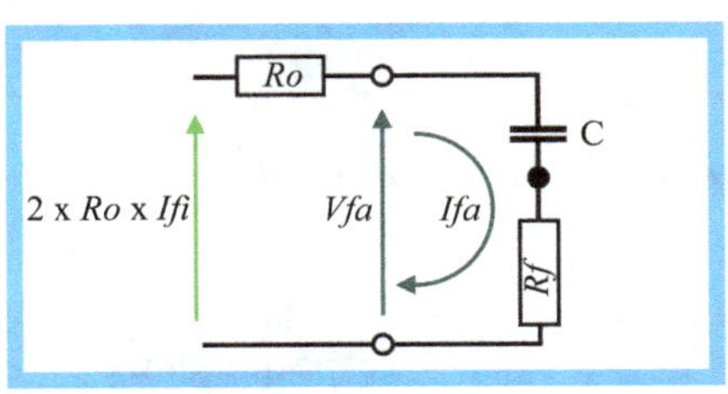

Figure 5.15 Circuit model used to calculate the response at the far end

$$far(Ifi, Qfa) := \left| \begin{array}{l} Ifa \leftarrow \dfrac{1}{Rf + Ro} \cdot \left(2 \cdot Ro \cdot Ifi - \dfrac{Qfa}{C} \right) \\ Qfa \leftarrow Qfa + Ifa \cdot dt \\ Ifr \leftarrow Ifa - Ifi \\ (Ifr \;\; Ifa \;\; Qfa) \end{array} \right.$$

Figure 5.16 Mathcad function which calculates the response at the far end

$$Vg := 100 \qquad Ro := 100 \qquad Rn := 1 \qquad C := 500 \cdot 10^{-9} \qquad Rf := 0$$

$$n := 10 \quad l := 15 \quad v := 3 \cdot 10^{8} \qquad T := \frac{l}{v} = 5 \times 10^{-8} \quad dt := \frac{T}{n} = 5 \times 10^{-9}$$

$$near(Ini, Vg) := \left| \begin{array}{l} Ina \leftarrow \dfrac{2 \cdot Ro \cdot Ini + Vg}{Ro + Rn} \\ Inr \leftarrow Ina - Ini \\ (Inr \;\; Ina) \end{array} \right.$$

$$forward(\boldsymbol{F}, In) := \left| \begin{array}{l} \boldsymbol{F}_1 \leftarrow In \\ for \;\; x \in n + 1 \,..\, 1 \\ \quad \boldsymbol{F}_{x+1} \leftarrow \boldsymbol{F}_x \\ If \leftarrow \boldsymbol{F}_{n+2} \\ (\boldsymbol{F} \;\; If) \end{array} \right.$$

$$far(Ifi, Qfa) := \left| \begin{array}{l} Ifa \leftarrow \dfrac{1}{Rf + Ro} \cdot \left(2 \cdot Ro \cdot Ifi - \dfrac{Qfa}{C} \right) \\ Qfa \leftarrow Qfa + Ifa \cdot dt \\ Ifr \leftarrow Ifa - Ifi \\ (Ifr \;\; Ifa \;\; Qfa) \end{array} \right.$$

$$back(\boldsymbol{B}, If) := \left| \begin{array}{l} \boldsymbol{B}_{n+1} \leftarrow If \\ for \;\; x \in 2 \,..\, n + 1 \\ \quad \boldsymbol{B}_{x-1} \leftarrow \boldsymbol{B}_x \\ In \leftarrow \boldsymbol{B}_1 \\ (\boldsymbol{B} \;\; In) \end{array} \right.$$

Figure 5.17 First page of Mathcad worksheet

length of the line is 15 m, the velocity of propagation is that of light in a vacuum, the source resistance is 1 Ω, and that the capacitor at the far end is 500 nF. For this computation, it is assumed that there is no resistance in series with the capacitor. Hence $Rf = 0$.

$Toff := 5 \cdot 10^{-5}$ $\qquad Noff := \dfrac{Toff}{dt} = 1 \times 10^{4}$

$N := 20 \cdot n - 1 \qquad i := 1..N \qquad t_i := i \cdot dt$

```
Ina := | F[n+2] ← 0
       | B[n+1] ← 0
       | for i ∈ 1..N − 1
       |   | Vgen ← Vg  if i > 1
       |   | if i > Noff
       |   |   | Vgen ← 0
       |   |   | Rn ← 200
       |   | (Inr  Ina) ← near(Ini, Vgen)
       |   | (F  Ifi) ← forward(F, Inr)
       |   | (Ifr  Ifa  Qfa) ← far(Ifi, Qfa)
       |   | (B  Ini) ← back(B, Ifr)
       |   | Ina[i] ← Ina
       | Ina
```

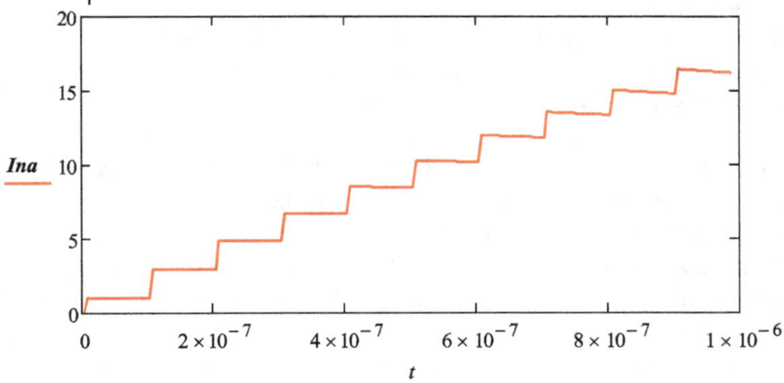

$T = 5 \times 10^{-8} \quad Vg = 100 \quad Ro = 100 \quad Rn = 1 \quad Rf = 0 \quad C = 5 \times 10^{-7}$

Figure 5.18 Second page of Mathcad worksheet

The time taken for a signal to make one traverse of the line is 50 ns. It is assumed that the line is divided into ten segments. So the calculations are carried out at 5 ns intervals.

The functions defined on this first page of the worksheet calculate the response at the near end of the line, the shifting of the signal from the near to the far end, the response at the far end, and the shifting of the reflected signal back to the near end.

The control parameters for the computation are defined at the top of Figure 5.18. It is assumed that the switch SW1 closes at 5 ns and that it opens at

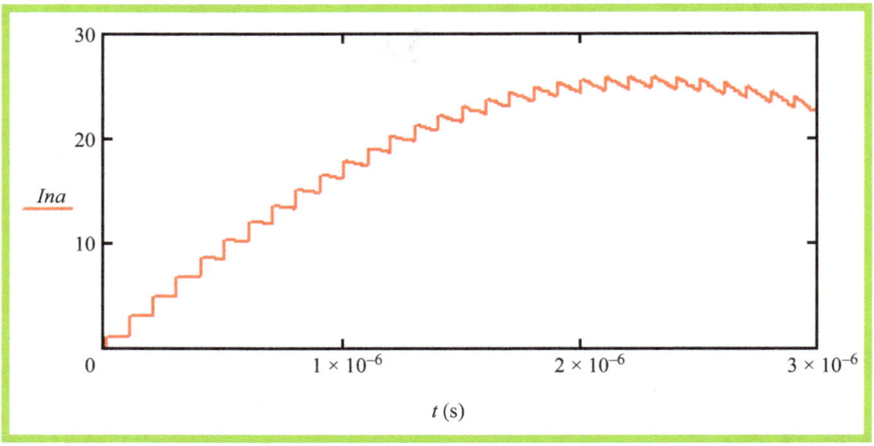

Figure 5.19 Waveform of current at the near end over a period of 3 μs

50 μs. This means that the voltage applied to line drops to zero when the number of iterations *Noff* is 10,000.

The main program simply takes the output of each subroutine and supplies these to the input of the next. The value of the current at the near end at each step is recorded in the vector ***Ina***, and this is the final output.

The waveform of the input current *Ina* over a period of 1 μs is given by the graph at the bottom of Figure 5.18. This shows that the current delivered to the line is constant at 1A for 100 ns; that is, for the time, it takes for the step to propagate to the far end and then back to the near end.

Then the size of the step changes to 2A. This is because the capacitor is acting as a short circuit and sending the current back to the near end where it acts as a voltage source in series with the generator. Thereafter, the waveform looks like a series of steps. Since the capacitor is gradually charging up, it starts to act as a voltage source at the far end. This causes a downward slope on each step.

Eventually, the amplitude of the downward slope is greater than that of the upward step, and the current reaches peak amplitude, as shown by Figure 5.19.

The result is a ringing waveform which rapidly dies down as the voltage on the capacitor settles at 100 V. Since SW1 switches to the open-circuit condition at 50 μs, the energy stored in the capacitor flows back into the cable and out into the environment, as illustrated by Figure 5.20.

In this model, it is assumed that the radiation resistance of the cable is 100 Ω. So the voltage appearing at the near end of the cable would be as shown by Figure 5.21.

5.3.3 Inductive load

Figure 5.22 shows the power supply line, this time with an inductive termination. When the switch closes, a current step will propagate along the cable assembly, just

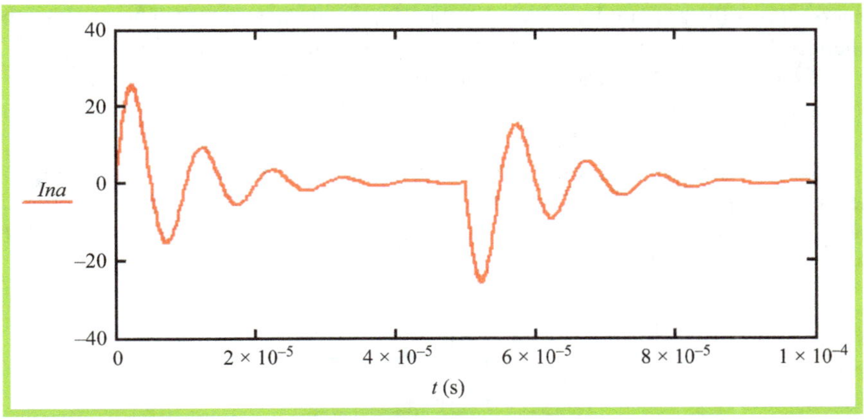

Figure 5.20 Waveform of current at the near end over a period of 100 μs

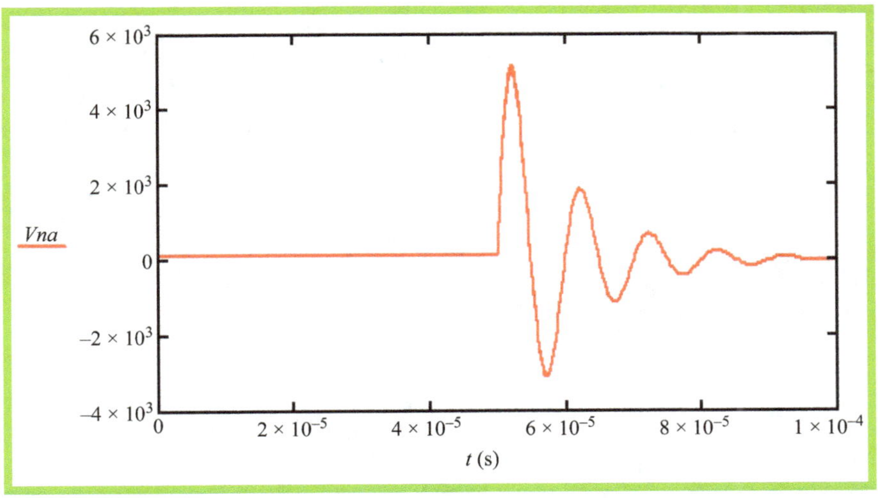

Figure 5.21 Voltage appearing at the near end of the cable when power is switched off

as with a capacitive load. But when it reaches the far end, the inductance acts as an open-circuit and the voltage doubles. This voltage propagates back to the near end, which doubles the current.

To simulate the reaction of the system, it is necessary to redefine the far end equations. Figure 5.23 shows the circuit model.

The mesh equation derived from Figure 5.23 is as follows:

$$2 \cdot Ro \cdot Ifi = (Ro + R\mathbf{1}) \cdot Ifa + L\mathbf{1} \cdot \frac{dIfa}{dt} \tag{5.35}$$

Where *dIfa* is the amount by which *Ifa* changes during the time *dt*. This enables the function defining the response to be formulated. Figure 5.24 is a copy.

Modifying the Mathcad worksheet to incorporate the changes necessary to simulate the response of the system illustrated by Figure 5.22 and running the program leads to the simulation of Figure 5.25.

The waveform of the input current over a period of 1 μs is given by the graph at the bottom of Figure 5.25. This shows that the current delivered to the line is constant at 1A for 100 ns; that is, for the time, it takes for the step to propagate to the far end and then back to the near end.

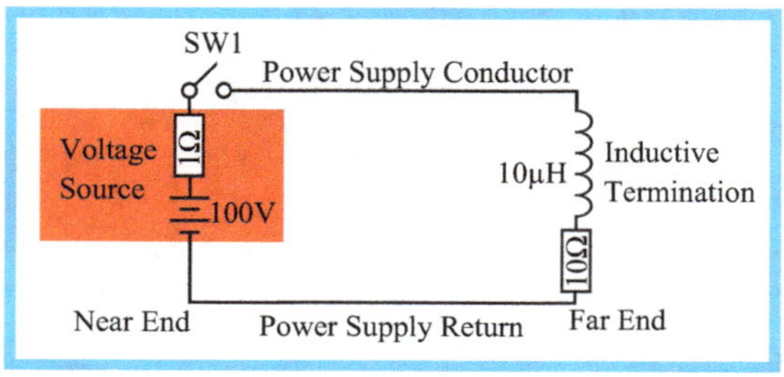

Figure 5.22 Power line wiring diagram, inductive termination

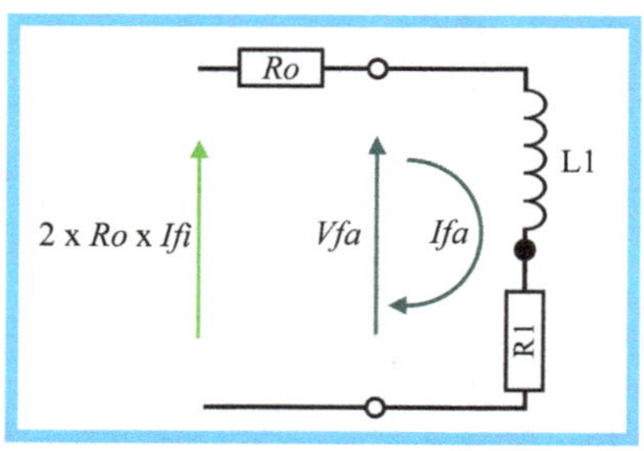

Figure 5.23 Circuit model which defines the response at the far end

$$far(Ifi, Ifa) := \left| \begin{array}{l} dIfa \leftarrow \dfrac{dt}{L} \cdot [2 \cdot Ro \cdot Ifi - (Ro + Rf) \cdot Ifa] \\ Ifa \leftarrow Ifa + dIfa \\ Ifr \leftarrow Ifa - Ifi \\ (Ifr \quad Ifa) \end{array} \right.$$

Figure 5.24 Mathcad function which defines the response at the far end

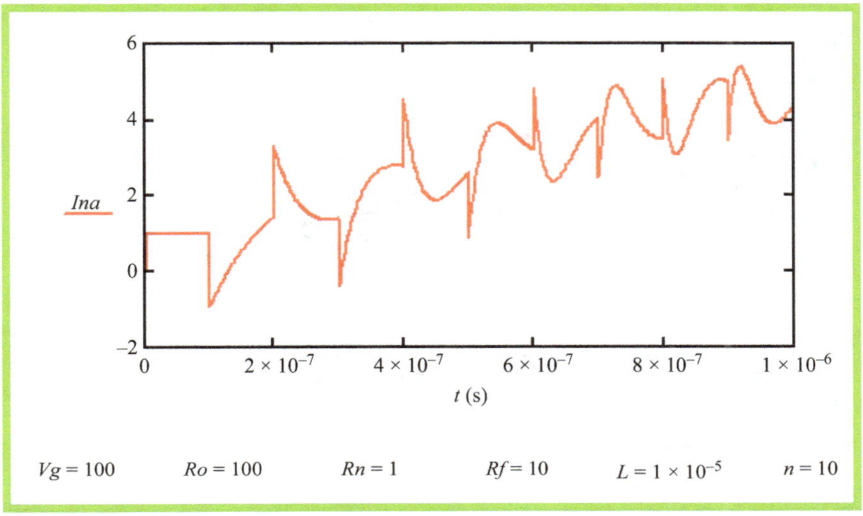

Figure 5.25 Waveform of current at the near end: Inductive load

The inductance at the far end acts as an open-circuit. When the step arrives at the far end, the direction of the current reverses and the step change propagates back to the near end. In the meantime, the constant voltage applied to the inductor causes the current to rise exponentially. So the amplitude of the reflected current drops exponentially. When the pulse arrives back at the near end and is again reflected, the waveform during the first 200 ns, and the next 800 ns, is as displayed in Figure 5.25.

The data listed under the graph of Figure 5.25 defines the parameters for the circuit of Figure 5.22. It was necessary to increase the number of segments n to 100 because the current was changing more rapidly than with a capacitive termination. Extending the period of the simulation leads to the waveform of Figure 5.26.

The curve rises exponentially, flattening out at about 9A. In this simulation, the switch opens at 10 μs, the surplus energy radiates out into the environment as a burst of high-frequency radiation, and the current in the load drops exponentially towards zero.

Figure 5.27 simulates the waveform of the voltage *Vna* at the near end. In radiating out into the environment, the current creates a high voltage across the terminals of the switch; of the order of 1 kV.

5.3.4 Assessment

A capacitor at the far end of the cable acts as a short circuit to any transient current. The amplitude of the current will double. Since the source resistance at the near end is less than the characteristic impedance of the line, the current increases again and the resultant waveform is rather like a staircase. The current delivered to the capacitor builds up until it is greater than that of the generator. A ringing waveform

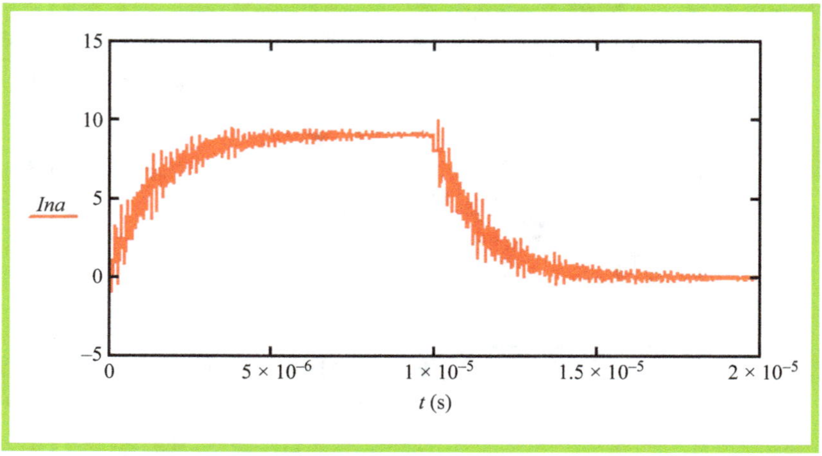

Figure 5.26 Waveform of the current at the near end, showing the effect of switch-off

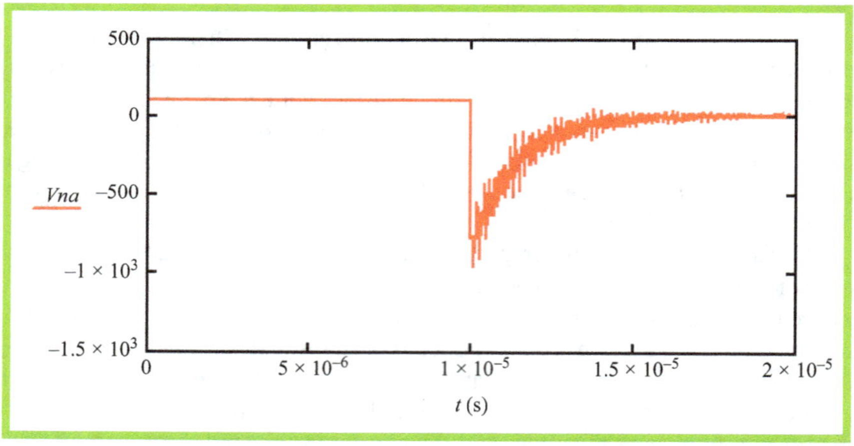

Figure 5.27 Waveform of voltage at the near end, showing effect of switch-off

is created. When the source is switched off, the energy stored in the system departs in the form of a high voltage ringing pulse.

An inductor at the far end acts as an open-circuit. So any transient voltage will double in magnitude and the reflected voltage propagates back. Reflections at the near end create a pulsed waveform which gradually increases the current delivered to the load. Current in the cable contains a wide variety of high-frequency components. When the source is switched off, the energy departs explosively in the form of a step voltage applied along the length of the cable.

During switch-on, in both cases, current build up is in the form of a series of pulses along the line. It is inevitable that the line will act as a radiating antenna. After switch-off, in both cases, the departure of the stored energy creates a high voltage pulse at the near end.

5.3.5 Conclusion

A filter with reactive components will interact with the supply cable to create high-frequency resonance, and this resonance will be a rich source of EMI.

5.4 Power line filter

5.4.1 Introduction

A filter is described which absorbs transient energy in the supply line as well as minimising the effect of those transients on the load. Traditional filters simply reflect that energy backs into the line where the only place for it to go is into the environment. This manifests itself as cross-coupled interference as well as radiated interference.

A model is created which simulates the interaction between the filter and the supply line. It is shown that the device performs its intended function in protecting the load and in preventing load transients from reaching the line. It also absorbs any transient energy, whatever the source. Any system which includes such filters will experience much reduced levels of EMI.

5.4.2 Basic power line

At least two conductors are involved in routing power from one location in a system to another; the send conductor and the return conductor. Figure 5.28 provides a schematic diagram of such a link. At the near end, the battery simulates a DC source of power, the 0.1 Ω resistor simulates the source impedance, and SW1 is initially open. The line itself is 15 m long. The load at the far end is a 10 Ω resistor in series with an isolating switch, SW2, which is initially closed.

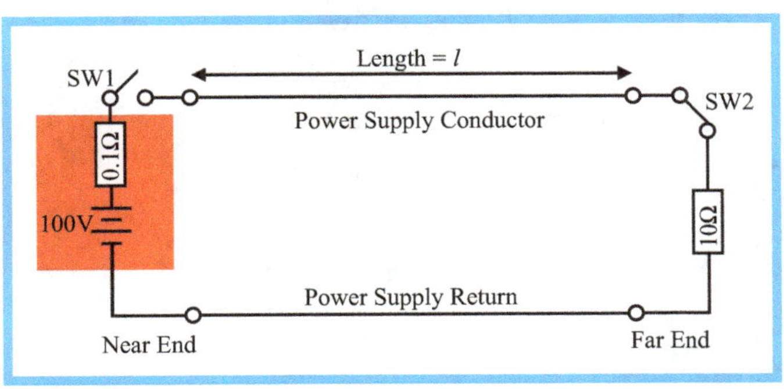

Figure 5.28 Wiring assembly of power supply line

The response of the line can be simulated using the technique described in Section 5.3. If it is assumed that SW1 closes at the start of the simulation and SW2 opens after 10 μs, then the waveform of the current at the near end of the line would be as illustrated by Figure 5.29.

At switch-on, the current rises in a series of small steps to settle at a steady 10 A. The shape on the curve is that of an L–R series circuit; the inductance of the line in series with the 10 Ω load.

When the load is switched off, the energy stored in the line is trapped by a short circuit at the near end and an open-circuit at the far end. So current oscillates back and forth. If the velocity of propagation v is 300 m/μs, the time T for a single traverse is as follows:

$$T = \frac{l}{v} = 50 \text{ ns} \tag{5.36}$$

Since it takes four traverses of a signal to complete a cycle, the frequency f is as follows:

$$f = \frac{1}{4 \cdot T} = 5 \text{ MHz} \tag{5.37}$$

If there were no radiation losses, a 5 MHz square wave would exist on the line, decaying slowly due to energy losses in the 100 mΩ resistance at the near end. However, each conductor acts as an antenna, and current emanates from the line in the form of a burst of radiation. This is illustrated by the response following the switch-off at 10 μs.

There is no doubt that this energy will depart into the environment in the form of EMI.

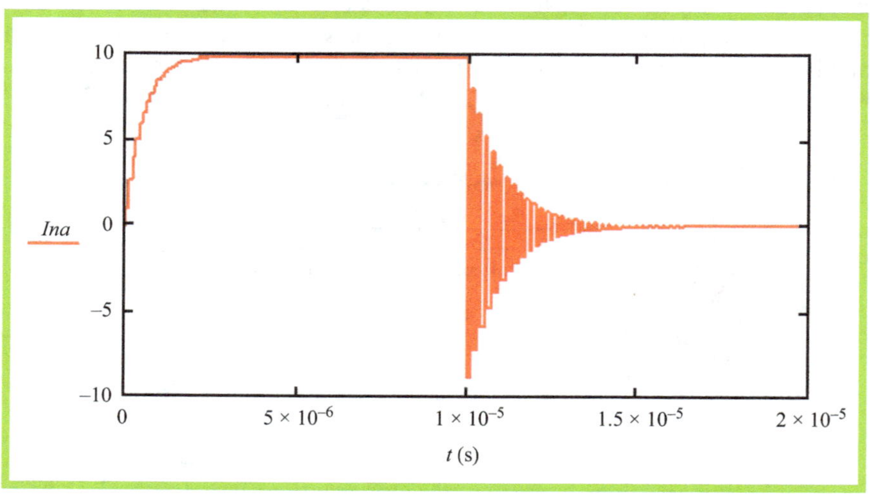

Figure 5.29 Waveform of current in power line

5.4.3 Adding a filter

This ringing pulse can be eliminated by the filter circuit of Figure 5.30. This assembly differs from conventional power line filters in that it includes resistive components. There is a resistor in parallel with each inductor and a resistor in series with the capacitor.

Under steady-state conditions, the inductor provides zero impedance and the capacitor acts as an open-circuit. So the current delivered to a 10 Ω resistor at the far end by a 100 V source at the near end is just less than 10 A.

When a transient step arrives at the far end, the inductors present a high value of impedance and the capacitor provides zero impedance. So the load at this end is effectively two 50 Ω resistors in series. If the characteristic impedance of the line is 100 Ω, all of the transient energy will be absorbed. There will be no reflections and no high-frequency interference.

5.4.4 Block diagram

Analysis of this power link can be divided into four functional entities; the interface circuit at the near end, the forward propagation of energy, the interface circuitry at the far end, and the propagation of energy back down the line. This is illustrated by the block diagram of Figure 5.31. Signal processing in each block is defined by a unique set of equations.

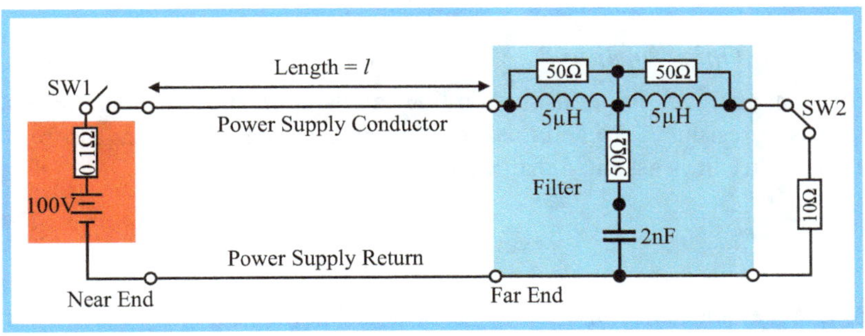

Figure 5.30 Wiring assembly of line and filter

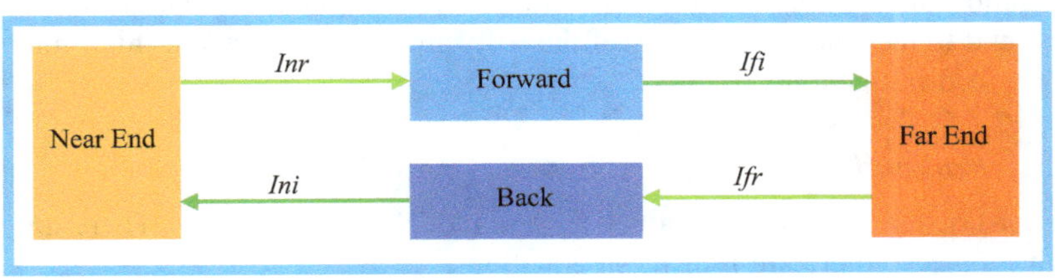

Figure 5.31 Block diagram: identifying signal processing functions

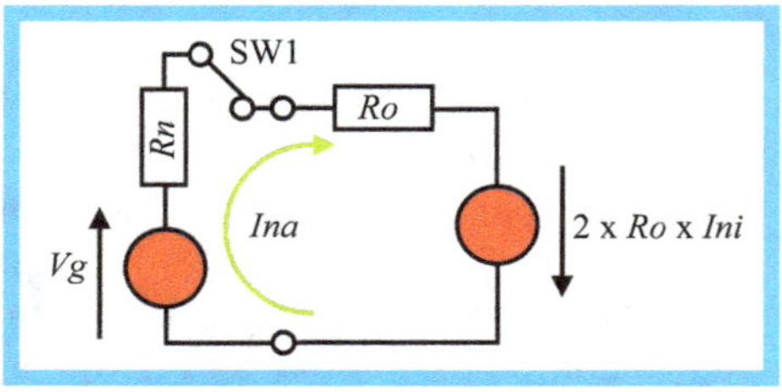

Figure 5.32 Circuit model of interface at the near end

5.4.5 Near end

Equations at the near end are defined by the circuit model of Figure 5.32.

$$Vg + 2 \cdot Ro \cdot Ini = (Ro + Rn) \cdot Ina \tag{5.38}$$

This leads to

$$Ina = \frac{Vg + 2 \cdot Ro \cdot Ini}{Ro + Rn} \tag{5.39}$$

Since the current *Ina* at the near end is defined as the sum of the incident and reflected currents, the value of the reflected current is as follows:

$$Inr = Ina - Ini \tag{5.40}$$

5.4.6 Transmission line

For transient analysis, any transmission line can be defined by two parameters, the characteristic resistance *Ro* and the time *T* it takes a signal to propagate from one end to the other. It is assumed that the line is lossless, so

$$Ro = \sqrt{\frac{L}{C}} \tag{5.41}$$

where *L* is the inductance of the loop as measured at the near end when the far end terminals are shorted together and *C* is the capacitance between the conductors when the terminals at the far end are open-circuit.

The line can be represented as two shift registers, one simulating the forward propagation of charges, the other simulating the propagation of charges back from the far end to the near end. Details of the software used to carry out this simulation are provided in Section 5.2.

5.4.7 Far end

Equations at the far end are defined by the model of Figure 5.33. There are four circuit loops. So there are four mesh equations:

$$2 \cdot Ro \cdot Ifi = (Ro + \mathbf{R1} + \mathbf{R2}) \cdot Ifa + \frac{Qfa}{C1} - \frac{Qfc}{C\mathbf{1}} - \mathbf{R1} \cdot Ifb - \mathbf{R2} \cdot Ifc \tag{5.42}$$

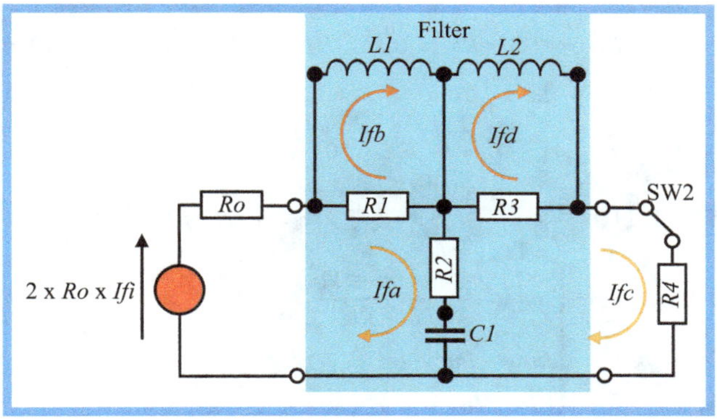

Figure 5.33 Circuit model of the interface at the far end: filter, switch, and load

$$0 = -\mathbf{R1} \cdot Ifa + \mathbf{R1} \cdot Ifb + \mathbf{L1} \cdot \frac{dIfb}{dt} \tag{5.43}$$

$$0 = -\frac{Qfa}{\mathbf{C1}} - \mathbf{R2} \cdot Ifa + \frac{Qfc}{C1} - (\mathbf{R2} + \mathbf{R3} + \mathbf{R4}) \cdot Ifa - \mathbf{R3} \cdot Ifd \tag{5.44}$$

$$0 = -R3 \cdot Ifc + R3 \cdot Ifd + L2 \cdot \frac{dIfd}{dt} \tag{5.45}$$

Let

$$Qf = (Qfa - Qfc) = (Ifa - Ifc) \cdot dt \tag{5.46}$$

5.4.8 First page of worksheet

Having defined the equations associated with each block of Figure 5.31, there is now enough information to allow a Mathcad worksheet to be compiled.

The two pages are replicated by Figures 5.34 and 5.35.

The first two lines of Figure 5.34 assign values to the parameters of the circuit models of Figures 5.32 and 5.33 using the components defined by Figure 5.30.

The third line defines the values used in the simulation of the cable link.

The fourth line defines the time *Toff* when SW2 opens.

The function *near* ( ) is derived from (5.39) and (5.40).

The functions *forward* ( ) and *back* ( ) are copied from Figure 5.9 of Section 5.2.

Each simulates the actions of a shift register.

The function *far* ( ) is derived from (5.42) to (5.46).

5.4.9 Second page of worksheet

This page defines the main program and creates a graph which simulates the waveform of the current *Ina* at the near end of the line. Figure 5.35 is a copy of the text.

$$Vg := 100 \quad Ro := 100 \quad Rn := 0.1 \quad R1 := 50 \quad R2 := 50 \quad R3 := R1$$

$$L1 := 5\cdot 10^{-6} \quad L2 := L1 \quad C1 := 2\cdot 10^{-9} \quad l := 15 \quad v := 3\cdot 10^{8}$$

$$n := 10 \qquad T := \frac{l}{v} = 5\times 10^{-8} \qquad dt := \frac{T}{n} = 5\times 10^{-9}$$

$$Toff := 10\cdot 10^{-6} \qquad Noff := \frac{Toff}{dt} = 2\times 10^{3}$$

$$near(Ini, Vg) := \left| \begin{array}{l} Ina \leftarrow \dfrac{2\cdot Ro\cdot Ini + Vg}{Ro + Rn} \\ Inr \leftarrow Ina - Ini \\ (Inr \;\; Ina) \end{array} \right.$$

$$forward(\boldsymbol{F}, In) := \left| \begin{array}{l} \boldsymbol{F}_1 \leftarrow In \\ for\ x \in n+1..1 \\ \quad \boldsymbol{F}_{x+1} \leftarrow \boldsymbol{F}_x \\ If \leftarrow \boldsymbol{F}_{n+2} \\ (\boldsymbol{F} \;\; If) \end{array} \right. \qquad back(\boldsymbol{B}, If) := \left| \begin{array}{l} \boldsymbol{B}_{n+1} \leftarrow If \\ for\ x \in 2..n+1 \\ \quad \boldsymbol{B}_{x-1} \leftarrow \boldsymbol{B}_x \\ In \leftarrow \boldsymbol{B}_1 \\ (\boldsymbol{B} \;\; In) \end{array} \right.$$

$$far(Ifi, Ifa, Ifb, Ifc, Ifd, Qf, R4) := \left| \begin{array}{l} dIfb \leftarrow \dfrac{dt}{L1}\cdot R2\cdot(Ifa - Ifb) \\ Ifb \leftarrow Ifb + dIfb \\ Ifa \leftarrow \dfrac{2\cdot Ro\cdot Ifi - \dfrac{Qf}{C1} + R1\cdot Ifb + R2\cdot Ifc}{Ro + R1 + R2} \\ dIfd \leftarrow \dfrac{dt}{L2}\cdot R3\cdot(Ifc - Ifd) \\ Ifd \leftarrow Ifd + dIfd \\ Ifc \leftarrow \dfrac{1}{R2 + R3 + R4}\left(\dfrac{Qf}{C1} + R2\cdot Ifa + R3\cdot Ifd\right) \\ Qf \leftarrow Qf + (Ifa - Ifc)\cdot dt \\ Ifr \leftarrow Ifa - Ifi \\ (Ifr \;\; Ifa \;\; Ifb \;\; Ifc \;\; Ifd \;\; Qf) \end{array} \right.$$

Figure 5.34 First page of worksheet: definition of functions

The number N defines the number of transits of the leading edge of the waveform along the line. Since the time taken for one transit is 50 ns, 400 such transits take 20 μs. This defines the timescale of the simulation and the time variable t.

$N := 400 \cdot n - 1 \qquad i := 1..N \qquad t_i := i \cdot dt$

```
Ina :=  | F[n+2] ← 0
        | B[n+1] ← 0
        | R4 ← 10
        | for i ∈ 1..N − 1
        |     | Vgen ← Vg  if i > 1
        |     | R4 ← 10^6  if i > Noff
        |     | (Inr  Ina) ← near(Ini, Vgen)
        |     | (F  Ifi) ← forward(F, Inr)
        |     | (Ifr  Ifa  Ifb  Ifc  Ifd  Qf) ← far(Ifi, Ifa, Ifb, Ifc, Ifd, Qf, R4)
        |     | (B  Ini) ← back(B, Ifr)
        |     | mon[i] ← Ina
        | mon
```

$Vg = 100 \quad Ro = 100 \quad Rn = 0.1 \quad R1 = 50 \quad R2 = 50 \quad L1 = 5 \times 10^{-6}$

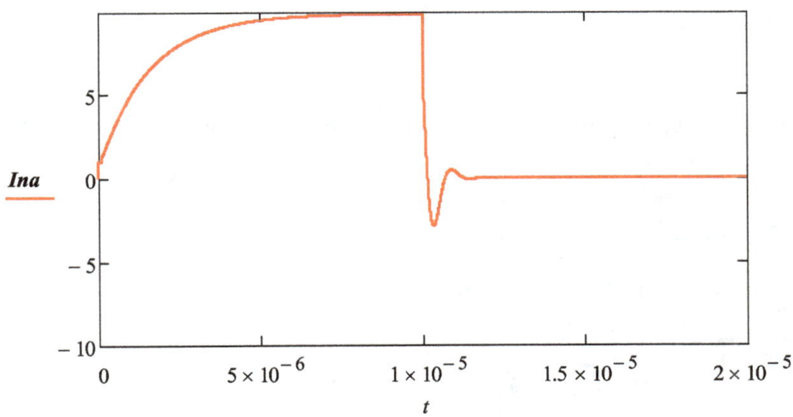

Figure 5.35 Second page of worksheet: main program and graph of current waveform

The main program takes the output of each function and uses it as the input of the next function. This means that any current (*Ina, Ini, Inr, Ifi, Ifa, Ifb, Ifc, Ifd,* or *Ifr*) can be selected as the output variable. In this case, the output is *Ina*.

5.4.10 Conclusion

Comparing the graph of Figure 5.35 with the waveform of Figure 5.29 shows that the simple filter of Figure 5.33 has a dramatic effect on the EMC of the system. The high amplitude, high-frequency current which occurs at switch-off has disappeared. This is the current which is the cause of most of the radiated emission from the system.

The design concept is simple. Just add resistors to the L and C components to absorb high-frequency energy. Select values which provide a load resistance equal to the characteristic impedance of the line, over the frequency range where radiated interference might be a problem.

At low frequencies, the filter is transparent.

Installing such filters in the electronic system could eliminate the vast majority of radiated interference problems.

5.5 Transient emission

5.5.1 Introduction

An experiment is described which enables transient radiation to be measured and analysed. It involves the application of a step pulse to one end a twin-conductor cable which is open at the far end and the use of a current transformer to monitor the differential-mode current. The waveform of this current is initially that of a square wave, but it quickly morphs into a sine wave. A circuit model of this configuration is created to replicate the observed waveform. Analysis of the model provides an insight into the mechanisms involved.

5.5.2 Setup

The setup used to investigate the transient response of a twin-conductor cable is shown in Figure 5.36.

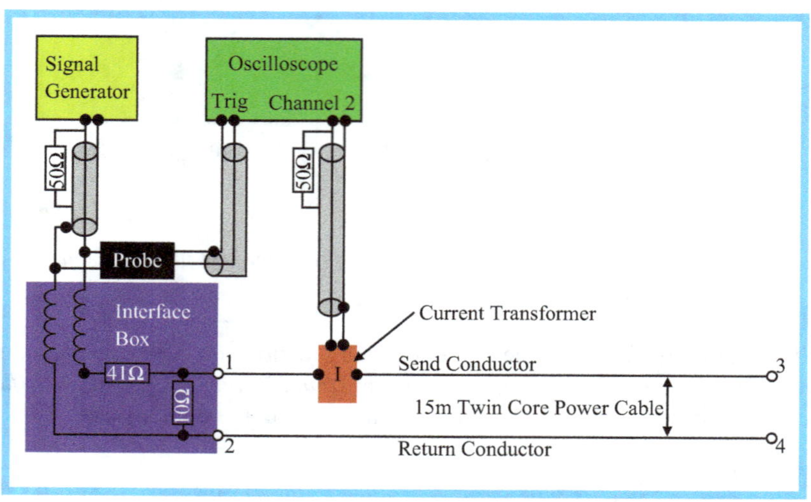

Figure 5.36 Setup used to investigate the transient response of twin-conductor cable

An interface box was first constructed in order to apply a step pulse to such a cable whilst maintaining isolation between the signal generator and the input terminals. This included a common-mode choke which provided high impedance to any antenna-mode current flow between the assembly under test and the ground terminals of the generator. The use of a current transformer to monitor the current ensured that there was complete isolation between the oscilloscope and the cable. Included in the interface box was a simple resistor network which provided a 50 Ω load to the signal generator and a low resistance signal source at the input to the cable.

5.5.3 Test method

Since the cable was open-circuit at the far end and supplied via a low resistance at the near end, reflections would occur at both terminations. The response to a step voltage input was expected to be a ringing pulse which decayed relatively slowly. So the signal generator was set to provide a square wave pulse at 100 kHz. This gave a mark time of 20 μs and a mark/space ratio of unity. This delay provided more than enough time for the reverberations of the line to die down before the occurrence of the next step.

The current transformer was a clamp-on device allowed the voltage waveform on channel 2 of the oscilloscope to replicate the waveform of the current in the send conductor. A times-ten voltage probe was used to provide a trigger pulse to the oscilloscope to eliminate any jitter in the monitored waveform.

A photograph of the current waveform is reproduced by Figure 5.37.

5.5.4 Vertical scaling

The objective now was to create a circuit model which replicated the waveform of Figure 5.37. This meant that the response of the test equipment needed to be

Vertical scale = 10 mV/div
Horizontal scale = 100 ns/div

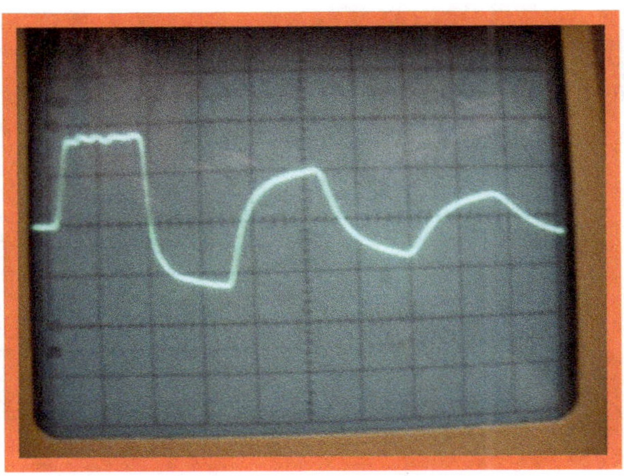

Figure 5.37 Waveform of current in the send conductor

simulated as well as that of the cable. So the first task was to measure the amplitude of the voltage delivered to the input terminals (terminals 1 and 2 of Figure 5.36).

In Figure 5.36, the 41 Ω resistor is in series with the 50 Ω output resistor of the signal generator. So the source resistance of the supply to the input terminals was as follows:

$$Rg = \frac{10 \cdot (41 + 50)}{10 + 41 + 50} = 9.01 \ \Omega \tag{5.47}$$

The relationship between Figure 5.38(a) and (b) illustrates this. The amplitude of the voltage step *Vin,* as monitored by the voltage probe, was as follows:

$$Vin = 0.84 \ \text{V} \tag{5.48}$$

Calibration of the current probe in the frequency domain enabled the circuit model of Figure 5.39 to be created. This related the current in the primary winding, *Iprim*, to the amplitude voltage *Vch2* monitored at the oscilloscope, over the frequency range 50 kHz to 20 MHz. The number of turns on the secondary winding was as follows:

$$Turns = 10 \tag{5.49}$$

A full description of the calibration process is provided in Section 2.5. For transient analysis, it can be assumed that the inductor *L1* is open-circuit, and the

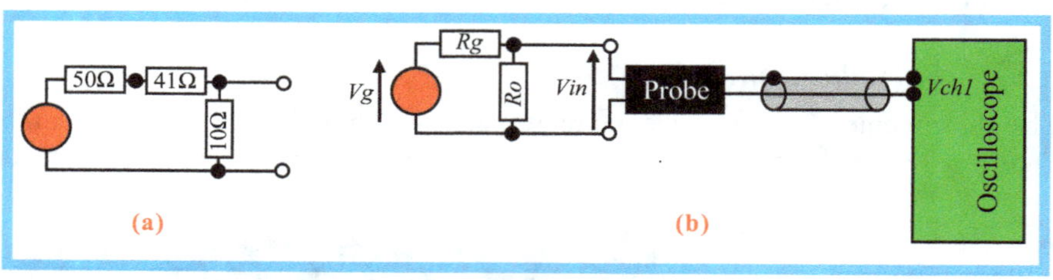

Figure 5.38 Circuit model of input circuitry: (a) model used to derive source resistance and (b) relating voltages Vch1, Vin, and Vg

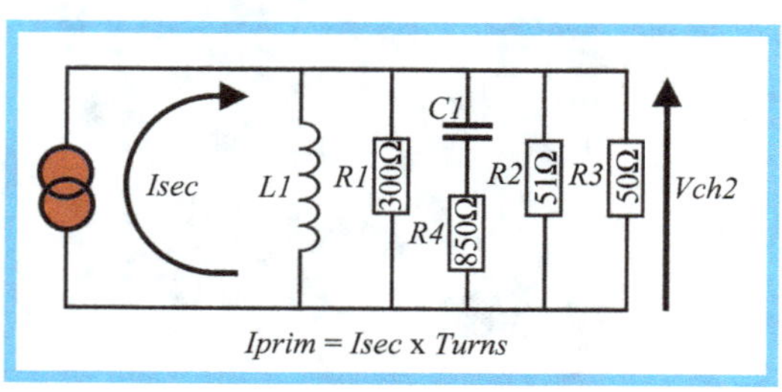

Figure 5.39 Circuit model of the current transformer

capacitor *C1* is short circuit. So the ratio between *Isec* and *Vch2* is as follows:

$$\frac{Isec}{Vch\mathbf{2}} = \frac{1}{R1} + \frac{1}{R2} + \frac{1}{R3} + \frac{1}{R4} = 0.044 \tag{5.50}$$

And the transfer resistance *RT* is as follows:

$$RT = \frac{Vch\mathbf{2}}{Iprim} = \frac{Vch2}{Isec \cdot Turns} = 2.27\ \Omega \tag{5.51}$$

The amplitude *Vch2* of the first step in the waveform monitored on channel 2 of the oscilloscope was derived from examination of the waveform of Figure 5.37:

$$Vch\mathbf{2} = 17\text{ mV} \tag{5.52}$$

This means that the measured value of the differential-mode current flowing in the cable changes during the first transit was:

$$Idiff = \frac{Vch2}{RT} = 7.489\text{ mA} \tag{5.53}$$

So, the characteristic resistance *Ro* of the cable was as follows:

$$Ro = \frac{Vin}{Idiff} = 112\ \Omega \tag{5.54}$$

From Figure 5.38(b)

$$Vg = Vin \cdot \frac{Rg + Ro}{Ro} = 0.907\text{ V} \tag{5.55}$$

5.5.5 Horizontal scaling

Each horizontal division of the graticule shown on Figure 5.37 represents 100 ns. If the vertical line on the left-hand side of the display represents zero time, the time *T1* of the first rising edge of the waveform is as follows:

$$T1 = 20\text{ ns} \tag{5.56}$$

In simulating the response, *T1* is the time that the input voltage *Vgen* switches from zero to *Vg*.

The first trailing edge occurs at 185 ns. The interval between the rising and falling edges represents the time taken for the leading edge to propagate from the near end of the cable to the far end and then back to the near end. If *T* is the time for one traverse, then

$$T = \frac{185 - 20}{2} = 82.5\text{ ns} \tag{5.57}$$

The time taken for the scan to sweep from the first to the tenth graticule line is as follows:

$$T\mathbf{2} = 1000 \text{ ns} \tag{5.58}$$

5.5.6 Modelling

Initial observation was that it was not a square waveform as would have been predicted using the textbook analysis of electromagnetic theory. Even so, the response for the first 165 ns did approximate that of a square wave. The finite rise time was an indication of the response time of the oscilloscope. The small ripple at the top of the waveform was probably due to reflections at the current transformer. Textbook analysis indicates that the time duration of 165 ns represents the time for the leading edge to propagate along the line, be reflected, and propagate back to the near end.

The first departure from the textbook response was the exponential decay of the first trailing edge. This indicated that current was departing from the line via capacitive coupling. The waveform then gradually morphed from a square wave to a sine wave, as well as decaying in amplitude. This decay was to be expected, since energy was departing from the line in the form of electromagnetic radiation. Losses on the source resistance would also contribute to this decay.

Initial attempts to model this response were reasonably successful, but the simulation failed to replicate the kinks at the top of the second and third peaks. It was then reasoned that the departure of current occurred during the time the pulse was propagating forward from the near end to the far end. The send conductor was acting as a transmitting antenna. But during the time the current was flowing back to the near end, the return conductor was acting as a transmitting antenna and delivering energy back to the send conductor.

The computations follow the same sequence as that described in Section 5.4. But this time the focus of attention is on the behaviour of the cable, rather than on that of the terminations.

Figure 5.40 illustrates the sequence of calculations. The parameters are as follows:

Ina is the amplitude of the current absorbed by the interface at the near end.
Inr is the amplitude of the current reflected back into the line at the near end.

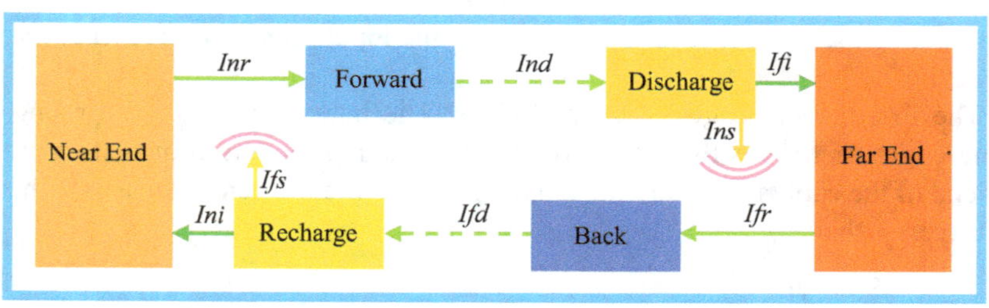

Figure 5.40 Block diagram which identifies the signal processing functions

Ind is the same amplitude as *Inr*, but delayed by *T* seconds.
Ins is the current needed to store charge *Qns* at the surface of the send conductor.
Ifi is the current delivered to the far end; the incident current.
Ifa is the current absorbed at the far end.
Ifr is the current reflected back into the line at the far end.
Ifd is the same amplitude as *Ifr*, but delayed by *T* seconds.
Ifs is the current needed to store charge *Qfs* on the surface of the return conductor.
Ini is the incident current arriving at the near end.

It is assumed that the energy involved in storing charge *Qns* on the send conductor departs from the system in the form of an electromagnetic wave.

5.5.7 Equations

The equations defining the responses at the near and far end are basically the reflection equations derived in any textbook on electromagnetic theory. The signal processing involved in simulating the time delays is the same as that used in Section 5.4. Equations involved in the simulation of the discharge current *Ins* can be derived from the circuit models of Figure 5.41.

The energy source in the line can be represented as a current *Ind* flowing in the characteristic resistance *Ro*. The voltage between the conductors is as follows:

$$Vnd = Ro \cdot Ind \tag{5.59}$$

The current flowing in the capacitor *Crad* can be derived from Figure 5.41(b)

$$Vnd = Ro \cdot Ins + \frac{Qns}{Crad} \tag{5.60}$$

From (5.59) and (5.60)

$$Ins = Ind - \frac{Qns}{Ro \cdot Crad} \tag{5.61}$$

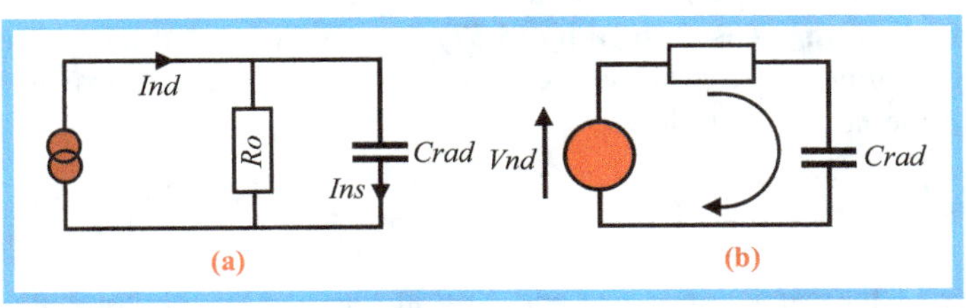

Figure 5.41 Deriving the amplitude of the discharge current Ins: (a) using a current source to simulate loop current in the cable and (b) converting the circuit model to one which uses a voltage source

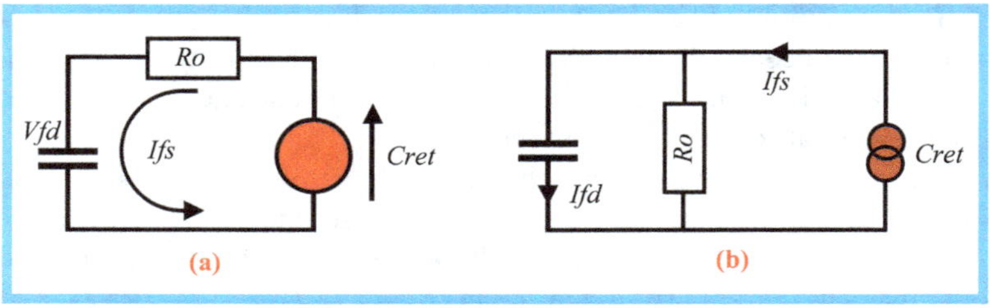

Figure 5.42 Deriving the amplitude of the recharge current Ifs: (a) using a current source to simulate loop current in the cable and (b) converting the circuit model to one which uses a voltage source

Since the current *Ins* has departed from the system, the current *Ifi* arriving at the far end must be

$$Ifi = Ind - Ins \tag{5.62}$$

The same relationships apply to the derivation of the current *Ifs* which represents the departure of energy from the return conductor. But this energy is not lost to the system. It is picked up by the send conductor, as illustrated by Figure 5.40.

From Figure 5.42

$$Ifs = Ifd - \frac{Qfs}{Ro \cdot Cret} \tag{5.63}$$

And the loop current delivered to the near end is assumed to be:

$$Ini = Ifd + Ifs \tag{5.64}$$

5.5.8 Computation

The computations are set out in a two-page Mathcad worksheet. The first page is illustrated by Figure 5.43.

The constants *Vg*, *Ro*, and *Rg* are defined by (5.55), (5.54), and (5.47). To simulate an open-circuit termination at the far end, the load *RL* was set at 100 MΩ. The propagation time *T* is defined by (5.57). It is assumed that the line is divided into $n = 100$ segments. So the time taken for a transient pulse to propagate from one segment to the next is as follows:

$$dt = \frac{T}{n} \tag{5.65}$$

The values of the capacitors *Crad* and *Cret* at the start of the computation were unknown. Initially, arbitrary values were assigned to these components.

The functions *near* ( ), *forward* ( ), *far* ( ), and *back* ( ) are essentially those used in Section 5.2.

$$Vg := 0.91 \quad Ro := 112 \quad Rg := 9 \quad RL := 100\cdot10^{6} \quad T := 82.5\cdot10^{-9} \quad n := 100$$

$$dt := \frac{T}{n} \qquad \begin{array}{ll} Crad := 280\cdot10^{-12} & \text{Radiating capacitor: Select-on-test variable} \\ Cret := 100\cdot10^{-12} & \text{Recharging capacitor: Select-on-test variable} \end{array}$$

$$near(Ini, Vg) := \left| \begin{array}{l} Ina \leftarrow \dfrac{2\cdot Ro\cdot Ini + Vg}{Ro + Rg} \\ Inr \leftarrow Ina - Ini \\ (Inr \;\; Ina) \end{array} \right. \qquad forward(\boldsymbol{F}, Inr) := \left| \begin{array}{l} \boldsymbol{F}_1 \leftarrow Inr \\ for\ x \in n+1..1 \\ \quad \boldsymbol{F}_{x+1} \leftarrow \boldsymbol{F}_x \\ Ind \leftarrow \boldsymbol{F}_{n+2} \\ (\boldsymbol{F} \;\; Ind) \end{array} \right.$$

$$discharge(Ind, Qns) := \left| \begin{array}{l} Ins \leftarrow Ind - \dfrac{Qns}{Ro\cdot Crad} \\ Ifi \leftarrow Ind - Ins \\ Qns \leftarrow Qns + Ins\cdot dt \\ (Ifi \;\; Ins \;\; Qns) \end{array} \right.$$

$$far(Ifi) := \left| \begin{array}{l} Ifa \leftarrow \dfrac{2\cdot Ro\cdot Ifi}{Ro + RL} \\ Ifr \leftarrow Ifa - Ifi \\ (Ifr \;\; Ifa) \end{array} \right. \qquad back(\boldsymbol{B}, Ifr) := \left| \begin{array}{l} \boldsymbol{B}_{n+1} \leftarrow Ifr \\ for\ x \in 2..n+1 \\ \quad \boldsymbol{B}_{x-1} \leftarrow \boldsymbol{B}_x \\ Ifd \leftarrow \boldsymbol{B}_1 \\ (\boldsymbol{B} \;\; Ifd) \end{array} \right.$$

$$recharge(Ifd, Qfs) := \left| \begin{array}{l} Ifs \leftarrow Ifd - \dfrac{Qfs}{Ro\cdot Cret} \\ Qfs \leftarrow Qfs + Ifs\cdot dt \\ Ini \leftarrow Ifd + Ifs \\ (Ini \;\; Ifs \;\; Qfs) \end{array} \right.$$

Figure 5.43 Defining input variables and functions used in the main program

The *discharge* ( ) function is developed from (5.61) and (5.62). The charge *Qns* accumulated on the capacitor *Crad* after *dt* seconds is as follows:

$$Qns \leftarrow Qns + Ins \cdot dt \tag{5.66}$$

And the incident current *Ifi* arriving at the far end is as defined by (5.62).

The recharge ( ) function is developed from (5.63) and (5.64). The charge *Qfs* in the capacitor *Cret* is as follows:

$$Qfs \leftarrow Qfs + Ifs \cdot dt \tag{5.67}$$

A copy of the second page of the Mathcad worksheet is provided by Figure 5.44.

The value of the transfer resistance *RT* is provided by (5.51). This constant relates the amplitude of the voltage observed on channel 2 of the oscilloscope, *Vch2,* to the amplitude of the current in the primary of the current monitor transformer. This is simulated by the current *Ina* flowing in the resistor *Rg* at the near end of the line.

$RT := 2.27$ Transfer resistance: Ratio of *Vch2* to current in the line, *Ina*.

$T1 := 20 \cdot 10^{-9}$ $N1 := floor\left(\frac{T1}{dt}\right) = 24$ The count, *i*, at leading edge

$T2 := 1000 \cdot 10^{-9}$ $N2 := floor\left(\frac{T2}{dt}\right) = 1212$ The count, *i*, at end of sweep

$i := 1..N2$ $t_i := i \cdot dt$ The time variable

```
Ina := | F[n+2] ← 0
       | B[n+1] ← 0
       | for i ∈ 1..N2
       |   | Vgen ← Vg  if i > N1
       |   | (Inr  Ina) ← near(Ini, Vgen)
       |   | (F  Ind) ← forward(F, Inr)
       |   | (Ifi  Ins  Qns) ← discharge(Ind, Qns)
       |   | (Ifr  Ifa) ← far(Ifi)
       |   | (B  Ifd) ← back(B, Ifr)
       |   | (Ini  Ifs  Qfs) ← recharge(Ifd, Qfs)
       |   | A[i] ← RT·Ina
       | A
```

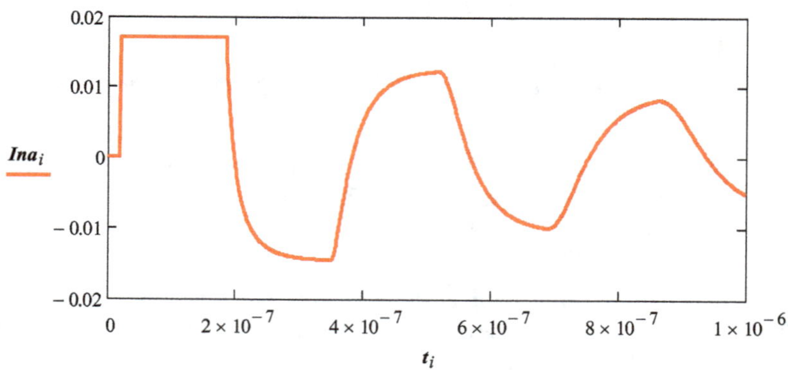

Figure 5.44 Main program and graph showing the response of the current at the near end

The constant *T1* is defined by (5.56). The constant *N1* defines the number of time steps between the start of the scan and the first leading edge of the waveform. The sweep time *T2* is defined by (5.58), and the constant *N2* is the number of time steps of the simulation.

The main program invokes the functions defined in the block diagram of Figure 5.40 and creates the vector ***A*** which records the value of the selected parameter at every time ***t***. In this particular case, it is the amplitude of the voltage which would be displayed at channel 2 of the oscilloscope and is proportional to the current *Ina* flowing in the resistor *Rg*.

The result of the simulation is displayed on the graph at the bottom of Figure 5.44. This shows the waveform of the current *Ina* at the near end of the cable. It took about half a dozen iterations of the program to select values for the constants *Crad* and *Cret*.

5.5.9 Assessment

Comparing this graph with the actual waveform recorded on Figure 5.37 shows that it is a fair representation of the actual response. This being so, it becomes possible to deduce the mechanisms involved in propagating a transient signal along a twin conductor cable. The basic model acts as a perfect transmission line. The inclusion of *Crad* allows the radiation losses to be simulated. The inclusion of *Cret* improves the similarity between the response of the model and the recorded waveform.

It can be reasoned that, during forward propagation of the wave, the send conductor acts as a transmitting antenna, whilst the return conductor acts as a receiving antenna. During the propagation of reflected energy back from the far end, the functions of the conductors are reversed. The return conductor acts as a transmitter, and the send conductor acts as a receiver.

This continuous interchange of energy causes most of that energy to flow along the path defined by the routing of the cable. Since current is also emitted from the outward-facing surfaces of the conductors, that current disappears into the environment and constitutes a loss of energy from the system.

5.5.10 Transient emission

Figure 5.45 illustrates the waveform of the current *Ins*. This is a measure of the transient emission from the cable. Since there is a delay of 83 ns between the time the current step is delivered to the near end and the time it arrives at the far end, the waveform of Figure 5.45 is a record of what has happened in the past. This means that during the first nanosecond, all the current delivered to the send conductor departs into the environment.

5.5.11 Surface charges

In departing into the environment, the current *Ins* in the send conductor leaves a charge of *Qns*. So the voltage *Vns* on the surface of the send conductor is as follows:

$$Vns = \frac{Qns}{Crad} \tag{5.68}$$

In departing from the send conductor, the current *Ifs* flowing back towards the near end deposits a charge *Qfs*. So the voltage *Vfs* on the surface of the return conductor

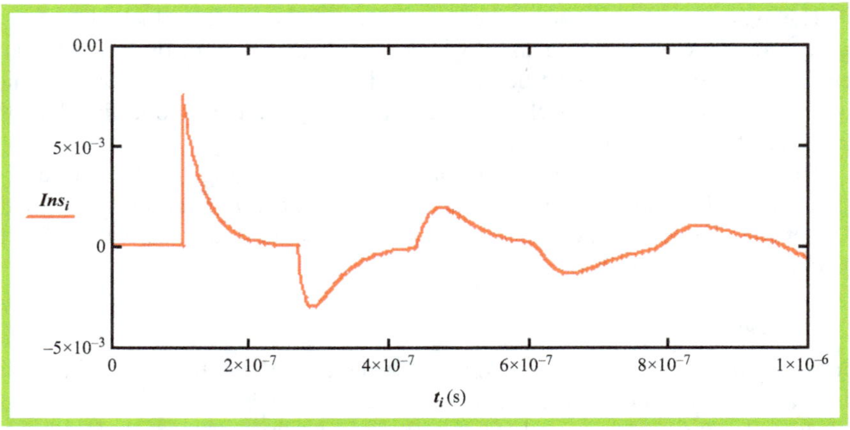

Figure 5.45 Waveform of the transient emission from the cable

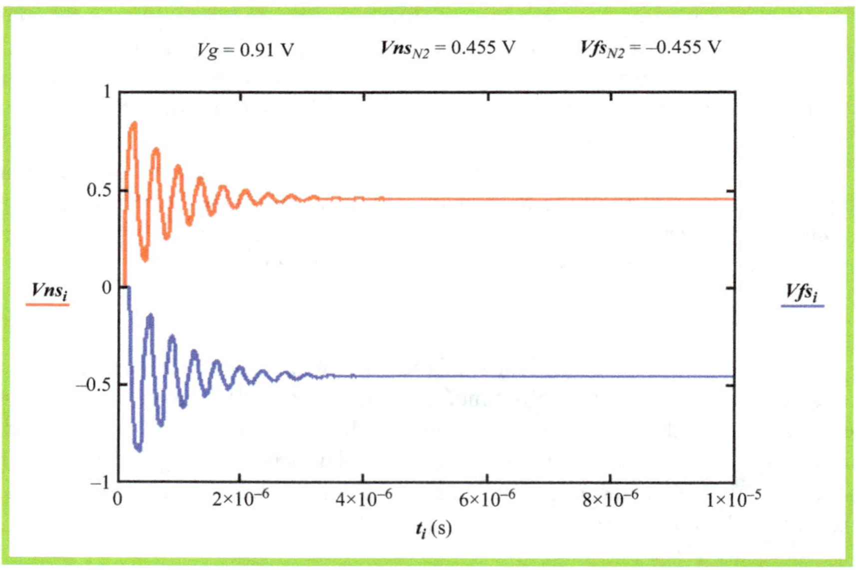

Figure 5.46 Waveforms of the voltages on the send and return conductors

is as follows:

$$Vfs = \frac{Qfs}{Cret} \tag{5.69}$$

Figure 5.46 shows the waveforms of these voltages over a period of 10 μs. As expected, each voltage settles down to a constant value of 455 mV. The figure also records the applied voltage, Vg.

5.5.12 Conclusion

This section has shown how a model can be created to simulate the transient emission of a twin-core cable. Since the response is a fair copy of the actual waveform and since the techniques and equations of circuit theory have been used exclusively in the creation of the model, it has been demonstrated that the method can be used to analyse the performance of any similar assembly.

That is, the transient emission of any twin conductor cable can be measured using simple test equipment and analysed using general purpose mathematical software.

Assessment of the model also provides an insight into the mechanisms involved in transient emission.

5.6 Answering the Catt Question

5.6.1 Introduction

Older readers of electronics world may remember articles by Ivor Catt in which he posed the Question *When the front edge of a voltage step is propagating along a transmission line, where does the negative charge on the bottom conductor come from*?

This brought to mind one of the lessons taught to final year students of Electrical Engineering at Glasgow University. It was explained that electrons could not possibly move at the speed of light. Nor did they need to, since there is an abundance of atoms in any copper conductor. If each atom has a free electron, these entities need only move at a snail's pace to carry the rated current. The question as to how power is transported from Glasgow to London in less time that it takes to blink was left unanswered.

Part of the answer can be found at Wikipedia. *The photon is a type of elementary particle. It is the quantum of the electromagnetic field including electromagnetic radiation such as light and radio waves, and the force carrier for the electromagnetic force. Photons are massless, and they always move at the speed of light in vacuum.*

Since photons apply force, they can cause electrons to be ejected from copper atoms. This causes current flow in the conductor, rather like tipping over a pack of cards. Developing this idea allows a mechanism to be postulated; one which provides an answer.

After posing his question to several eminent scientists, Ivor Catt poured scorn on each and every reply. From his point of view, the fact that charges are delivered to the bottom conductor brings the whole science of electromagnetic theory into question. Even so, he has made no attempt to provide an alternative explanation.

5.6.2 The Catt Anomaly

Traditionally, when a TEM step (i.e. logic transition from low to high) travels through a vacuum from left to right, guided by two conductors (the signal line and the 0v line), there are four factors which make up the wave:

- *Electric current in the conductors.*
- *Magnetic field, or flux, surrounding the conductors.*
- *Electric charge on the surface of the conductors.*
- *Electric field, or flux, in the vacuum terminating on the charge.*

The key to grasping the anomaly is to concentrate on the electric charge on the bottom conductor. During the next 1 ns, the step advances one foot to the right. During this time, extra negative charge appears on the surface of the bottom conductor in the next one foot length, to terminate the lines (tubes) of electric flux which now exist between the top (signal) conductor and the bottom conductor.

Where does this new charge come from? Not from the upper conductor, because by definition, displacement current is not the flow of real charge. Not from somewhere to the left, because such charge would have to travel at the speed of light in a vacuum. This last sentence is what those 'disciplined in the art' cannot grasp, although paradoxically it is obvious to the untutored mind. A central feature of conventional theory is that the drift velocity of electric current is slower than the speed of light.

5.6.3 Electromagnetic theory

Books on electromagnetic theory usually include a chapter on transmission lines. But only after the topics of static fields, conductors and charges, voltage and current, capacitance, energy and forces, magnetic field, magnetic force, vector identities, amperes law, inductance, Maxwell's equations, and reflection of electromagnetic waves have been fully discussed.

The deductions and analyses recorded in these preceding chapters came about as a result of many experiments and much puzzling as to how the observed results could be explained using scientific reasoning. It was then necessary for other engineers and scientists to review and agree the explanations. The work involved could be measured in terms of man-centuries.

Analysis of transmission lines is based on the knowledge that the currents, voltages, and charges in the conductors are the result of the action of the electromagnetic field.

The existence of charges on the bottom conductor referred to in the Catt Question does not disprove electromagnetic theory: It is the inevitable result of deductions made during the development of that theory.

5.6.4 Transmission line

An important feature of transmission line analysis is the fact that the return conductor plays an essential role in the propagation of energy. So it is useful to reprise the way it is introduced in the textbooks. Figure 5.47 illustrates a simple circuit model using lumped components.

Developing this model further, the line is assumed to consist of a series of minute sections, where the components are defined in terms of the length of the section, as illustrated by Figure 5.48.

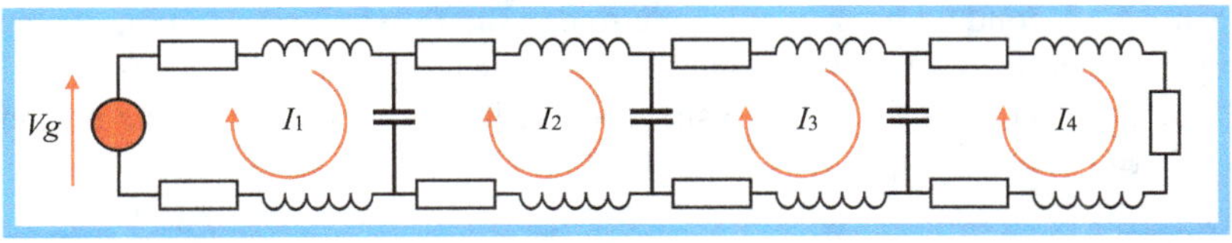

Figure 5.47 Lumped-parameter model of a twin-conductor line

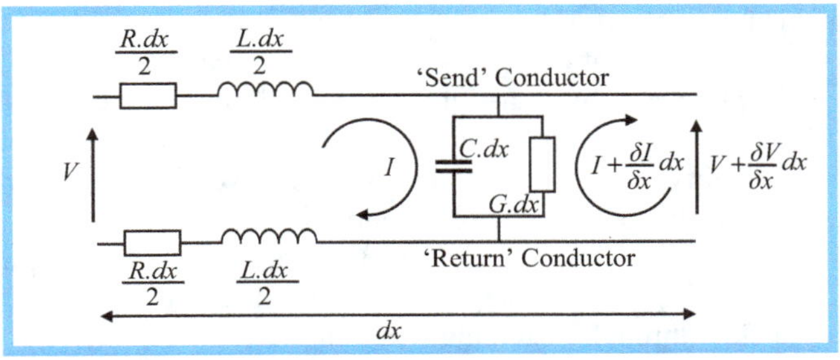

Figure 5.48 Circuit model of a single segment of the line

In Figure 5.48, R is defined as Ω per metre, L as Henry per metre, C as Farad per metre, and G as Siemens per metre.

Mesh analysis gives the equations:

$$(R + j \cdot \varpi \cdot L) \cdot dx \cdot I = V - \left(V - \frac{\delta V}{\delta x}\right) \cdot dx \tag{5.70}$$

$$(R + j \cdot \varpi \cdot C) \cdot dx \cdot V = I - \left(I + \frac{\delta I}{\delta x}\right) \cdot dx \tag{5.71}$$

Since these equations are the starting point of all subsequent analysis of transmission lines, it is important to note that they have been derived from Figure 5.48 and that inherent in this figure are unstated assumptions. These are as follows:

Current flows across the gap, from the send conductor to the return conductor.
Current flows back, from the return conductor to the send conductor.
Current flows in a series of tiny loops, as indicated by Figure 5.47.
Current consists of the flow of charges.

In books on electromagnetic theory, the deductions emanating from (5.70) and (5.71) lead to the conclusion that charges propagate at a speed of light in the medium. That is, they propagate along each conductor at a velocity slightly lower than that of light in a vacuum. They also propagate through the insulation.

These deductions are supported by the results of many tests conducted over the last century. The equations of transmission line analysis are used in the design of every piece of electronic equipment in existence. That is the justification for the above assumptions.

Since both conductors play an active role in the forward propagation of energy, the replacement of the return conductor with a surface at zero volts will eliminate any hope of visualising the mechanisms involved. This may be the basic reason for the Catt Anomaly.

5.6.5 Charges, electrons, and photons

It has been shown in this book that, by invoking the relationships derived by electromagnetic theory, the analytical tools of circuit theory can be used to simulate all the mechanisms involved in the propagation of EMI.

The basic assumptions of engineers who design circuits and electrical systems are that charges propagate at near-light velocity along conductors and through insulation. This assumption is inherent in the definitions of voltage, current, and charge in circuit theory, and these definitions lead inexorably to the circuit model of Figure 5.48 and to the deductions of transmission line theory.

On the other hand, scientists declare that the electron is the smallest particle which possesses the property of 'charge'. Yet electrons do not move at the speed of light and do not flow through insulating material.

Hence, there cannot be a one-to-one relationship between electrons and charges. Since scientists conclude that the photon is the *quantum of the electromagnetic field*, it is logical to reason that currents, voltages, charges, electric flux, and magnetic flux are manifestations of the interactions between photons and conductors.

What follows is a description of the mechanism involved in the propagation of charges along a transmission line. It is based on the relationships of Figure 5.48.

5.6.6 Coupling mechanism

When photons interact with copper atoms, they cause the release of the 'free' electrons, which depart in the opposite direction. An atom with an electron missing is positively charged. An atom with an extra electron is negatively charged. Electrons can move in any direction in a conductor, but they cannot depart from the surface of the conductor. The action of stopping at the surface causes the emission of a photon into the environment and for that surface to become positively charged.

When photons arrive at the bottom conductor, they cause that surface to become negatively charged and for photons to be released into the body of the conducting material. The process then continues in the bottom conductor. It can be likened to the movement of the steel balls in Newton's Cradle or to a pack of cards tipping over. Figure 5.49(a) illustrates this.

As well as driving positive charge into the body of the return conductor, photons can also move charged particles back along the surface. In turn, this causes the emission of photons back towards the send conductor. That is, the conducting surface acts as a reflector to some of the incoming photons. Figure 5.49(b) illustrates this.

When photons arrive back at the send conductor, the wave front will have moved forward. This means that they augment the number of photons being driven forward along that conductor. This action is the same as positive feedback.

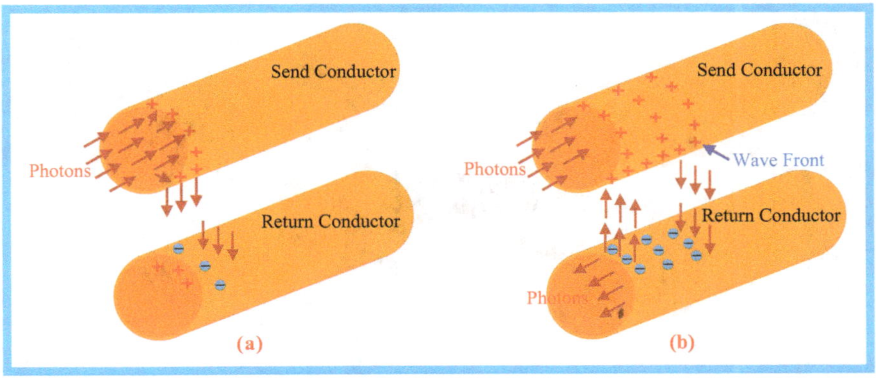

Figure 5.49 Action of photons in: (a) transferring charge between conductors and (b) returning energy to the send conductor

Current flows in a loop; along the send conductor, across the gap, and back via the return conductor.

5.6.7 Emission losses

Figure 5.49(a) and (b) only shows a few of the photons involved. In reality, photons are emitted from every point on each conductor, and depart in every direction. It is possible for many of those on the adjacent surfaces to return to the conductor opposite. The energy used to do this is represented by the voltage developed across the characteristic resistance.

Those which are emitted from the outward facing surfaces depart into the environment. If the cable-under-review is completely isolated, these photons depart into the environment and constitute a loss of energy. A voltage is developed on these outer surfaces due to the departing photons. As far as electromagnetic theory is concerned, current has departed from the system. In doing so it has left trapped charges on those surfaces. Section 5.5 describes a way of simulating this type of loss.

5.6.8 Magnetic field

Inside the body of the material, atoms can move in any direction. As well as moving axially and radially, they can propagate round the circumference of the conductor. This is illustrated by Figure 5.50. The action can be compared to a slingshot. Since this happens with both conductors, the flow of charges follows a figure-of-eight pattern.

This is possibly the start of an explanation as to how the lines of force in the magnetic field form a series of concentric circles.

5.6.9 Electromagnetic field

The action of the electromagnetic field can be explained in terms of the propagation of photons, both inside and outside a pair of conductors. It correlates with the

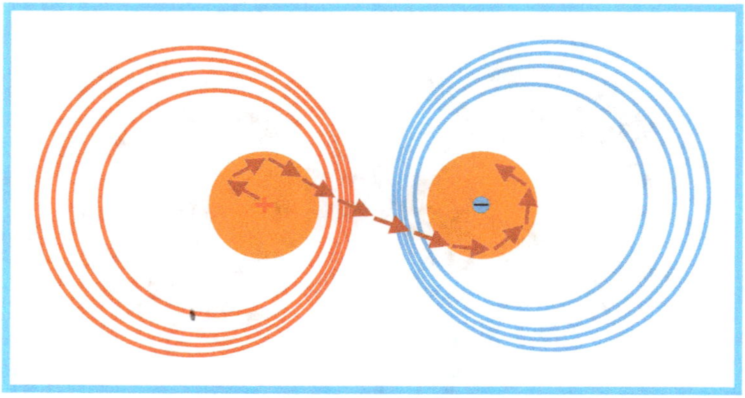

Figure 5.50 Propagation of photons round the surface creates magnetic fields of force

modelling technique used to analyse the properties of any conducting assembly, as described in Section 3.1.

Charges can propagate in any direction in the body of a conductor; axially, radially, and circumferentially. Photons have the same degree of freedom. This correlates with the fact that the analysis of the electromagnetic field emanating from a half wave transmitting dipole has longitudinal, radial, and latitudinal components of both the *H*- and the *E*-field.

Although photons do not carry charge, their interaction with atoms of conducting material allows the flow of these entities to be visualised in terms of moving charges.

5.6.10 Conclusion

This section has described a mechanism whereby photons interact with the atoms of conducting material to cause charges to appear to move at near-light velocity along a conductor, how the departure of photons from the surface of a conductor creates positive charge on that surface and how the arrival of photons at a conductor creates negative charge on that surface.

Such an explanation establishes a clear relationship between quantum and electromagnetic theories; an explanation which can be understood by circuit designers and system designers. It also answers the Catt Question.

5.7 Flow of partial currents

5.7.1 Introduction

The analysis of the drift velocity of electrons along a conductor conjures up a picture of water sluggishly flowing along a ditch in the countryside. This is misleading. A more appropriate visualisation of the flow of charges along a conductor is that of an avalanche hurtling down a mountainside.

A reprise of a lesson copied from a blackboard 60 years ago derives a figure for the drift velocity when a copper conductor is carrying its rated current. This was described as a snail's pace. The simulation of the flow of partial currents along a transmission line provides a dramatically different scenario.

If an electron is ejected from a copper atom, that atom becomes positively charged. If an electron is attached, the charge on the atom goes negative. If photons happen to illuminate a small patch on the surface of a copper conductor, this will initiate a chain reaction. Positive charges will depart along the wire in one direction with negative charges flowing in the other direction.

If a step voltage is applied at the near end of a twin conductor cable, a wave front will propagate along that cable, with positive charges flowing along one conductor and negative charges keeping pace along the other conductor.

With a short-circuited termination at the far end, the chain reactions continue unabated. A negative wave front arriving from the return conductor reappears as a positive wave front departing along the send conductor. A positive wave front arriving from the send conductor reappears as a negative wave front departing along the return conductor.

The total current at any point in a conductor is the sum of the partial currents.

5.7.2 Drift velocity

The atomic mass of any element is usually quoted in terms of the unified atomic mass and referred to as the Dalton, which has the symbol '*Da*':

$$Da = 1.6605 \times 10^{-27}\ \text{kg} \tag{5.72}$$

The mass of the copper atom, *m*, has been measured to be

$$\text{mass} = 63.546 \cdot Da\ \text{kg} \tag{5.73}$$

The density of copper is as follows:

$$\text{density} = 8930\ \text{kg/m}^3 \tag{5.74}$$

Hence, the number of atoms of copper per cubic metre is as follows:

$$n = \frac{\text{density}}{\text{mass}} = 8.466 \times 10^{28}\ \text{Atoms/m}^3 \tag{5.75}$$

The atomic number of copper is 29. That is, there are 29 electrons in orbit round a nucleus which has 29 protons and 35 neutrons. If it is assumed that there is one free electron per atom, *n* is the number of free electrons per cubic metre capable of drift through the atomic structure. The charge, *e*, on each electron is as follows:

$$e = 1.602 \times 10^{-19}\ C \tag{5.76}$$

The charge contained by one cubic metre of copper is as follows:

$$\text{Qdensity} = n \cdot e \, \text{C/m}^3 \tag{5.77}$$

If I amperes flows into a conductor for dt seconds, the magnitude of the charge delivered to the conductor will be

$$Q = I \cdot dt \, C \tag{5.78}$$

If the cross-sectional area of the conductor is A and the length of the conductor into which the charge flows is dx, the volume of the conducting material holding this charge will be

$$\text{Volume} = A \cdot dx \, \text{m}^3 \tag{5.79}$$

The charge density in this volume is as follows:

$$\text{Qdensity} = \frac{Q}{\text{Volume}} = \frac{I \cdot dt}{A \cdot dx} \, \text{C/m}^3 \tag{5.80}$$

Using (5.77) to substitute for Qdensity

$$n \cdot e = \frac{I}{A \cdot vel} \, \text{C/m}^3 \tag{5.81}$$

where *vel* is the drift velocity of the electrons.

The diameter *dia* of an American Wire Gauge (AWG 20) conductor is:

$$dia = 0.8125 \times 10^{-3} \text{m} \tag{5.82}$$

So the cross-sectional area is as follows:

$$A = \frac{\pi \cdot dia^2}{4} = 5.185 \times 10^{-7} \, \text{m}^2 \tag{5.83}$$

The rated current *Irated* is a function of the way the conductor is used. Typically,

$$Irated = 11 \, \text{A} \tag{5.84}$$

Rearranging (5.81) and assigning values to the parameters *n, e, A,* and *Irated*

$$vel = \frac{Irated}{A \cdot n \cdot e} = 1.565 \times 10^{-3} \, \text{ms}^{-1} \tag{5.85}$$

This drift velocity has been described as a snail's pace. However, the main point is that there is superabundance of charge-carrying entities; 8×10^{10} in a cubic μm.

5.7.3 Loop resistance

The resistivity of copper is as follows:

$$\rho = 1.7 \times 10^{-8} \, \Omega\text{m} \tag{5.86}$$

If an AWG 20 conductor is used in a 15 m length of twin-conductor cable, the loop resistance would be

$$R = \frac{\rho \cdot 15 \cdot 2}{A} = 0.984\ \Omega \tag{5.87}$$

5.7.4 Transmission line model

Figure 5.51 illustrates a configuration where the resistance at the near end is *Rn* and the resistance at the far end is zero. At the near end, the source voltage is *Vgen*, the incident current is *Ini*, the absorbed current is *Ina*, and the reflected current is *Inr*. At the far end, the incident current is *Ifi*, the absorbed current is *Ifa*, and the reflected current is *Ifr*.

All the currents are defined as loop currents. Positive current is defined as a clockwise flow. This means that positive current flows forward along the send conductor and back along the return conductor. Since *Ini*, *Inr*, *Ifi,* and *Ifr* are partial currents, there are two separate currents flowing along each conductor. Since *Ina* is defined as the sum of *Ini* and *Inr*, then the current flow at the near end can be treated as unidirectional. The same is true for *Ifa* at the far end.

The modelling technique assumes that the line is lossless. However, by setting the resistor at the near end equal in value to the series resistance of the conductors, copper losses can be included in the simulation.

5.7.5 Simulation

The loop resistance of a 15 m length of 20 AWG cable has been calculated to be 0.984 Ω. So, if the terminals at the far end are shorted together and an ohmmeter connected to the terminals at the near end, the ohmmeter reading would be about 1 Ω. So it is reasonable to assign this value to *Rn*. Assigning a value of 10 V to the generator *Vgen* and a value of 100 Ω to the characteristic resistance *Ro* creates a model which can be used to simulate the responses of the partial currents under steady-state conditions.

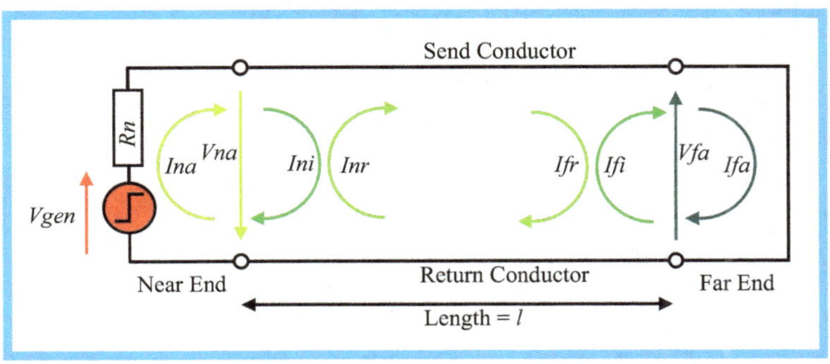

Figure 5.51 Transmission line

Figure 5.52 is a reproduction of the Mathcad worksheet used to do just that. Since examples have already been provided of the processing details in previous sections, it is only necessary to state that the vectors ***Ifi***, ***Ifr***, and ***Ifa*** hold values of these parameters over the range zero to *Tsweep*.

Figure 5.53 is a graph which simulates the response of these three currents over a period of 200 ns. The current at the far end, *Ifa*, is shown as the solid green curve.

$$Vg := 10 \qquad Ro := 100 \qquad Rn := 1 \qquad Rf := 0 \qquad n := 100$$

$$l := 15 \qquad v := 3\cdot 10^{8} \qquad T := \frac{l}{v} = 5\times 10^{-8} \qquad dt := \frac{T}{n} = 5\times 10^{-10}$$

$$near(Ini, Vg) := \left| \begin{array}{l} Ina \leftarrow \dfrac{2\cdot Ro\cdot Ini + Vg}{Ro + Rn} \\ Inr \leftarrow Ina - Ini \\ (Inr \;\; Ina) \end{array}\right. \qquad far(Ifi) := \left| \begin{array}{l} Ifa \leftarrow \dfrac{2\cdot Ro\cdot Ifi}{Rf + Ro} \\ Ifr \leftarrow Ifa - Ifi \\ (Ifr \;\; Ifa) \end{array}\right.$$

$$forward(\boldsymbol{F}, In) := \left| \begin{array}{l} \boldsymbol{F}_1 \leftarrow In \\ for\ x \in n+1 .. 1 \\ \quad \boldsymbol{F}_{x+1} \leftarrow \boldsymbol{F}_x \\ If \leftarrow \boldsymbol{F}_{n+2} \\ (\boldsymbol{F} \;\; If) \end{array}\right. \qquad back(\boldsymbol{B}, If) := \left| \begin{array}{l} \boldsymbol{B}_{n+1} \leftarrow If \\ for\ x \in 2 .. n+1 \\ \quad \boldsymbol{B}_{x-1} \leftarrow \boldsymbol{B}_x \\ In \leftarrow \boldsymbol{B}_1 \\ (\boldsymbol{B} \;\; In) \end{array}\right.$$

$$N := 4\cdot n \qquad i := 1 .. N \qquad t_i := i\cdot dt \qquad Tsweep := N\cdot dt = 2\times 10^{-7}$$

$$(\boldsymbol{Ifi} \;\; \boldsymbol{Ifr} \;\; \boldsymbol{Ifa}) := \left| \begin{array}{l} \boldsymbol{F}_{n+2} \leftarrow 0 \\ \boldsymbol{B}_{n+1} \leftarrow 0 \\ for\ i \in 1 .. N-1 \\ \quad \left| \begin{array}{l} Vgen \leftarrow Vg \;\; if\ i > 1 \\ (Inr \;\; Ina) \leftarrow near(Ini, Vgen) \\ (\boldsymbol{F} \;\; Ifi) \leftarrow forward(\boldsymbol{F}, Inr) \\ (Ifr \;\; Ifa) \leftarrow far(Ifi) \\ (\boldsymbol{B} \;\; Ini) \leftarrow back(\boldsymbol{B}, Ifr) \\ \boldsymbol{Ifi}_i \leftarrow Ifi \\ \boldsymbol{Ifr}_i \leftarrow Ifr \\ \boldsymbol{Ifa}_i \leftarrow Ifa \end{array}\right. \\ (\boldsymbol{Ifi} \;\; \boldsymbol{Ifr} \;\; \boldsymbol{Ifa}) \end{array}\right.$$

$$L := T\cdot Ro = 5\times 10^{-6} \qquad \boldsymbol{I}_i := \frac{Vg}{Rn}\left(1 - e^{\frac{-Rn}{L}\cdot t_i}\right)$$

Figure 5.52 Mathcad worksheet which analyses the response of a short-circuited line

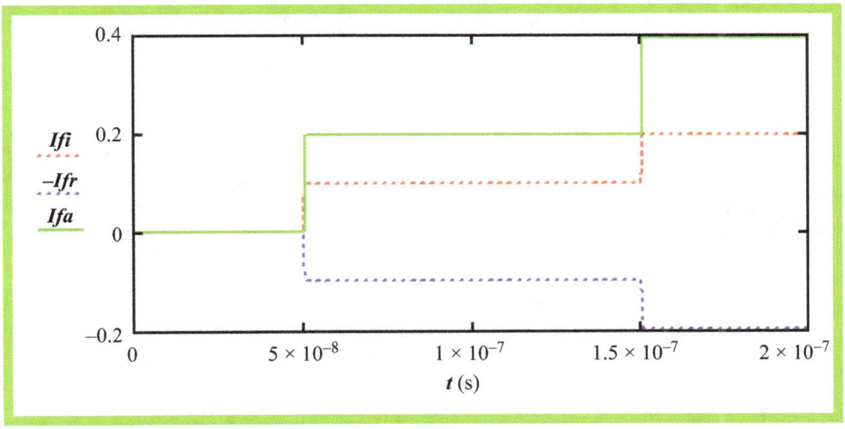

Figure 5.53 Response of a short-circuited line over a period of four traverses of the step

The incident current, *Ifi*, is shown as the dotted red curve, and the reflected current, *Ifr*, is shown as a dotted blue curve. The reflected current is shown as a negative current because it is flowing backwards.

5.7.6 Mechanism

The graph of Figure 5.53 and an assessment of the computations involved in its creation provide an insight into the mechanism involved in the actual system.

Since a step voltage of 10 V is applied and the characteristic resistance of the line is 100 Ω, the current *Inr* delivered to the line will be 0.1 A. This manifests itself as a series of tiny charges which propagate forward at velocity *v*. In the model, a shift register is used to simulate this action. The number of stages *n* in the register is 100. So each section represents a time duration *dt* of 500 ps. So, the charge delivered is $Inr \cdot dt = 50$ pF. During the next 50 ns, this charge will propagate forward along the length of the line.

It takes *T* seconds for the leading edge to arrive at the terminals at the far end. During this time, nothing happens at the far end. Then, at $t = 50$ ns, the electromagnetic field manifests itself as a voltage of 10 V applied via a resistance of 100 Ω. Hence, the current *Ini* changes from 0 to 0.1 A. Then it stays constant at this value for the next 100 ns.

Since the load *Rf* is zero, the current *Ifa* at the far end is twice the value of *Ini*. Hence, the reflected current *Ifr* is equal in value to *Ifi*.

At the far end terminals, this reflected current manifests itself as a series of 50 pF discharges which then propagate back towards the near end at velocity *v*. Figure 5.54 illustrates the responses over a period of 20 μs. As expected, the current delivered to the far end rises exponentially towards 10 A.

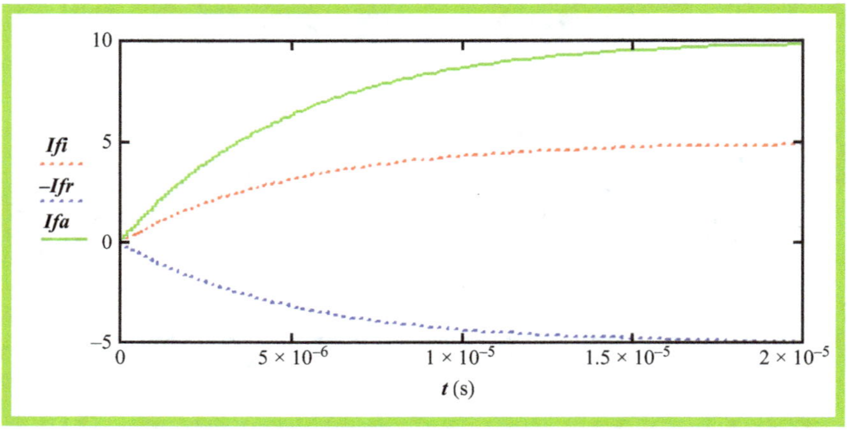

Figure 5.54 Response of a short-circuited line over a period of 20 μs

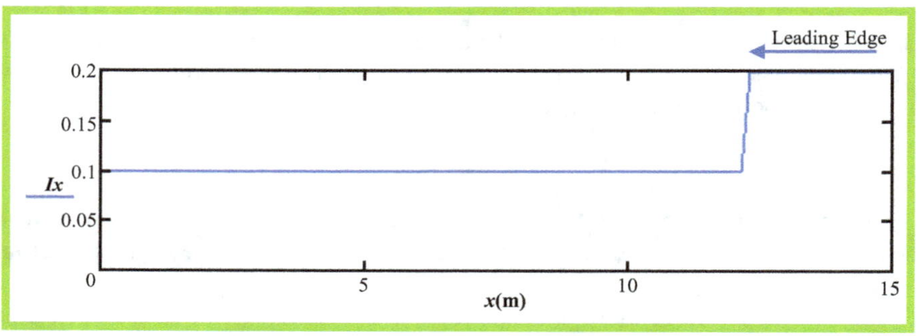

Figure 5.55 Snapshot of the current distribution along the send conductor at 60 ns

However, the fact that the partial currents, *Ifi* and *Ifr*, only reach a value of 5 A under steady-state conditions need some explanation.

5.7.7 Partial currents

The relationship between partial current and total current is best explained by analysing the current distribution in the cable at a particular instant. This is illustrated by Figure 5.55. The worksheet used to create this figure is a slight modification of that recorded in Section 5.2.

It is assumed that a snapshot is taken of the distribution of current 60 ns after the step voltage is applied at the near end. That is, 10 ns after the step has been reflected at the far end. At this instant, the leading edge is propagating back towards the near end, but the current *Ifi* still arriving at the far end is 0.1 A. This is the incident current. The step of 0.1 A propagating back is the reflected current *Ifr*.

This reflected current is really the current in the return conductor continuing its journey back: along the send conductor. The incoming charges at the near end do not 'know' that they have been routed through an angle of 180 degrees at the far end. A negative current flowing in the opposite direction is a positive current.

5.7.8 Charge carriers

Given an insight into the relationship between the propagation of charges and current, it becomes possible to relate these parameters to electrons and atoms.

The analysis of drift velocity has shown that there is a superabundance of charge carriers in the assembly.

If an electron is dislodged from a copper atom, that atom will become positively charged. It will attract an electron from a neighbouring atom. The neighbouring atom will become positively charged.

Similarly, the dislodged electron will attach itself to a neighbour on the opposite side, causing that neighbour to become negatively charged.

If photons were to illuminate an area on the surface of a copper wire, many electrons would be displaced from the region at that surface. This would initiate a chain reaction with positive charges propagating in one direction along the conductor and negative charges flowing away in the opposite direction.

Since the cable is short-circuited at the far end, a chain reaction with a positive leading edge moving forward along the send conductor will continue unchecked through the terminals and continue to propagate back along the return conductor with a negative leading edge.

If there is also a chain reaction moving forward along the return conductor with a negative leading edge, this will also continue through the shorted terminals and propagate back along the send conductor as a step with a positive leading edge.

There is no conflict between the two chain reactions. At any point in the conductor, they support each other. Figure 5.55 illustrates this.

5.7.9 Photon action

Such a depiction of charges propagating back and forth along the conductors is only half the picture. The real source of energy is the flow of photons. These propagate to and fro between the conductors and create charges on the surface of both conductors; positive charges on the send conductor and negative charges on the return conductor. It is these induced charges which sustain the current flow.

Since this interchange of photons takes place at the surface of the conductors, the flow of charges along the conductors will be concentrated at the surfaces.

A step voltage applied at the near end of the cable will cause a step current to propagate along each conductor. There will be positive charges at the front end of the send conductor and negative charges at the front end of the return conductor.

As described above, the flow of these charges will continue unchecked at the far end, and the leading edge will propagate back to the near end.

During the transit of energy back down the line, the return conductor acts as a transmitting antenna, delivering energy back to the send conductor. During this time, the send conductor continues to deliver energy to the return conductor. That is, each conductor acts as a transmitting antenna and as a receiving antenna. Both conductors contribute to the kinetic energy being stored in the cable.

5.7.10 Resistance and inductance

Figure 5.53 provides a picture of the current increasing as a series of discrete current steps. If the source resistance *Rn* in the model is set at zero, the amplitude of every step would be 0.1 A and the current would increase linearly with time without limit.

As the current flowing in each conductor increases, more and more atoms become involved in the propagation of charge. So, more and more of them are forced to flow below the surface of each conductor. Eventually, the movement of charges along the cable involves the entire cross section.

When the conductor is acting as a transmitter, there is radial flow of sub-atomic particles up to the surface, where the charges are trapped and photons are emitted. When the conductor is acting as a receiver, the arriving photons cause similar particles to be driven down onto the conducting material.

There will be impacts between particles moving in opposite directions, whether they are moving laterally, radially, or circumferentially. Particles moving in opposite directions are bound to cause friction. This causes heating. In circuit analysis, this effect is defined by the voltage developed across a resistance.

This resistance causes the amplitude of the steps to decrease with time. Eventually, the steps become imperceptible. But that does not mean that the propagation to and fro of the photons ceases. Under steady-state conditions, this propagation manifests itself as the magnetic field.

The stored energy in the field is the same as that of the dynamic energy of the charges. An analogy would be that of a flywheel which has picked up speed due to tiny impulses at every rotation.

Electromagnetic theory uses the concept of inductance to relate the energy in the magnetic field to the voltage developed by a step change in current.

5.7.11 Simple model

Since the far end is short-circuited, the line behaves as an inductor, where

$$L = T \cdot Ro = 5 \times 10^{-8} \cdot 100 = 5 \times 10^{-6}\ \text{H} \tag{5.88}$$

The series resistance of the conducting loop has been assumed to be 1 Ω.

If a 10 V step were to be applied to a series L–R circuit, the model used to analyse the response would be as shown by Figure 5.56.

The loop equation for transient analysis is as follows:

$$Vg = R \cdot I + L \cdot \frac{dI}{dt} \tag{5.89}$$

Analysing this using Heaviside transformations gives the relationship between current and time:

$$I = \frac{Vg}{R} \cdot \left(1 - e^{-\frac{R}{L} \cdot t}\right) \tag{5.90}$$

Plotting this relationship on the same graph as *Ifa* leads to Figure 5.57. The response *Ifa* at the end terminals of the transmission is illustrated by the solid green curve. The response *I* of the circuit model of Figure 5.56 is shown as a dotted black curve. The fact that they are identical shows that the modelling technique invoked in the Mathcad worksheet of Figure 5.52 produces an accurate simulation. Since the modelling technique is based on the relationships of transmission line theory, the simulation of the flow of partial currents must also be accurate.

The description of the mechanisms involved in relating the movement of electrons and the action of photons is consistent with both the simulation of Figure 5.54 and with the known properties of photons and electrons.

The properties of the photon are defined by Wikipedia as: *The photon is a type of elementary particle. It is the quantum of the electromagnetic field including*

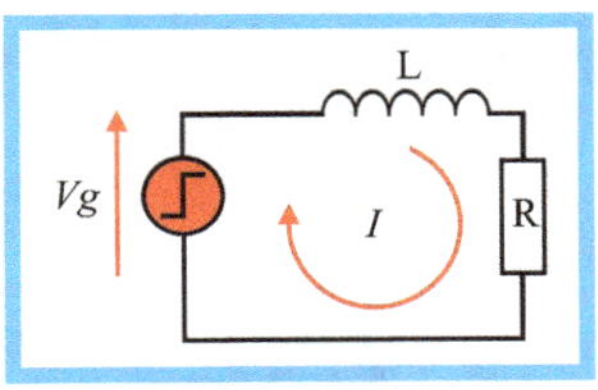

Figure 5.56 Lumped-parameter model of configuration of Figure 5.51

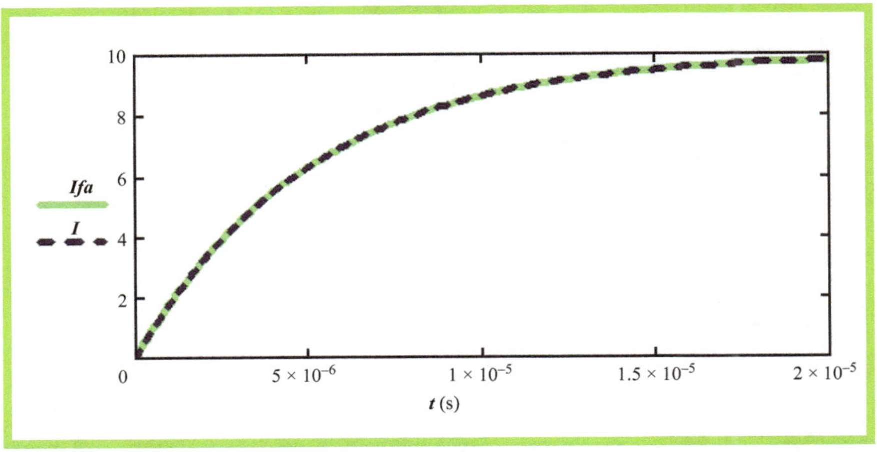

Figure 5.57 Waveforms, analysed using transmission line and Heaviside equations

electromagnetic radiation such as light and radio waves, and the force carrier for the electromagnetic force. Photons are massless, and they always move at the speed of light in vacuum. It has also been stated by the scientific community that photons do not carry charge.

The properties of atoms and electrons can be found in most science data books.

5.7.12 Conclusion

A plausible description has been provided of the mechanism underlying the behaviour of photons, electrons, and atoms in a transmission line. It provides a solid basis for further analysis as well as an improved understanding of the phenomenon.

Chapter 6
Three conductor transient models

6.1 Transient cross-coupling

6.1.1 Introduction

Propagation of charge along a twin-conductor line can be simulated by a pair of shift registers, and the transient response at each termination can be analysed by time-step analysis. Simulating the response of a cable routed along the structure is more complex, since the differential-mode and common-mode signals propagate at different velocities. There is also inter-mode coupling at each end.

The creation of a model which simulates such an assembly requires a high degree of confidence; confidence which can only be achieved by comparing the simulation with the results of measurements carried out on a test rig. This brings with it the requirement to simulate the test equipment as well as the assembly-under-test.

Fortunately, there existed a rig which had previously been characterised using frequency response analysis. The model derived from this exercise was transformed into one which could handle transient signals, and the test equipment re-configured to deliver and monitor square wave signals.

Photos were taken of the waveforms observed on the screen of the oscilloscope.

The task then was to create a model which simulated these waveforms. Since the setup included modules which did not form part of the equipment-under-test, these modules affected the overall response of the system to the output of the signal generator. So, it was necessary to simulate the effects of the interface module at the input to the test rig, the current transformer, and to correlate the sweep time of the simulation with that of the oscilloscope.

The waveforms appearing on the screen were then compared with the response of the model. Close correlation was achieved. This established a fair degree of confidence in the technique.

It was then possible to identify that the part of the simulation which represents the assembly-under-test and define this as an accurate model of that assembly. That is, as a circuit diagram with values assigned to all the components.

A description of every aspect of the simulation is provided and a copy of the Mathcad worksheet which performs the calculations is included. Any electronics engineer can replicate the process, since it involves general-purpose test equipment and general-purpose software.

6.1.2 Frequency response test

Figure 6.1 defines the setup previously used to carry out frequency response measurements of a three-conductor signal link. This link consisted of a twin-conductor cable routed along a 15 mm copper pipe. The terminals at the far end were shorted together. A signal generator was used to provide a voltage across a 10 Ω resistor in the interface module. This allowed the voltage *Vin* to be monitored by channel 1 of the oscilloscope and enabled the cable terminals at the near end to be driven by a low impedance source. A current transformer was used to allow channel 2 of the scope to monitor the differential-mode current in the cable.

The signal generator was used to apply an alternating signal to the near end terminals, the voltage *Vch1* at channel 1 of the oscilloscope was used to monitor the input voltage, and *Vch2* was used to monitor the output current. This was done over a set of spot frequencies *F1* and a graph created of the variation of cable admittance with frequency.

Then the current transformer was clamped round both conductors of the cable, and the test was repeated, this time providing a curve defining the transfer admittance of the assembly. That is, the ratio of the common-mode current to the differential-mode voltage.

6.1.3 Definitive circuit model

Then a circuit model was used to replicate these two curves. This was done by comparing the response of a general circuit model of the assembly with the test results, and varying the component values until close correlation was achieved between theoretical results and measured results. The resultant model, with values assigned to all the components, can be described as the definitive circuit model. This is illustrated in Figure 6.3.

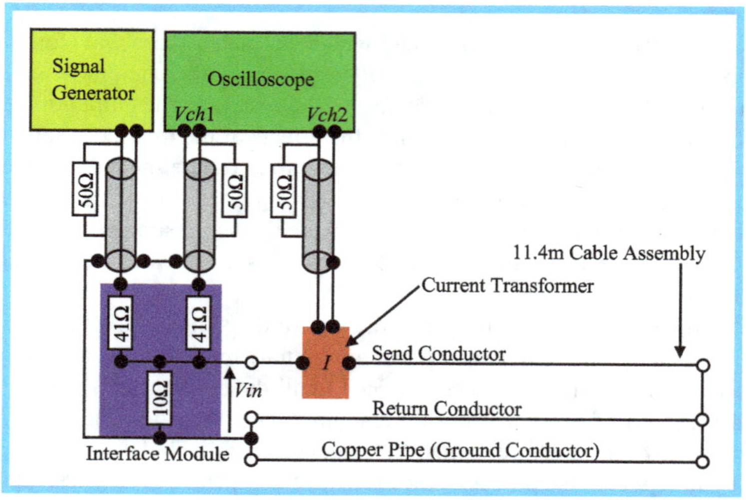

Figure 6.1 Test setup for characterisation of signal link

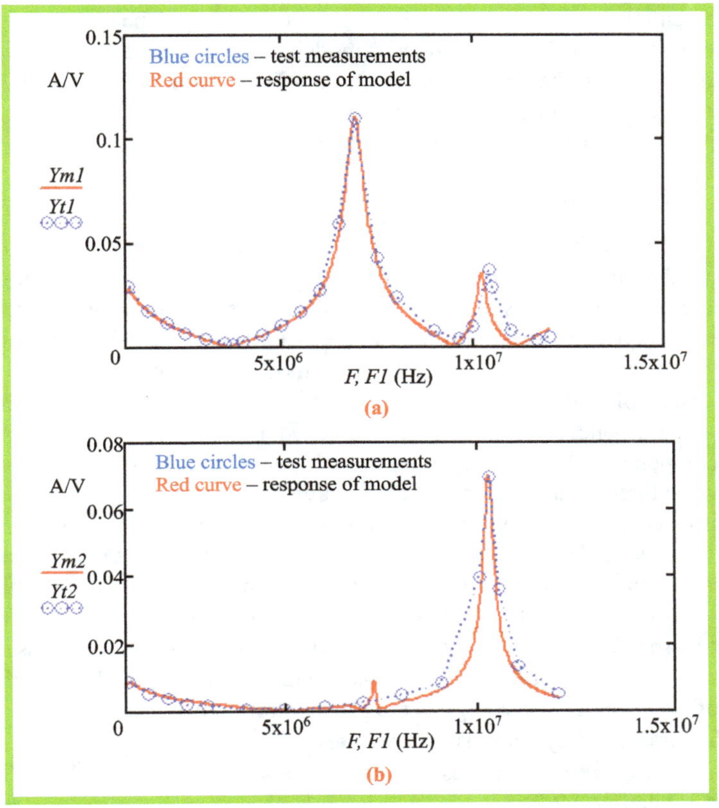

Figure 6.2 Correlation between test results and circuit model: (a) ratio of differential-mode current to input voltage and (b) ratio of common-mode current to input voltage

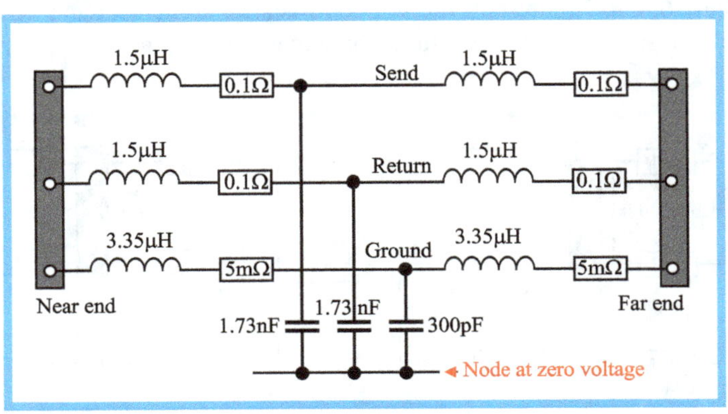

Figure 6.3 Definitive circuit model of the signal link of Figure 6.1

An essential feature of the process was that the lumped parameter model was converted into a distributed parameter model using a set of transfer equations before carrying out the main computations. These equations used hyperbolic functions to define values for the impedances of a Triple-T network. A full description of this process is provided in Section 2.6.

6.1.4 Transient model

The Mathcad worksheet used to create Figure 6.2(a) and (b) was based on mesh analysis of the circuit model. Such an approach resulted in fewer equations than would have been created by nodal analysis. It allowed a one-to-one correlation to be achieved between loop currents in the simulation and the current modes involved in assessing test results; differential-mode and common-mode.

The choice of a star configuration rather than a delta configuration for the model enabled a one-to-one correlation to be achieved between each conductor and its reactive components. The fact that the star configuration resulted in the creation of a node between three capacitors was pure serendipity. This node represents a location in the far distance at which the energy level of the electromagnetic field is zero. It removes the need to define a location on the ground as a 'zero-volt node', enables this conductor to be treated in the same way as the send and return conductors, and makes it possible to use circuit theory to simulate EMI coupling mechanisms (if a delta configuration had been chosen, this development would not have been possible).

Section 5.7 demonstrates the fact that the driving force involved in the transmission of electromagnetic energy is the propagation of photons to and fro between the conductors of the system. This correlates with the deductions of transmission line theory that charge propagates forward at the same velocity as the voltage and current (see Section 2.2.2).

Replacing each T-Network of Figure 6.3 with a single resistor results in a circuit model where the junction of *Ro1, Ro2,* and *Ro3* is at zero volts at each end of the link. This allows the transient current in each conductor to be calculated using mesh analysis. The zero-volt terminals at each end are independent. So, the processing of signals at one end is not a function of what is happening at the other end.

In Figure 6.4, *Rs1* represents the source resistance at the near end. *Rs2* and *Rs3* simulate the resistances of the return conductor and the ground conductor. *Vg*

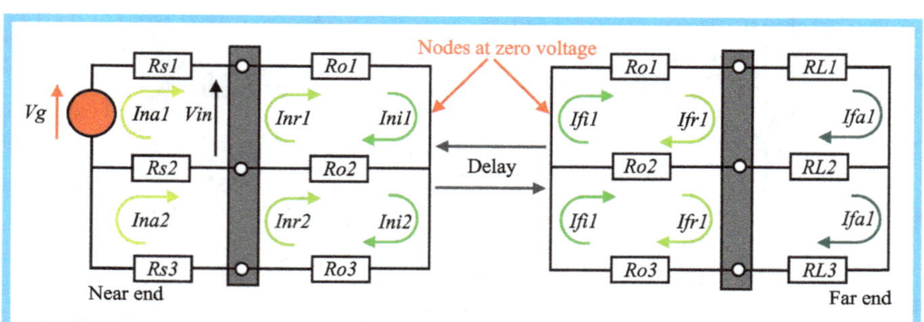

Figure 6.4 Cross-coupling model for transient analysis

simulates the source voltage. In the case of the setup of Figure 6.1, resistors *RL1, RL2,* and *RL3* are all zero.

Ini1, Inr1, and *Ina1* are the incident, reflected, and absorbed currents in the differential-mode loop at the near end. *Ini2, Inr2,* and *Ina2* are the incident, reflected, and absorbed currents in the common-mode loop.

Ifi1, *Ifr1*, and *Ifa1* are the incident, reflected, and absorbed currents in the differential-mode loop at the far end. *Ifi2*, *Ifr2*, and *Ifa2* are the incident, reflected, and absorbed currents in the common-mode loop.

6.1.5 Transient response test

Having constructed a model which can simulate transient cross-coupling, the task then was to carry out a set of transient tests on the assembly of Figure 6.1 and to replicate the resultant waveforms.

Such an approach takes advantage of the facts that the signal generator can create rectangular waveforms as well as sinusoidal signals; and that the oscilloscope can display sinusoidal signals or transient waveforms with equal ease.

So, a set of transient tests was carried out on the setup of Figure 6.1 and photographs taken of the waveforms. A square wave at 200 kHz was found to give a good approximation to a series of step functions. Then the current transformer was clamped round both conductors to monitor the common-mode signal, and more photographs were taken.

The next step was to assign component values to the model to simulate its performance. It was also necessary to include the response of the test equipment in the simulation. This was done, and a direct comparison made between the photographic records and the simulated waveforms.

6.1.6 Resistor values

The characteristic resistances *Ro1, Ro2,* and *Ro3* are as follows:

$$Ro1 = \sqrt{\frac{Lc1}{Cc1}} = \sqrt{\frac{(1.5 + 1.5) \times 10^{-6}}{1.73 \times 10^{-9}}} = 41.643\ \Omega \tag{6.1}$$

$$Ro2 = \sqrt{\frac{Lc2}{Cc2}} = \sqrt{\frac{Lc1}{Cc1}} = 41.643\ \Omega \tag{6.2}$$

$$Ro3 = \sqrt{\frac{Lc3}{Cc3}} = \sqrt{\frac{(3.35+3.35) \times 10^{-6}}{300 \times 10^{-12}}} = 149.4\ \Omega \tag{6.3}$$

Where the values of *Lc1*, *Lc2*, *Lc3*, *Cc1*, *Cc2*, and *Cc3* are defined by the circuit model of Figure 6.3.

The value of the source resistance *Rsource* at the near end was derived from Figure 6.1:

$$Rsource = \frac{1}{\left(\frac{1}{41+50}\right) + \left(\frac{1}{41+50}\right) + \left(\frac{1}{10}\right)} = 8.2\ \Omega \tag{6.4}$$

Since the propagation of charges in the model is assumed to be lossless, it was necessary to treat the series resistances of the send and return conductors as a load at the source:

$$Rs1 = Rsource + 0.1 + 0.1 = 8.4 \ \Omega \tag{6.5}$$

Similarly

$$Rs2 = 0.1 + 0.1 = 0.2 \ \Omega \tag{6.6}$$

$$Rs3 = 0.005 + 0.005 = 0.01 \ \Omega \tag{6.7}$$

The series resistors are defined by the model of Figure 6.3. The resistors *RL1*, *RL2*, and *RL3* at the far end are each defined to be zero.

6.1.7 Source voltage

The ratio K of the voltage *Vch1* monitored at the oscilloscope to the voltage *Vin* appearing the terminals of the send and return conductors can be derived from Figure 6.1.

$$K = \frac{Vch1}{Vin} = \frac{50}{50+41} = 0.549 \tag{6.8}$$

During the test of the setup of Figure 6.1, the amplitude of the output of the square waveform was set at a value such that the step voltage *Vch1* monitored at channel 1 of the oscilloscope was 0.8 V. This meant that the voltage *Vin* applied between the input terminals at the near end was as follows:

$$Vin = \frac{Vch1}{K} = 1.456 \text{ V} \tag{6.9}$$

At the instant the step voltage was applied, the load presented by the cable was as follows:

$$R2 = Ro1 + \frac{Ro2 \cdot Ro3}{Ro2 + Ro3} = 74.209 \ \Omega \tag{6.10}$$

So the voltage *Vg* of the model of Figure 6.4 had to be

$$Vg = \frac{Rs + R2}{R2} \cdot Vin = 1.62 \text{ V} \tag{6.11}$$

6.1.8 Propagation times

Figure 6.2 identifies the fact that the current in the common-mode loop resonates at a higher frequency than that in the differential-mode loop. Since the terminals at the far end are shorted together, the peak amplitudes correspond to the half-wave frequencies:

With differential-mode

$$fh1 = 6.93 \times 10^{-6} \text{ Hz} \tag{6.12}$$

With common-mode

$$fh2 = 10.26 \times 10^6 \text{ Hz} \tag{6.13}$$

This means that the propagation times are as follows:

$$T1 = \frac{1}{2 \cdot fh1} = 7.215 \times 10^{-8} \text{ s} \tag{6.14}$$

$$T2 = \frac{1}{2 \cdot fh2} = 4.873 \times 10^{-8} \text{ s} \tag{6.15}$$

Charges propagate along the common-mode loop (loop 2) at a higher velocity than those in the differential-mode loop (loop 1) because the relative permittivity is less. If loop 1 is represented by $n1$ segments and loop 2 by $n2$ segments and

$$\frac{n1}{n2} = \frac{T1}{T2} \tag{6.16}$$

Then the time dt taken for a charge to propagate along any segment can be defined:

$$dt = \frac{T1}{n1} = \frac{T2}{n2} \text{ s} \tag{6.17}$$

If loop 1 is simulated by 15 segments, the time taken for a charge to propagate along one segment would be

$$dt = \frac{T1}{n1} = \frac{7.215 \times 10^{-8}}{15} = 4.81 \times 10^{-9} \text{ s}$$

If loop 2 is represented by 10 segments, the time taken for a charge to propagate along this loop will be

$$n2 \cdot dt = 10 \cdot dt = 4.81 \times 10^{-8} \text{ s} \tag{6.18}$$

This is close enough to $T2$ to allow a reasonably accurate simulation.

6.1.9 Reflection equations

The reflection equations at the near end are as follows:

$$\begin{aligned} Vni1 + Vnr1 &= Vna1 - Vg \\ Vni2 + Vnr2 &= Vna2 \end{aligned} \tag{6.19}$$

$$\begin{aligned} Ini1 + Inr1 &= Ina1 \\ Ini2 + Inr2 &= Ina2 \end{aligned} \tag{6.20}$$

$$\begin{aligned} Vni1 &= (Ro1 + Ro2) \cdot Ini1 - Ro2 \cdot Ini2 \\ Vni2 &= -Ro2 \cdot Ini1 + (Ro2 + Ro3) \cdot Ini2 \end{aligned} \tag{6.21}$$

$$\begin{aligned} Vnr1 &= -(Ro1 + Ro2) \cdot Inr1 + Ro2 \cdot Inr2 \\ Vnr2 &= Ro2 \cdot Inr1 - (Ro2 + Ro3) \cdot Inr2 \end{aligned} \tag{6.22}$$

$$\begin{aligned} Vna1 &= (Rs1 + Rs2) \cdot Ina1 - Rs2 \cdot Ina2 \\ Vna2 &= -Rs2 \cdot Ina1 + (Ro2 + Ro3) \cdot Ina2 \end{aligned} \tag{6.23}$$

In vector-matrix form, these are as follows:

$$\mathbf{Vni} + \mathbf{Vnr} = \mathbf{Vna} - \mathbf{Vg} \tag{6.24}$$

$$\mathbf{Ini} + \mathbf{Inr} = \mathbf{Ina} \tag{6.25}$$

$$\mathbf{Vni} = \mathbf{Ro} \times \mathbf{Ini} \tag{6.26}$$

$$\mathbf{Vnr} = -(\mathbf{Ro} \times \mathbf{Inr}) \tag{6.27}$$

$$\mathbf{Vna} = \mathbf{Rs} \times \mathbf{Ina} \tag{6.28}$$

Subtracting (6.27) from (6.26) and invoking (6.25)

$$\mathbf{Vni} - \mathbf{Vnr} = \mathbf{Ro} \times \mathbf{Ina} \tag{6.29}$$

Adding (6.24) to (6.29) and invoking (6.28)

$$2 \cdot \mathbf{Vni} = (\mathbf{Ro} + \mathbf{Rs}) \times \mathbf{Ina} - \mathbf{Vg} \tag{6.30}$$

This leads to

$$\mathbf{Ina} = (\mathbf{Ro} + \mathbf{Rs})^{-1} \times (2 \cdot \mathbf{Ro} \times \mathbf{Ini} + \mathbf{Vg}) \tag{6.31}$$

Similarly

$$\mathbf{Ifa} = (\mathbf{Ro} + \mathbf{R}L)^{-1} \times (2 \cdot \mathbf{Ro} \times \mathbf{Ifi}) \tag{6.32}$$

6.1.10 Transient emission

As a step current *Inr* travels along the send conductor, some of the energy departs into the environment as transient emission. This can be simulated by the model of Figure 6.5.

The voltage developed across the characteristic resistance is as follows:

$$Vnr = Ro \cdot Inr = Ro \cdot Ins + \frac{Qns}{Crad} \tag{6.33}$$

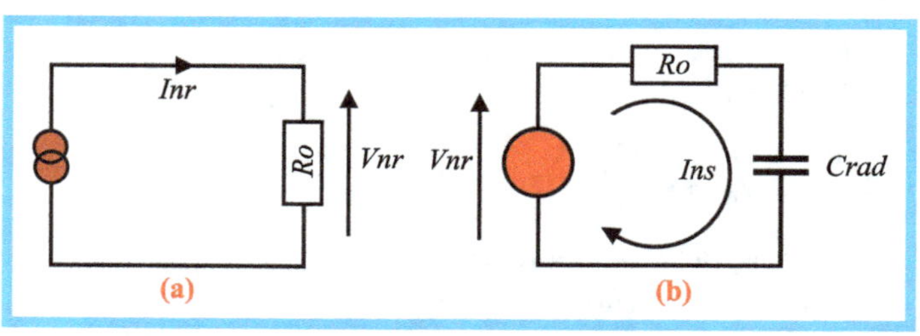

Figure 6.5 Deriving the amplitude of the discharge current Ins: (a) voltage developed across characteristic resistance and (b) current delivered to radiation capacitance

Hence,

$$Ins = Inr - \frac{Qns}{Ro \cdot Crad} \tag{6.34}$$

where *Crad* is the radiation capacitor (this is the only select-on-test parameter).

The amplitude of the current *Ifi* arriving at the far end is that which emanated from the near end minus that which was used to store charge in *Crad*:

$$Ifi = Inr - Ins \tag{6.35}$$

6.1.11 Current transformer

Since the response of the current transformer is inherent in the overall response of the system depicted in Figure 6.1, it was necessary to include that response in the simulation. Figure 6.6 is the circuit model derived during calibration testing of the device.

The number of *Turns* on the secondary winding is 10. A full description of the calibration process is provided in Section 2.5.

This model can be converted to one which can simulate the transient response of the transformer. In Figure 6.7, the resistor *R1* is as follows:

$$R1 = \frac{1}{\left(\frac{1}{300}\right) + \left(\frac{1}{850}\right) + \left(\frac{1}{51}\right) + \left(\frac{1}{50}\right)} = 22.7\ \Omega \tag{6.36}$$

The loop equations for Figure 6.7 are as follows:

$$\begin{aligned} Vout &= R1 \cdot I\sec - R1 \cdot Ind \\ 0 &= -R1 \cdot I\sec + R1 \cdot Ind + L1 \cdot \frac{dIind}{dt} \end{aligned} \tag{6.37}$$

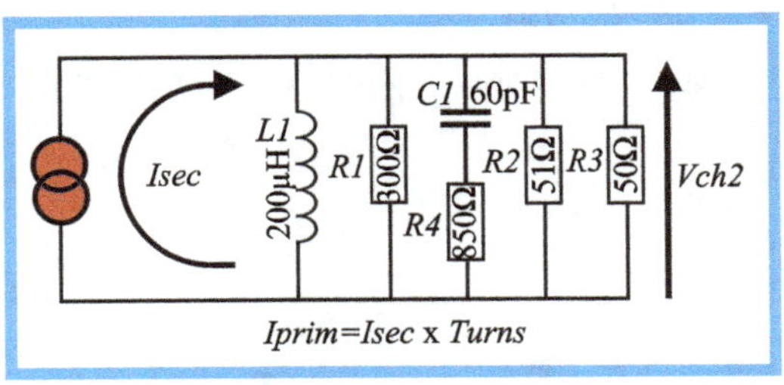

Figure 6.6 Circuit model to simulate frequency response of current transformer

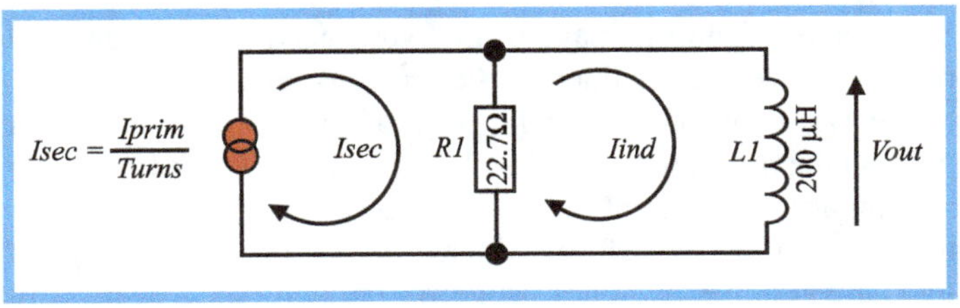

Figure 6.7 Circuit model to simulate transient response of current transformer

Hence,

$$dIind = \frac{dt}{L1} \cdot R1 \cdot (I\sec - Iind) \tag{6.38}$$

Section 5.1 describes how a Mathcad function can be created to analyse the transient response of Figure 6.7, that is to relate the voltage *Vout* monitored on channel 2 of the oscilloscope to the current *Iprim* in the current transformer. *Iprim* could be the current in the differential-mode loop or in the common-mode loop.

6.1.12 Waveform

The signal generator was set to give a square wave output at

$$f = 200 \times 10^3 \text{ Hz} \tag{6.39}$$

The mark/space ratio was unity. So, the mark time was as follows:

$$Tmark = \frac{1}{2 \cdot f} = 2.5 \times 10^{-6} \text{ s} \tag{6.40}$$

The number of steps *M* of the simulation during this time was as follows:

$$M = floor\left(\frac{Tmark}{dt}\right) = 519 \tag{6.41}$$

It was necessary to invoke the Mathcad function '*floor*' to round the result down to the nearest whole number to keep the source waveform in the simulation synchronised with the step changes.

6.1.13 Oscilloscope display

The scan time of the oscilloscope was 5 μs. So, the time duration for the simulation was set at

$$Tscope = 5 \times 10^{-6} \tag{6.42}$$

Giving

$$N = floor\left(\frac{Tscope}{dt}\right) = 1039 \tag{6.43}$$

6.1.14 Mathcad worksheet

Figure 6.8 is a copy of the first page of the worksheet. This assigns values to the parameters which have been defined in the previous text.

Figure 6.9 is a copy of the second page of the worksheet. This defines the functions used in the main program. They have been developed from functions used in the preceding articles on transient analysis; apart from the function *sqwv*(*n*,*M*). This simulates a voltage waveform which is zero for the first two time steps, switches to $-Vg/2$ and then switches between $-Vg/2$ and $Vg/2$ every 2.5 μs.

$Lc1 := 3.0 \cdot 10^{-6}$ $\quad Lc2 := Lc1$ $\quad Lc3 := 6.7 \cdot 10^{-6}$

$Cc1 := 1730 \cdot 10^{-12}$ $\quad Cc2 := Cc1$ $\quad Cc3 := 300 \cdot 10^{-12}$

$Rs1 := 8.4 \quad Rs2 := 0.2 \quad Rs3 := 0.01 \quad RL1 := 0 \quad RL2 := 0 \quad RL3 := 0$

$Crad1 := 300 \cdot 10^{-12} \quad Crad2 := 300 \cdot 10^{-12} \quad Turns := 10 \quad R1 := 22.7$

$L1 := 200 \cdot 10^{-6} \quad K := \dfrac{50}{50+41} = 0.549$ Vch1/Vin

$Ro1 := \sqrt{\dfrac{Lc1}{Cc1}} = 41.643 \quad Ro2 := Ro1 \quad Ro3 := \sqrt{\dfrac{Lc3}{Cc3}} = 149.443$

$Vg := 1.62$ Peak to peak amplitude of voltage source

$$\mathbf{Ro} := \begin{pmatrix} Ro1 + Ro2 & -Ro2 \\ -Ro2 & Ro2 + Ro3 \end{pmatrix}$$

$$\mathbf{Rs} := \begin{pmatrix} Rs1 + Rs2 & -Rs2 \\ -Rs2 & Rs2 + Rs3 \end{pmatrix}$$

$$\mathbf{RL} := \begin{pmatrix} RL1 + RL2 & -RL2 \\ -RL2 & RL2 + RL3 \end{pmatrix}$$

$$\mathbf{Rn} := \mathbf{Ro} + \mathbf{Rs} \quad \mathbf{Gn} := \mathbf{Rn}^{-1} \quad \mathbf{Rf} := \mathbf{Ro} + \mathbf{RL} \quad \mathbf{Gf} := \mathbf{Rf}^{-1}$$

$fh1 := 6.93 \cdot 10^{6} \quad T1 := \dfrac{1}{fh1 \cdot 2} = 7.215 \times 10^{-8}$

$fh2 := 10.26 \cdot 10^{6} \quad T2 := \dfrac{1}{fh2 \cdot 2} = 4.873 \times 10^{-8}$

$n1 := 15 \quad dt := \dfrac{T1}{n1} = 4.81 \times 10^{-9} \quad n2 := floor\left(n1 \cdot \dfrac{T2}{T1}\right) = 10$

$f := 200 \cdot 10^{3} \quad Tmark := \dfrac{1}{2 \cdot f} = 2.5 \times 10^{-6} \quad Tscope := 2 \cdot Tmark$

$M := floor\left(\dfrac{Tmark}{dt}\right) \quad N := floor\left(\dfrac{Tscope}{dt}\right)$

$n := 1 .. N \quad t_n := n \cdot dt$

Figure 6.8 First page of Mathcad worksheet

Figure 6.10 is a copy of the third page of the worksheet. This shows the function *ItoV(Iprim,Iind)* and the main program.

The function *ItoV(Iprim,Iind)* converts the value of the current being monitored to a voltage *Vout* which simulates that appearing at channel 2 of the oscilloscope.

The main program is a development of that described in Section 5.5. Since it simulates the action of two modes of propagation, it is about twice as long. The iterative part of the computation calculates a value for every current defined in Figure 6.4, at every time-step. Following normal mathematical convention, the variables in bold font are vectors.

```
near(Ini1 ,Ini2 ,Vs) :=  | Ini ← (Ini1)
                         |       (Ini2)
                         | Ina ← Gn·(2·Ro·Ini + Vs)
                         | Inr ← Ina − Ini
                         | (Inr1  Inr2  Ina)

far (Ifi1 ,Ifi2) :=      | Ifi ← (Ifi1)
                         |       (Ifi2)
                         | Ifa ← Gf·(2·Ro·Ifi)
                         | Ifr ← Ifa − Ifi
                         | (Ifr1  Ifr2  Ifa)

rad(Inr ,Qns ,Ro ,Crad) := | Int ← Qns / (Ro·Crad)
                           | Ins ← Inr − Int
                           | Qns ← Qns + Ins·dt
                           | (Int  Qns)

back(B ,Ifr ,n) :=       | B_(n+1) ← Ifr
                         | for i ∈ 2.. n + 1
                         |    B_(i−1) ← B_i
                         | Ini ← B_1
                         | (B  Ini)

forward (F ,Int ,n) :=   | F_1 ← Int
                         | for i ∈ n + 1.. 1
                         |    F_(i+1) ← F_i
                         | Ifi ← F_(n+2)
                         | (F  Ifi)

sqwv(n ,M) :=            | x ← trunc((n − 2)/M)
                         | sgn ← | −1  if 2·floor(x/2) = x
                         |       | 0  if n < 2
                         |       | 1  otherwise
                         | (Vg/2)·sgn
```

Figure 6.9 Second page of Mathcad worksheet

To allow the simulation to coincide with the waveform monitored on the oscilloscope, it includes a delay of 2.5 μs between the application of the falling edge of the generator waveform and the first rising edge. This is achieved by the statement:

$$j \leftarrow i - M \ \text{if} \ i > M.$$

```
ItoV(Iprim ,Iind) :=  | Isec ← Iprim / Turns
                      | dI ← (dt / L1)·R1·(Isec − Iind)
                      | Iind ← Iind + dI
                      | Vout ← R1·(Isec − Iind)
                      | ( Vout  Iind )

( Vch1  Vch2 ) :=     | F1[n1+2] ← 0
                      | F2[n2+2] ← 0
                      | B1[n1+1] ← 0
                      | B2[n2+1] ← 0
                      | for  i ∈ 1 .. M + N
                      |    | Vgen[1] ← sqwv(i ,M)
                      |    | Vgen[2] ← 0
                      |    | ( Inr1  Inr2  Ina ) ← near(Ini1 ,Ini2 ,Vgen)
                      |    | ( Int1  Qns1 ) ← rad(Inr1 ,Qns1 ,Ro1 ,Crad1)
                      |    | ( Int2  Qns2 ) ← rad(Inr2 ,Qns2 ,Ro2 ,Crad2)
                      |    | ( F1  Ifi1 ) ← forward(F1 ,Int1 ,n1)
                      |    | ( F2  Ifi2 ) ← forward(F2 ,Int2 ,n2)
                      |    | ( Ifr1  Ifr2  Ifa ) ← far(Ifi1 ,Ifi2)
                      |    | ( B1  Ini1 ) ← back(B1 ,Ifr1 ,n1)
                      |    | ( B2  Ini2 ) ← back(B2 ,Ifr2 ,n2)
                      |    | ( Vout1  Iind1 ) ← ItoV(Ina[1] ,Iind1)
                      |    | ( Vout2  Iind2 ) ← ItoV(Ina[2] ,Iind2)
                      |    | j ← 1
                      |    | j ← i − M  if  i > M
                      |    | Vch2[j] ← Vout2
                      |    | Vch1[j] ← K·(Vgen[1] − Rs1·Ina[1])
                      | ( Vch1  Vch2 )
```

Figure 6.10 Third page of Mathcad worksheet

6.1.15 Results

Figure 6.11 shows the results of the tests. The current transformer was clamped round the cable to measure the common-mode current.

That is, the current flowing along the return conductor and back via the structure. The input voltage was a square waveform of 5 μs duration, and the amplitude measured by channel 1 of the oscilloscope. This is illustrated by Figure 6.11(a). The

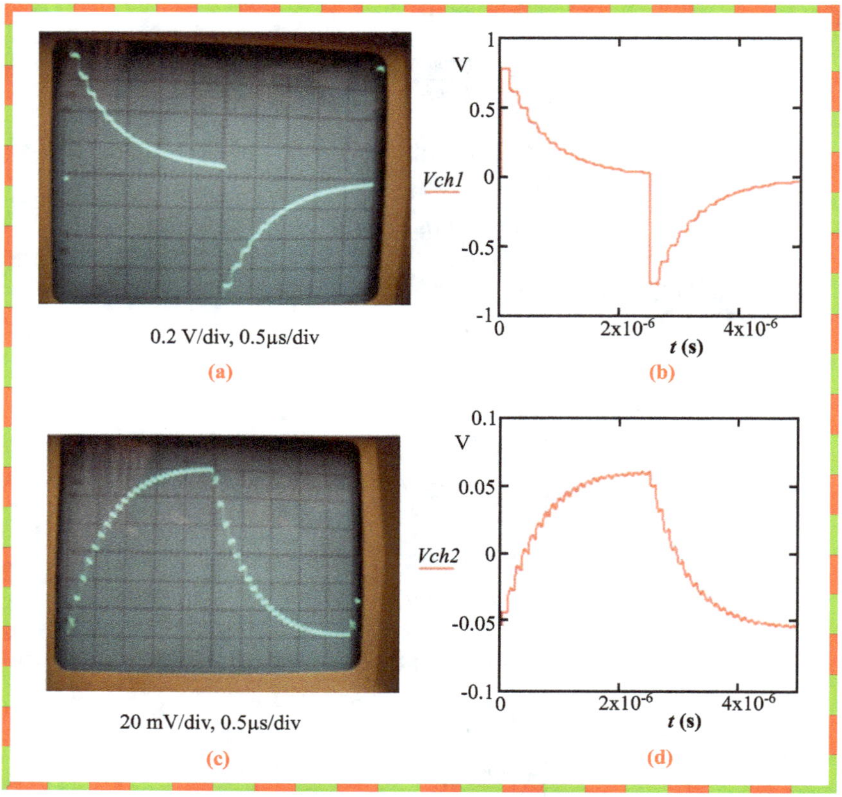

Figure 6.11 Comparing the test results with the simulation: (a) waveform of the input voltage, as monitored by channel 1 of the oscilloscope, (b) simulated waveform at channel 1, Vch1, (c) waveform of the current in the common-mode loop, as displayed by channel 2, and (d) simulated waveform at channel 2, Vch2

simulation provided by the model is shown by Figure 6.11(b). The current waveform is shown by Figure 6.11(c) and the simulated current by Figure 6.11(d).

6.1.16 Assessment

The close correlation between the test and simulated waveforms provides a high degree of confidence in the modelling technique. Of particular significance is the saw tooth appearance of the waveforms in Figure 6.11(c) and (d). This is because the step change in the common-mode current arrives back at the near end before that of the differential-mode current.

Including the parameters *Crad1* and *Crad2* in the model allowed the effect of radiation losses to be simulated (the rounded corners). Only a few iterations of the program were needed to assign values to these components.

6.1.17 Conclusion

It is possible to create models which simulate the cross-coupling between the differential-mode and common-mode currents in a test rig, in both the frequency and the time domain, using the analytical tools of circuit theory. In both cases, the radiated emission was included in the simulation.

Since the equations of circuit theory are used exclusively and all program details are visible in the Mathcad worksheet, it follows that any electrical engineer can acquire the ability to analyse EMI cross-coupling requirements; and this can be done with the same accuracy and reliability as any other set of design requirements.

6.2 Suppressing transient emission

6.2.1 Introduction

Current transients in the conducting structure are a prolific source of interference in vehicles, aircraft, spacecraft, and ships. A simple way of reducing the threat from these sources is to route a return conductor alongside each power supply conductor and connect it to the conducting structure at each end. Most of the steady-state current will flow along the structure, but the return conductor will carry the transient spikes.

An analysis was carried out on a representative assembly; a test rig. This simulates the response of a setup where a 10 V step is applied to one end of a power line which delivers current to a 10 Ω load. The return conductor carries most of the return current during the first 2 μs. After about 20 μs, the return current is shared equally between the return conductor and the ground. After 200 μs, the ground conductor carries most of the current.

Since the highest voltage spikes are due to the highest rate of change of current and it is these spikes which affect the most sensitive circuitry, the use of dedicated return conductors will provide a significant reduction in the EMI suffered by the system.

6.2.2 Testing and modelling

The test rig consists basically of a 15 mm water pipe fixed round one room of the house. A wooden batten assembled on the pipe provides a support for a cable to be routed along the length of the assembly. Various terminations can be connected at each end. Test equipment can then be used to monitor currents and voltages at one end of the line. It is possible to create a circuit model to simulate the response of any such assembly.

This rig is a fair representation of the wiring of a signal link or a of a power link assembled on a conducting structure. The copper pipe represents the structure, or 'ground'. The cable represents any wiring harness assembled on the framework. The wooden batten provides spacing between wiring and structure. The terminations represent the interface circuitry of the equipment units installed at either end of the link.

For the configuration analysed here, a twin-core cable was assembled on the rig, as shown in Figure 6.12. This defines the physical dimensions of the cross section and the length of the assembly.

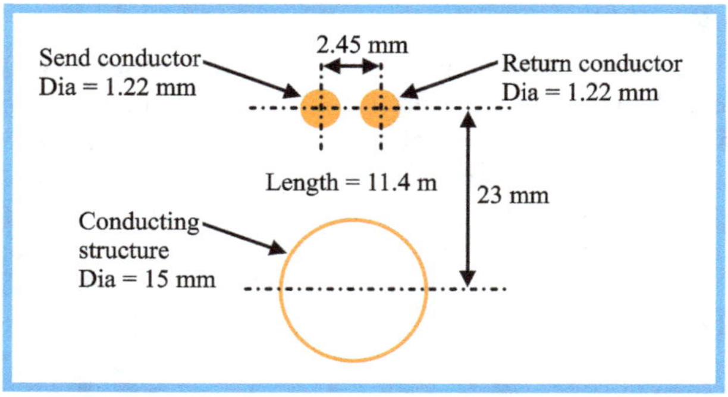

Figure 6.12 Cross section of conductors of test rig

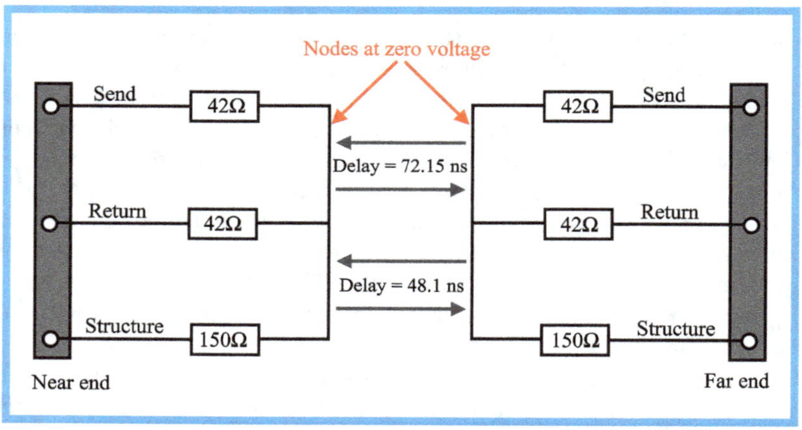

Figure 6.13 Transient circuit model of assembly-under-test

Tests were carried out to characterise the transient response, and a definitive model was derived from the results. Section 6.1 describes the setup, the tests, the analysis, the assessment, and the resultant model. This model is shown in Figure 6.13.

The unusual feature of this model is the lack of any direct connection between the near and far ends.

At the near end, the currents delivered to the link are independent of the circuitry at the far end. Since these currents are a function of the voltages delivered to the output terminals by reflections coming back from the far end (as well as voltage sources at the near end), they are functionally dependent on the response at the far end. An essential feature of the model is the definition of the time delay inherent in each circuit loop.

Each component in the model of Figure 6.13 represents the characteristic resistance of a conductor. That is, the characteristic impedance minus the parameters which simulate losses due to series resistance. Only the components associated with the test rig are identified. It is possible to add circuit components to represent any interface circuits at both ends to complete the circuit of the link.

Such an approach is similar to that used in the component libraries of SPICE (*Simulation Programs with Integrated Circuit Emphasis*) software. These packages contain circuit models of commercially available integrated circuits, transistors, and diodes, etc. These library models can be connected up to simulate a subsystem mounted on a printed circuit board. Then the functional performance of that assembly can be simulated.

The difference between the two techniques is that the three-conductor model can simulate all the mechanisms involved in electromagnetic interference. SPICE models cannot do this, because they are based on the concept of the 'equipotential ground'.

6.2.3 Analysis

The assembly under review is shown in Figure 6.14. This represents a configuration where a 10 V DC power supply is connected to a 10 Ω load via the conductors of the test rig. The structure provides a low resistance return path for the supply. The return conductor is connected to the structure at both ends.

Transient behaviour can be simulated by the general circuit model of Figure 6.15. This defines all the variables. The power supply at the near end is represented by a voltage source *Vg* which applies a step input of 10 V. The resistors

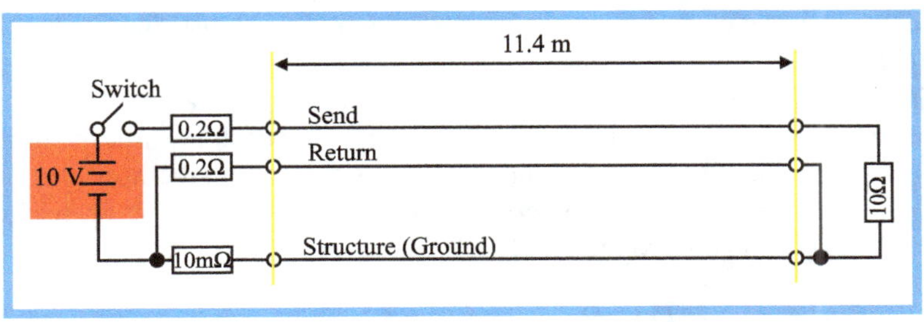

Figure 6.14 Assembly under review

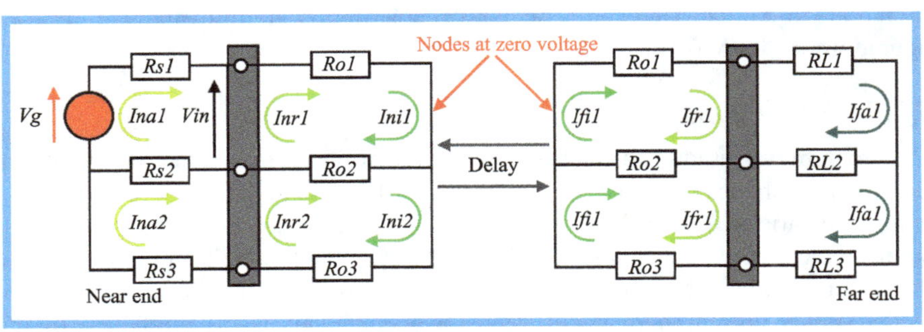

Figure 6.15 General circuit model of the assembly

Ro1, Ro2, and Ro3 simulate the characteristic resistance of each conductor. Values for the test rig are provided by Figure 6.13. All the currents are defined as loop currents; forward along the upper conductor and back along the lower conductor.

The three resistors *Rs1*, *Rs2*, and *Rs3* at the near end represent the series resistances of the send, return, and ground conductors. It is necessary to simulate these components external to the cabling because the basic model assumes that there are no losses.

Ini1, *Inr1*, and *Ina1* are the incident, reflected, and absorbed currents in the differential-mode loop at the near end. *Ini2*, *Inr2*, and *Ina2* are the incident, reflected, and absorbed currents in the common-mode loop. Initially, the incident currents at the near end are zero.

When the step voltage is applied to the differential-mode terminals, current *Inr1* flows forward along *Rs1* and *Ro1*, then back along *Ro2* and *Rs2*. This creates a voltage in the common-mode loop, thus inducing a current *Inr2*.

The cable is represented as a series of sections. Currents *Inr1* and *Inr2* manifest themselves as discrete charges which propagate from section to section along the cable; and this can be simulated using two shift registers. The differential-mode loop is represented by 15 sections, the common-mode loop by 10. This caters for the fact that the velocity of propagation in the common-mode loop is higher. The time taken for each charge to propagate along the link is defined by Figure 6.13.

At the far end, *Ifi1*, *Ifr1*, and *Ifa1* are the incident, reflected, and absorbed currents in the differential-mode loop. *Ifi2*, *Ifr2*, and *Ifa2* are the incident, reflected, and absorbed currents in the common-mode loop. *RL1* is assumed to be 10 Ω, whilst *RL2* and *RL3* are set at zero.

The normal reflection equations can be invoked to calculate the values of the currents incident at the far end and the currents reflected back towards the near end, at each instant. Then the reflected currents propagate back to the near end.

This process will continue indefinitely, with charges propagating backwards and forwards at near-light velocity.

6.2.4 Mathcad worksheet

A copy of the two-page worksheet which performs the computations is provided by Figures 6.16 and 6.17. This is basically a copy of that defined in Section 6.1, with alterations to the interface components to enable the program to simulate the configuration of Figure 6.14.

6.2.5 Simulation

The program was used to calculate the amplitude of the current in the supply, return, and ground conductor when a step voltage of 10 V is applied to the near end. By setting the variable *Tsweep* to the desired sweep time, the computations were carried out for three time periods: 2 μs, 20 μs, and 200 μs. Figure 6.18 illustrates the results.

The red curve at the top of the graph of Figure 6.18(a) simulates the waveform of the supply current ***Isup*** in the send conductor. This is the amplitude of the differential-mode current. The dashed green curve at the bottom of the graph

$Rs1 := 0.2 \qquad Rs2 := 0.2 \qquad Rs3 := 0.01$

$RL1 := 10 \qquad RL2 := 0 \qquad RL3 := 0$

$Ro1 := 42 \qquad Ro2 := Ro1 \qquad Ro3 := 150$

$Vg := 10 \qquad Crad1 := 300 \cdot 10^{-12} \qquad Crad2 := 300 \cdot 10^{-12}$

$$\boldsymbol{Rs} := \begin{pmatrix} Rs1 + Rs2 & -Rs2 \\ -Rs2 & Rs2 + Rs3 \end{pmatrix} = \begin{pmatrix} 0.4 & -0.2 \\ -0.2 & 0.21 \end{pmatrix}$$

$$\boldsymbol{Ro} := \begin{pmatrix} Ro1 + Ro2 & -Ro2 \\ -Ro2 & Ro2 + Ro3 \end{pmatrix} = \begin{pmatrix} 84 & -42 \\ -42 & 192 \end{pmatrix}$$

$$\boldsymbol{RL} := \begin{pmatrix} RL1 + RL2 & -RL2 \\ -RL2 & RL2 + RL3 \end{pmatrix} = \begin{pmatrix} 10 & 0 \\ 0 & 0 \end{pmatrix}$$

$$\boldsymbol{Rn} := \boldsymbol{Ro} + \boldsymbol{Rs} = \begin{pmatrix} 84.4 & -42.2 \\ -42.2 & 192.21 \end{pmatrix} \qquad \boldsymbol{Gn} := \boldsymbol{Rn}^{-1}$$

$$\boldsymbol{Rf} := \boldsymbol{Ro} + \boldsymbol{RL} = \begin{pmatrix} 94 & -42 \\ -42 & 192 \end{pmatrix} \qquad \boldsymbol{Gf} := \boldsymbol{Rf}^{-1}$$

$T1 := 7.2 \cdot 10^{-8}$

$T2 := \dfrac{T1}{1.5} = 4.8 \times 10^{-8}$

$n1 := 15 \qquad n2 := floor\left(n1 \cdot \dfrac{T2}{T1}\right) = 10$

$Tsweep := 2 \cdot 10^{-6} \quad dt := \dfrac{T1}{n1} = 4.8 \times 10^{-9} \quad N := floor\left(\dfrac{Tsweep}{dt}\right) = 416$

$n := 1 .. N \qquad \boldsymbol{t}_n := n \cdot dt$

$$near(Ini1, Ini2, Vs) := \left| \begin{array}{l} \boldsymbol{Ini} \leftarrow \begin{pmatrix} Ini1 \\ Ini2 \end{pmatrix} \\ \boldsymbol{Ina} \leftarrow \boldsymbol{Gn} \cdot (2 \cdot \boldsymbol{Ro} \cdot \boldsymbol{Ini} + \boldsymbol{Vs}) \\ \boldsymbol{Inr} \leftarrow \boldsymbol{Ina} - \boldsymbol{Ini} \\ (\boldsymbol{Inr}_1 \ \ \boldsymbol{Inr}_2 \ \ \boldsymbol{Ina}) \end{array} \right.$$

$$far(Ifi1, Ifi2) := \left| \begin{array}{l} \boldsymbol{Ifi} \leftarrow \begin{pmatrix} Ifi1 \\ Ifi2 \end{pmatrix} \\ \boldsymbol{Ifa} \leftarrow \boldsymbol{Gf} \cdot (2 \cdot \boldsymbol{Ro} \cdot \boldsymbol{Ifi}) \\ \boldsymbol{Ifr} \leftarrow \boldsymbol{Ifa} - \boldsymbol{Ifi} \\ (\boldsymbol{Ifr}_1 \ \ \boldsymbol{Ifr}_2 \ \ \boldsymbol{Ifa}) \end{array} \right.$$

Figure 6.16 First page of worksheet: defining the circuit parameters

simulates the current ***Ignd*** in the ground conductor, the 15 mm copper tube. ***Ignd*** is the amplitude of the common-mode current. The blue curve simulates the current ***Iret*** in the return conductor. This is the difference between that in the differential-mode and the common-mode loop.

It can be seen that the amplitude of the current in all three conductors increases as a series of steps which morph into a ripple at the top of an exponential rise. This rise is indicative of that in an L–R series circuit, as indeed it is. However, the most significant feature of the response is that the current in the ground conductor is

```
rad(Inr, Qns, Ro, Crad) :=  | Int ← Qns / (Ro·Crad)
                            | Ins ← Inr − Int
                            | Qns ← Qns + Ins·dt
                            | (Int  Qns)

forward(F, Int, n) :=  | F_1 ← Int                 back(B, Ifr, n) :=  | B_(n+1) ← Ifr
                       | for i ∈ n + 1 .. 1                            | for i ∈ 2 .. n + 1
                       |   F_(i+1) ← F_i                               |   B_(i−1) ← B_i
                       | Ifi ← F_(n+2)                                 | Ini ← B_1
                       | (F  Ifi)                                      | (B  Ini)

(Isup  Iret  Ignd) :=  | F1_(n1+2) ← 0
                       | F2_(n2+2) ← 0
                       | B1_(n1+1) ← 0
                       | B2_(n2+1) ← 0
                       | Vgen_2 ← 0
                       | for i ∈ 1 .. N
                       |   | Vgen_1 ← Vg  if i > 1
                       |   | (Inr1  Inr2  Ina) ← near(Ini1, Ini2, Vgen)
                       |   | (Int1  Qns1) ← rad(Inr1, Qns1, Ro1, Crad1)
                       |   | (Int2  Qns2) ← rad(Inr2, Qns2, Ro2, Crad2)
                       |   | (F1  Ifi1) ← forward(F1, Int1, n1)
                       |   | (F2  Ifi2) ← forward(F2, Int2, n2)
                       |   | (Ifr1  Ifr2  Ifa) ← far(Ifi1, Ifi2)
                       |   | (B1  Ini1) ← back(B1, Ifr1, n1)
                       |   | (B2  Ini2) ← back(B2, Ifr2, n2)
                       |   | Isig_i ← Ina_1
                       |   | Iret_i ← Ina_1 − Ina_2
                       |   | Ignd_i ← Ina_2
                       | (Isig  Iret  Ignd)
```

Figure 6.17 Second page of worksheet: the main program

about one third of that in the send conductor. Since it is the current in the ground conductor that causes all of the conducted emission in any electronic system, this modification represents an improvement of about 10 dB.

Figure 6.18(b) illustrates the same response over a period of 20 μs. This shows how the current is shared between the return conductor and the conducting structure. Current in the ground continues to rise. This is at the expense of a fall in the

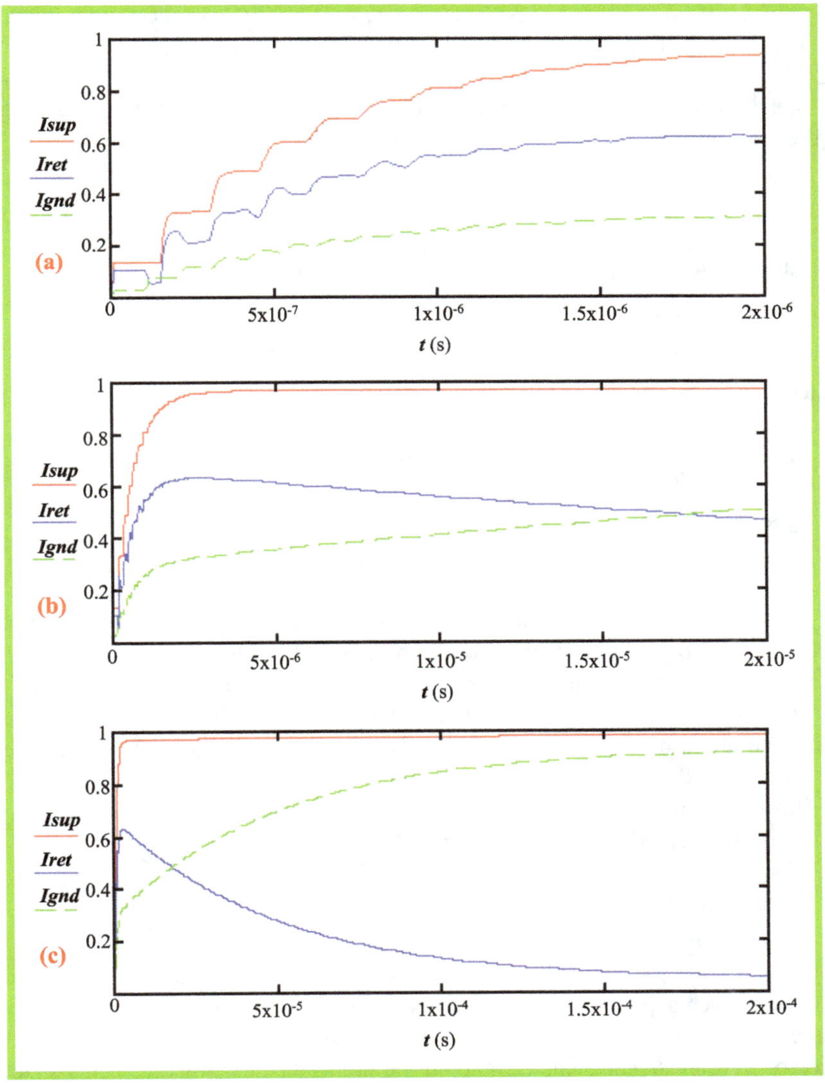

Figure 6.18 Amplitude of current in the supply, return and ground conductors (amperes): (a) over a period of 2 μs, (b) over a period of 20 μs, and (c) over a period of 200 μs

current in the return conductor, which reaches a peak at about 2 μs. After about 17 μs, currents in the return and ground conductors are equal in amplitude.

Figure 6.18(c) shows that the current in the structure continues to rise exponentially. After 200 μs, it is almost equal to that in the send conductor. This is matched by an exponential fall in the amplitude of current in the return conductor.

6.2.6 Accuracy

The worksheet computations result in a current of 0.979 A in the send conductor, 0.933 A in the ground conductor, and 0.047 A in the return conductor, after a period of 2 ms.

Steady-state analysis of currents in the assembly of Figure 6.14 gives a value for the load on the battery when the switch is closed:

$$Rtotal = 10 + 0.2 + \frac{0.2 \times 0.01}{0.2 + 0.01} = 10.21\ \Omega \tag{6.44}$$

This gives a value for the current in the send conductor:

$$Isend = \frac{Vg}{Rtotal} = \frac{10}{10.21} = 0.979\ \ \text{A} \tag{6.45}$$

The current in the return conductor is as follows:

$$Ireturn = \frac{0.01}{0.2 + 0.01} \times 0.979 = 0.047\ \ \text{A} \tag{6.46}$$

Giving a value for the current in the structure:

$$Iground = 0.979 - 0.047 = 0.932\ \ \text{A} \tag{6.47}$$

The fact that the simulation results in values for currents which are exactly the same as those derived from steady-state analysis provides a high level of confidence of its accuracy. Under steady-state conditions, the model of Figure 6.13 is transparent.

In Section 6.1, it is shown that the analysis of this same model provides a faithful simulation of the waveform displayed on the screen of an oscilloscope. This gives added assurance.

However, the highest level of confidence emanates from the fact that all the equations are derived from electromagnetic theory.

6.2.7 Application

The vast majority of power supplies to electrical equipment on ships, vehicles, aircraft, and spacecraft use the conducting structure as the return path for the current delivered to the load. Such supplies need to be turned on and off, and this creates high rates of change of current. Many supplies use switched-mode converters, which are a rich source of transients. Transient currents in the inductance of structure will inevitably create high voltages between equipment units. In turn, these voltages will appear as unwanted signals in series with signal conductors routed along that same structure. Power supplies are the main source of intra-system interference.

By routing a return conductor alongside the supply conductor, a significant reduction in conducted emission can be achieved. This can be done for supplies to electronic equipment, relays, fuel valves, hydraulic valves, motors, heaters, and lights. The main advantage of this design is that the structure remains the preferred conducting path for low frequency supplies: DC, 50, 60, and 400 Hz. So, there are no losses due to increased conductor resistance. In fact, there is a slight reduction in conductor resistance.

The modelling technique described here can be utilised to analyse the EMI characteristic of any of these power links. This can be done by the designer who is responsible for the performance of the system under review, since all the analytical tools are available in Circuit Theory.

6.2.8 Conclusion

A method of analysing the cross-coupling of transients between the differential-mode and the common-mode loop in a power supply cable has been described.

It has been shown that the use of a dedicated return conductor will provide a significant reduction in the conducted emission of the cable.

6.3 Optimum common-mode rejection

6.3.1 Introduction

There have been many designs of common-mode rejection circuits in the pages of EW and WW over the last 60 years, but this is the first which takes into account the characteristics of the interconnecting wires. Another unique feature is the use of transient, rather than frequency, analysis. Transient analysis covers the entire range, from DC to GHz.

A systematic approach is applied to the design process, similar to the use of library models in SPICE software to simulate the behaviour of selected integrated circuits on printed circuit board assemblies.

Given the availability of a circuit model of the signal link, it becomes possible to tailor the design of the interface circuitry of the signal receiver to maximise common-mode rejection. The design is basically the same as that adopted for a twin-conductor transmission line; terminate the link with resistors equal to the characteristic resistances of the conductors.

The simulation shows how the receiver performs in the presence of a high level of transient interference.

6.3.2 General circuit model

The starting point to the design process is the general circuit model of the configuration under review. For transient analysis, this is as illustrated by Figure 6.19. It consists of three sections:

The interface circuitry of the signal source at the near end of the link.
The signal link, comprising the send, return, and ground conductors.
The interface circuitry of the signal receiver at the far end.

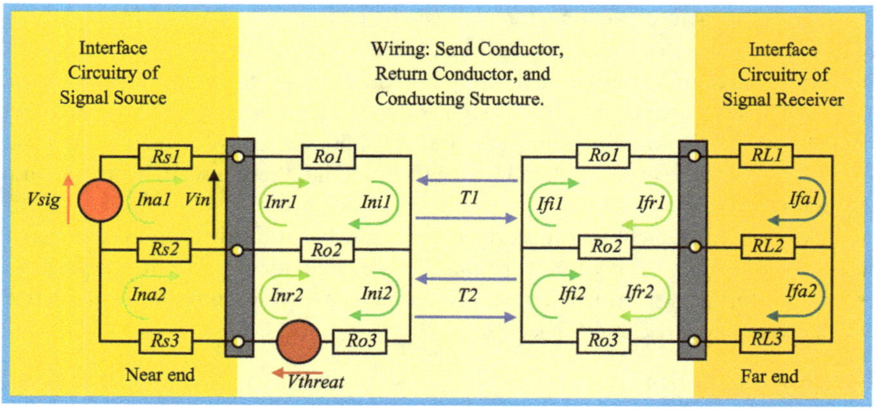

Figure 6.19 General circuit model for transient analysis of a signal link

The resistors *Ro1*, *Ro2*, and *Ro3* simulate the characteristic resistance of each conductor. The resistors *Rs1*, *Rs2*, and *Rs3* represent the series resistances of the circuitry at the signal source interface.

All the currents are defined as loop currents; forward along the upper conductor and back along the lower conductor: that is, clockwise. *Ini1*, *Inr1*, and *Ina1* are the incident, reflected, and absorbed currents in the differential-mode loop at the near end. *Ini2*, *Inr2*, and *Ina2* are the incident, reflected, and absorbed currents in the common-mode loop. Initially, the incident currents at the near end are set at zero.

When a step voltage is applied to the differential-mode terminals, current *Inr1* flows forward along *Rs1* and *Ro1*, then back along *Ro2* and *Rs2*. This creates a voltage in the common-mode loop, thus inducing a current *Inr2*.

The cable is represented as a series of sections. Currents *Inr1* and *Inr2* manifest themselves as discrete charges which propagate from section to section along the cable; and this can be simulated using two shift registers. The differential-mode loop is represented by *n1* sections, the common-mode loop by *n2* sections. This caters for the fact that the velocities of propagation in the differential-mode and common-mode loops are different.

At the far end, *Ifi1*, *Ifr1*, and *Ifa1* are the incident, reflected, and absorbed currents in the differential-mode loop. *Ifi2*, *Ifr2*, and *Ifa2* are the incident, reflected, and absorbed currents in the common-mode loop. Resistors in the interface circuitry of the signal receiver at the far end are *RL1*, *RL2*, and *RL3*.

There are two voltage sources:

> *Vsig* represents the signal voltage applied between the send and return conductors; *Vthreat* represents the threat voltage which exists in series with the ground conductor.

The significance of this model is that none of the component values and neither of the voltage sources have been defined. It can be used to simulate the response of any part of an electronic system which incorporates such a link.

6.3.3 Optimum loading

Section 6.1 defines the physical dimensions of a particular signal link and describes a set of tests which assign values to the parameters defining the interconnecting wiring. Assigning component values to the interface circuitry at each end completes the definition of the signal link under review: Figure 6.20.

Under worst case conditions, there are no losses. No losses mean maximum interference. So, it is assumed that *Rs1*, *Rs2*, and *Rs3* have zero resistance. The voltage source *Vsig* at the near end is assumed to be a square wave which switches between zero and 10 V. It is assumed that there is a threat voltage *Vthreat* due to transient currents in the structure and that this voltage switches between minus 5 V and plus 5 V.

Transmission line theory teaches us that if the value of the load resistance of a twin-conductor line is equal to the characteristic resistance, then there will be no reflections at the far end. Translating that recommendation into one for a three-conductor line leads to the values assigned to *RL1*, *RL2*, and *RL3* at the far end.

6.3.4 Mathcad worksheet

The Mathcad worksheet which carries out the computations is shown in Figures 6.21–6.23. This is basically a copy of that used in Sections 6.1 and 6.2.

Notes have been added to help follow the process.

6.3.5 Simulation

In this simulation, the signal voltage *Vsig* at the near end is defined as a square wave where the mark time is 4 μs and the amplitude switches from 0 to 10 V. The output *Vdiff* at the far end follows this waveform very accurately, as illustrated by the red curve in the graph at the top of Figure 6.24.

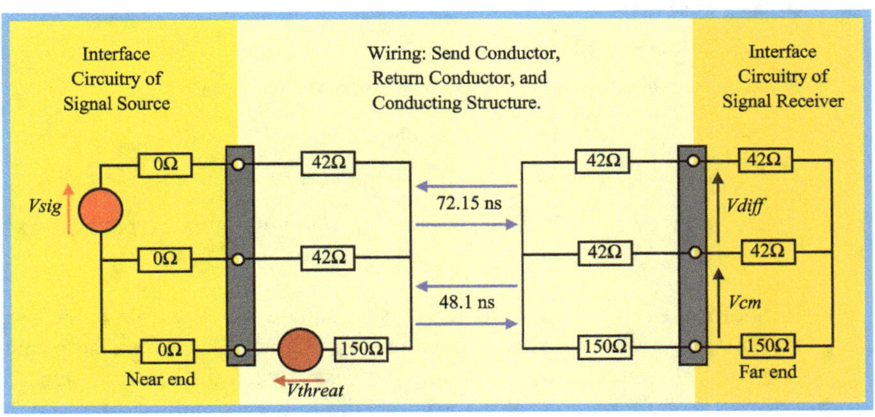

Figure 6.20 Definitive circuit model of signal link

$Rs1 := 0 \quad Rs2 := 0 \quad Rs3 := 0$

$RL1 := 42 \quad RL2 := 42 \quad RL3 := 150$ Circuit Components

$Ro1 := 42 \quad Ro2 := 42 \quad Ro3 := 150$

$Vg := 10$ Amplitude of voltage sources

$$\mathbf{Ro} := \begin{pmatrix} Ro1 + Ro2 & -Ro2 \\ -Ro2 & Ro2 + Ro3 \end{pmatrix}$$

$$\mathbf{Rs} := \begin{pmatrix} Rs1 + Rs2 & -Rs2 \\ -Rs2 & Rs2 + Rs3 \end{pmatrix}$$ Resistance matrices

$$\mathbf{RL} := \begin{pmatrix} RL1 + RL2 & -RL2 \\ -RL2 & RL2 + RL3 \end{pmatrix}$$

$\mathbf{Rn} := \mathbf{Ro} + \mathbf{Rs} \quad \mathbf{Rf} := \mathbf{Ro} + \mathbf{RL}$

$\mathbf{Gn} := \mathbf{Rn}^{-1} \quad \mathbf{Gf} := \mathbf{Rf}^{-1}$ Matrices of conductance values

$T1 := 7.2 \cdot 10^{-8}$ Differential-mode propagation time

$T2 := \frac{T1}{1.5} = 4.8 \times 10^{-8}$ Common-mode propagation time

$n1 := 15$ Number of segments in differential mode loop

$dt := \frac{T1}{n1} = 4.8 \times 10^{-9}$ Time to propagate along one segment

$n2 := floor\left(n1 \cdot \frac{T2}{T1}\right) = 10$ Number of segments in common-mode loop

$Tsweep := 2 \cdot 10^{-5}$ Duration of simulation

$N := floor\left(\frac{Tsweep}{dt}\right) = 4.166 \times 10^{3}$ Total number of steps in the simulation

$m1 := floor(0.2 \cdot N) = 833$

$Tsignal := m1 \cdot dt = 3.998 \times 10^{-6}$ Mark time of signal square wave

$m2 := floor(0.15 \cdot N) = 624$

$Tthreat := m2 \cdot dt = 2.995 \times 10^{-6}$ Mark time of threat square wave

$n := 1..N \quad t_n := n \cdot dt$ Time vector

Figure 6.21 First page of Mathcad worksheet: defining the basic parameters

The waveform of the threat voltage is also a square wave, with a mark time of 3 μs and a voltage which switches between −5 V and +5 V. This also propagates along the link and appears as an output *Vcm* between the return conductor and ground at the far end. The blue curve on the bottom graph of Figure 6.24 illustrates this.

$$near(Ini1, Ini2, Vs) := \left| \begin{array}{l} \boldsymbol{Ini} \leftarrow \begin{pmatrix} Ini1 \\ Ini2 \end{pmatrix} \\ \boldsymbol{Ina} \leftarrow \boldsymbol{Gn}\cdot(2\cdot\boldsymbol{Ro}\cdot\boldsymbol{Ini} + \boldsymbol{Vs}) \\ \boldsymbol{Inr} \leftarrow \boldsymbol{Ina} - \boldsymbol{Ini} \\ (\boldsymbol{Inr}_1 \;\; \boldsymbol{Inr}_2 \;\; \boldsymbol{Ina}) \end{array} \right.$$

Calculating the response at the Near end for each time step

$$far(Ifi1, Ifi2) := \left| \begin{array}{l} \boldsymbol{Ifi} \leftarrow \begin{pmatrix} Ifi1 \\ Ifi2 \end{pmatrix} \\ \boldsymbol{Ifa} \leftarrow \boldsymbol{Gf}\cdot(2\cdot\boldsymbol{Ro}\cdot\boldsymbol{Ifi}) \\ \boldsymbol{Ifr} \leftarrow \boldsymbol{Ifa} - \boldsymbol{Ifi} \\ (\boldsymbol{Ifr}_1 \;\; \boldsymbol{Ifr}_2 \;\; \boldsymbol{Ifa}) \end{array} \right.$$

Calculating the response at the Far end for each time step

$$forward(\boldsymbol{F}, Int, n) := \left| \begin{array}{l} \boldsymbol{F}_1 \leftarrow Int \\ for \;\; i \in n+1..1 \\ \quad \boldsymbol{F}_{i+1} \leftarrow \boldsymbol{F}_i \\ Ifi \leftarrow \boldsymbol{F}_{n+2} \\ (\boldsymbol{F} \;\; Ifi) \end{array} \right.$$

Shift register which simulates the forward propagation of charges along the loop from segment to segment. Can be used for both differential mode and common-mode transmission

$$back(\boldsymbol{B}, Ifr, n) := \left| \begin{array}{l} \boldsymbol{B}_{n+1} \leftarrow Ifr \\ for \;\; i \in 2..n+1 \\ \quad \boldsymbol{B}_{i-1} \leftarrow \boldsymbol{B}_i \\ Ini \leftarrow \boldsymbol{B}_1 \\ (\boldsymbol{B} \;\; Ini) \end{array} \right.$$

Shift register which can simulate the propagation of charges back along each loop

$$sqwv(n, m, Voffset) := \left| \begin{array}{l} x \leftarrow floor\left(\frac{n}{m}\right) \\ sgn \leftarrow \left| \begin{array}{ll} 0 & if \;\; 2\cdot floor\left(\frac{x}{2}\right) = x \\ 1 & otherwise \end{array} \right. \\ sgn\cdot Vg + Voffset \end{array} \right.$$

Simulating a square wave
n = total number of steps
m = number of steps during the mark time
It allows an offset voltage to be defined

Figure 6.22 Second page of Mathcad worksheet: defining the functions

This configuration acts as two transmission lines in parallel. There is minimal interaction between the two signals.

The only cross coupling visible in this simulation is the existence of rectangular pulses of about 24 ns duration. These are due to the fact that the time taken for the threat current to traverse the length of the link is less than the time taken for the signal current.

By terminating each conductor with its characteristic resistance, the cross-coupling at the far end will ensure that the voltages appearing at the far end are independent of each other (apart from the 24 ns glitches which can easily be filtered out).

```
(Vdiff  Vcm) := | F1[n1+2] ← 0
                | F2[n2+2] ← 0
                | B1[n1+1] ← 0                         Defining the shift registers
                | B2[n2+1] ← 0                         Two forward & two back
                | Vgen[2] ← 0
                | for i ∈ 1..N
                |   | Vgen[1] ← sqwv(i,m1,0)                    Signal square wave
                |   | Vgen[2] ← sqwv(i,m2,−5)                   Threat square wave
                |   | (Inr1  Inr2  Ina) ← near(Ini1,Ini2,Vgen)
                |   | (F1  Ifi1) ← forward(F1,Inr1,n1)
                |   | (F2  Ifi2) ← forward(F2,Inr2,n2)
                |   | (Ifr1  Ifr2  Ifa) ← far(Ifi1,Ifi2)
                |   | (B1  Ini1) ← back(B1,Ifr1,n1)
                |   | (B2  Ini2) ← back(B2,Ifr2,n2)
                |   | Vfa ← RL·Ifa
                |   | Vdiff[i] ← Vfa[1]                       Voltages at Far End
                |   | Vcm[i] ← Vfa[2]                         Compiling the output
                | (Vdiff  Vcm)                                vectors
```

Figure 6.23 Third page of Mathcad worksheet: the main program

Hence, it is possible to pick up the differential-mode signal simply by monitoring the voltage between the signal and return conductors at the far end. But there is a basic requirement for this interface; the current flowing forward along the send conductor must be balanced by the currents flowing back along the return and ground conductors.

6.3.6 Far end interface

The requirement for the differential-mode and common-mode currents to be balanced can be met by interface circuit of the analogue signal receiver illustrated by Figure 6.25. This is basically a non-inverting, unity gain, and differential-mode amplifier.

The signal current flows through the 42 Ω resistor into the dummy load of a 42 Ω resistor in parallel with a 150 Ω resistor, and the voltage at the junction is applied to the non-inverting input of the operational amplifier. This voltage is balanced at the inverting input by current fed back from the amplifier output flowing into the return conductor via a 42 Ω resistor and into the ground conductor via a 150 Ω resistor.

The voltage at the output of the signal receiver at the far end will be the same as the input of the signal transmitter at the near end, independent of any voltage *Vcm* in the common-mode loop.

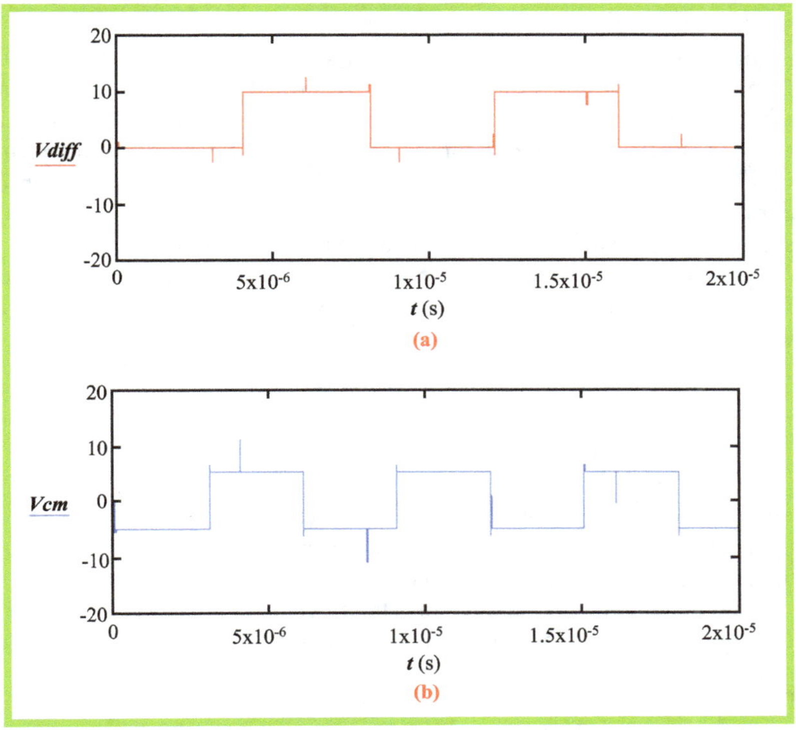

Figure 6.24 Waveforms of voltages appearing at the far end of the line: (a) waveform of differential-mode voltage and (b) waveform of common-mode voltage

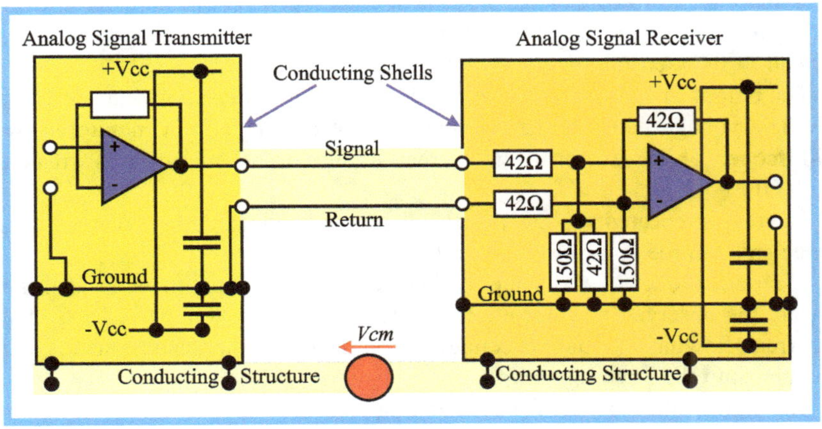

Figure 6.25 Interface providing optimum common-mode rejection

6.3.7 Conclusion

A systematic method of tailoring the design of the common-mode rejection interface in any electronic system has been described. Critical to the simulation is the fact that both the differential-mode and the common-mode loop have been completely defined. There are no input ports and output ports as would normally be expected in the definition of the wiring of an equipment unit or a printed circuit module.

Such a design can be implemented during the prototype design phase of a project, tested using bench-test methods, and subjected to a critical design review, long before a start is made on the drawings for the manufacture of the engineering model.

6.4 Ground bounce

6.4.1 Introduction

Ground bounce is due to supply current being switched off. All the energy stored in the supply wiring departs into the environment via the surface of the conducting structure, leaving charge on that surface, charge which manifests itself as a very high, short duration, voltage between the structure and any conductor routed along it. A test rig is used to demonstrate this, and a circuit model is used to simulate the response.

The construction of a switching unit which can deliver a current of about 400 mA and then switch that current OFF is described. This can be done at a repetition rate which allows the waveform to be monitored on a general-purpose oscilloscope. A signal generator set to give a square wave is used to control the times at which switching occurs.

The test rig consists of a twin-conductor cable routed along a copper pipe. The terminals are short-circuited at the far end. It is configured as two loops. In the culprit loop, current flows along the structure and back via one conductor. Current flowing in this loop develops a voltage between the open-circuit terminals at the near end of the victim loop.

The output of the switching unit is used to switch the supply current ON and OFF. The voltage between the terminals of the culprit loop is monitored by an oscilloscope. This waveform shows that a high transient pulse is created at switch OFF.

The test is repeated to show that the electromagnetic coupling creates a similar voltage spike in the victim loop.

A circuit model of the setup is then developed using time-step analysis to replicate the observed waveforms. It is reasoned that the voltage spike is created by current flowing out of the conducting surfaces of the scope and the signal generator, into the environment.

The significant feature of the modelling technique is that it uses the relationships of circuit theory to carry out the analysis. So, it can be carried out by any electrical engineer. There is no need to invoke any of the relationships of full field modelling.

Enough detailed information is provided for any circuit designer to replicate the experiment.

6.4.2 Test rig

The test rig described in Section 6.1 was used to demonstrate this phenomenon. An assessment of the results of the tests described in that section leads to the definition of parameter values which can be assigned to the transient coupling model of the rig. Figure 6.26 identifies these. The brown and blue conductors are part of a twin-conductor cable which is routed 11.4 m along a 15 mm copper pipe. Only the parameters associated with the rig are identified.

6.4.3 Switching module

Given the existence of a test rig which has been fully characterised, the first task was to create a power source which could deliver a relatively high current into a cable, wait until that current has settled down into its steady-state condition, and then switch it off. To do this repetitively, the switching device had to be a semi-conductor. This device had to be able to withstand the high voltage due to the transient developed across the inductance of the culprit loop. It was also necessary to decouple the module from the test equipment to minimise any interaction between that equipment and the cable assembly-under-test.

Figure 6.27 shows the circuit diagram of the module. To provide isolation from the instrumentation, it contains its own source of power; four AA batteries, supplying an open circuit voltage of about 6 V. It was designed to use a square wave input from a signal generator to create switched output of about 400 mA. The control signal is applied via the coaxial connection on the left-hand side and the switched output on the right-hand side.

When a positive voltage of about 4 V is applied via the coaxial connector, the pnp transistor BC327 switches off. No current flows in the 1 kΩ resistors. Since there is no current flowing in the base-emitter junction, the BC337 switches off. This means that

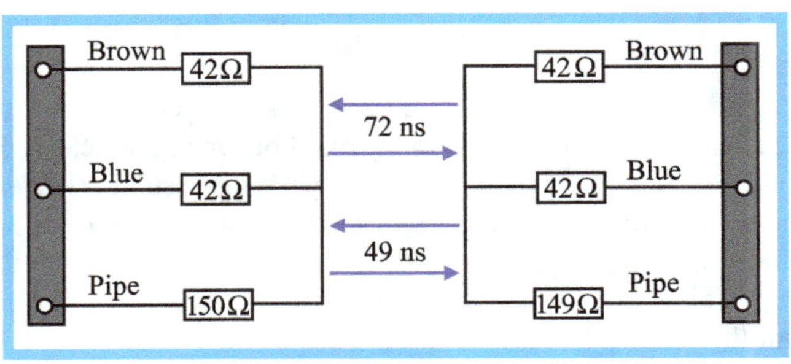

Figure 6.26 Definitive circuit model for transient analysis of the test rig

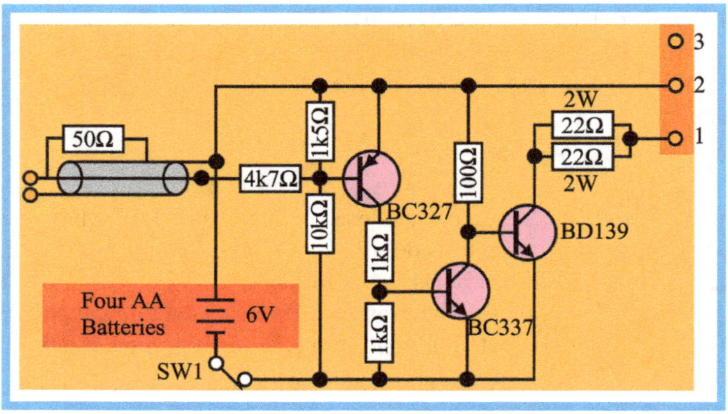

Figure 6.27 Switching module

current flows into the BD139 input via the 100 Ω resistor, switching it into the conducting state. If terminals 1 and 2 are connected together, current flows into the 22 Ω resistors. The effective resistance is about 11 Ω. When a negative voltage of about −4 V is applied via the input connector, BC327 and BC337 switch on, cutting off the drive to BD139, which switches off the output current.

The switch SW1 was used to conserve the batteries. It was only switched to the ON position when a test was in progress.

6.4.4 Test setup

The output terminals of the switching module were then connected to the near end of the test rig, as illustrated in Figure 6.28. The input terminals were connected to the signal generator. This was set to deliver a square waveform which switched between +4 V and −4 V at approximately 100 kHz.

This caused the BD139 transistor to switch between short-circuit and open-circuit at the same frequency as the signal generator. The time between switch ON and switch OFF was about 3.8 μs. The load consisted of the conducting path along the copper pipe and back along the brown conductor. This can be described as the culprit loop. The victim loop is defined as along the copper pipe and back via the blue conductor.

A ×-10 V probe was used to monitor the voltage at terminal 1 with respect to terminal 2; that is, to monitor the voltage applied between the near end of the line and the ground conductor. Figure 6.29(a) illustrates the resultant waveform on the screen of the oscilloscope. This shows that the voltage steps to about −4 V at switch ON and then decays exponentially towards zero as the current in the culprit loop builds up. This current is limited by the source voltage and the source resistance to about 0.4 A.

At the instant of switch OFF, the terminals 1 and 2 become open circuit, and a voltage of about 18 V appears across these terminals.

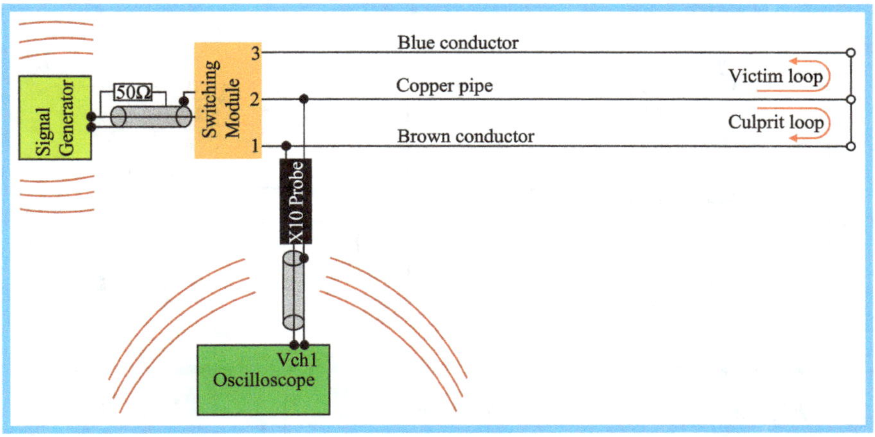

Figure 6.28 Test setup

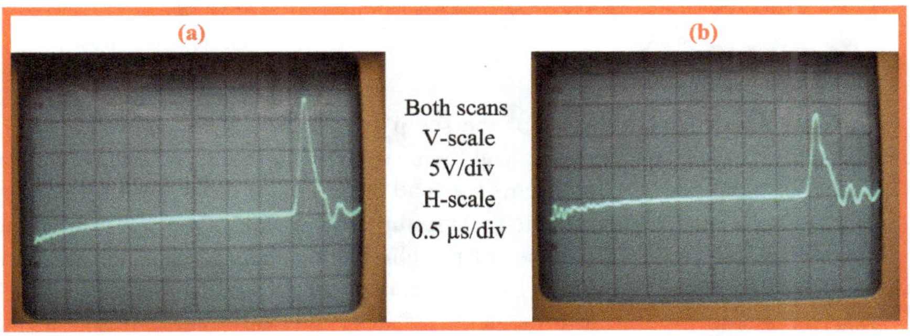

Figure 6.29 Voltage waveforms displayed on the oscilloscope: (a) waveform at input of the culprit loop and (b) waveform at output of the victim loop

The voltage probe was then connected between terminals 3 and 2 and the tests repeated. This gave the waveform of Figure 6.29(b).

6.4.5 Cross-coupling model

In the setup of Figure 6.28, the conductor which carries the current common to both the culprit loop and the victim loop is the copper pipe. That is, current flowing along the copper pipe induces current in both the blue and brown conductors. But the model of Figure 6.26 represents a configuration where it is the blue conductor which carries both the differential-mode and the common-mode current.

Modifying the model of the setup to simulate the new configuration leads to Figure 6.30. The values assigned to the propagation times for signals in both loops are the same, since the coupling between the brown conductor and the copper pipe is the same as that between the blue conductor and the pipe.

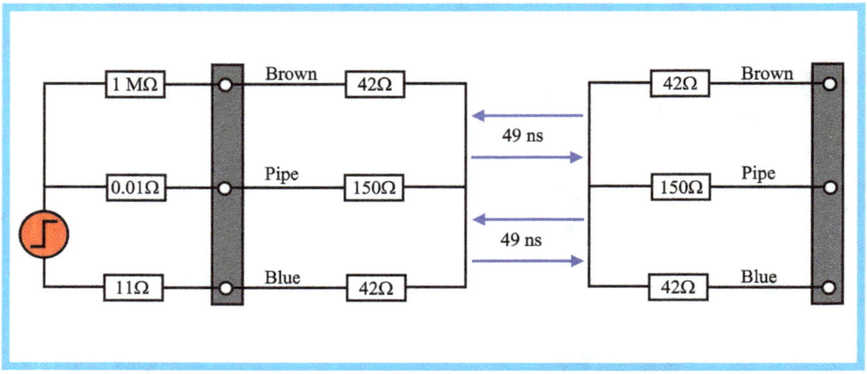

Figure 6.30 Circuit model for transient analysis of cross-coupling

The 1 MΩ resistor represents the resistance of the ×-10 V probe when it is connected to the oscilloscope. The 11 Ω resistance represents the parallel combination of the 22 Ω resistors of Figure 6.27.

6.4.6 Analysis

Figure 6.31(a) simulates the state of the setup during switch ON conditions. *Rs1* represents the resistance of the source at the near end, *Rs2* represents the resistance of the copper pipe, and *Rs3* represents the load presented to the input terminals by the voltage probe. *Vg1* represents the voltage supplied by the battery. *Ro1*, *Ro2*, and *Ro3* are the characteristic resistances of the blue conductor, pipe, and brown conductor. *RL1*, *RL2*, and *RL3* are the load resistors at the far end of the rig.

Ini1, *Inr1*, and *Ina1* are the incident, reflected, and absorbed currents at the near end of the culprit loop. *Ini2*, *Inr2*, and *Ina2* are the incident, reflected, and absorbed currents at the near end of the victim loop.

Ifi1, *Ifr1*, and *Ifa1* are the incident, reflected, and absorbed currents at the far end of the culprit loop. *Ifi1*, *Ifr2*, and *Ifa2* are the incident, reflected, and absorbed currents at the far end of the victim loop.

The parameter *T* is the time taken for a charge to propagate from one end of the line to the other. Shift registers are used to simulate the movement of charge from one segment to the next. If there are *n* segments, each time step *dt* takes *T/n* seconds.

Figure 6.31(b) defines the parameters involved when the transistor switch goes open-circuit. When this happens, the only source of energy in the system is the current flowing in the conductors of the test rig.

However, the rig is still connected to the conducting structure of the test equipment via the screens of the coaxial cables. One connects terminal 2 to the signal generator, the other links that terminal to the oscilloscope. When the switch goes open-circuit, all the stored energy flows out through the conducting surfaces of the test equipment. This action is simulated by the resistor *Rrad*.

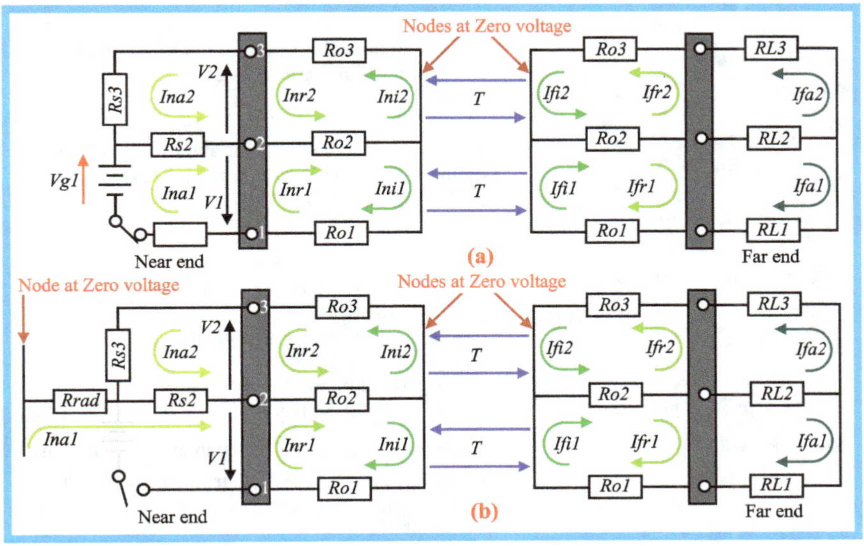

Figure 6.31 *Parameters involved in modelling the response of the setup of Figure 6.28: (a) definition of parameters for switch ON conditions and (b) definition of parameters for switch OFF conditions*

6.4.7 Computations

Figures 6.32 and 6.33 are copies of the first two pages of the Mathcad worksheet. The process is essentially the same as that described in Section 6.3.

Figure 6.32 is a copy of the first page of the Mathcad worksheet. This page defines the input parameters.

The values of the circuit components are those defined by Figure 6.30. Currents and voltages are those defined by Figure 6.31.

The only variable whose value was unknown was *Rrad*. So, this was initially treated as a 'select-on-test' component. It did not take many iterations of the program to correlate the simulated waveforms with those of the oscilloscope.

The number n of segments in the line is 10. Since the time T taken for a transient to complete one traverse of the line is 49 ns, the duration dt of each time step must be 4.9 ns. The number of traverses of the line N is set at 100. So, the simulation is for a period of 4.9 μs.

Measurements of the waveforms of Figure 6.29 indicate that the time between switch ON and switch OFF was 4 μs. Setting the number M to represent 82 traverses enabled the simulation to replicate this time period. That is, $M \cdot dt = 4 \times 10^{-6}$.

Figure 6.33 is a copy of the second page of the worksheet. This defines the functions and the main program. This utilises the reflection equations which are derived in Section 6.1.9.

The output of the main program is two vectors. The vector ***V1*** records the amplitude of the voltage at the near end of the culprit loop at each time step. The

$Ro1 := 41.6 \quad Ro2 := 149.4 \quad Ro3 := Ro1$ $Rs1 := 11 \quad Rs2 := 0.01 \quad Rs3 := 10^{6}$ $RL1 := 0 \quad RL2 := 0 \quad RL3 := 0$	Components of the circuit model are as defined in Figure 6.31(a)
$\mathbf{Ro} := \begin{pmatrix} Ro1 + Ro2 & Ro2 \\ Ro2 & Ro2 + Ro3 \end{pmatrix}$ $\mathbf{Rs} := \begin{pmatrix} Rs1 + Rs2 & Rs2 \\ Rs2 & Rs2 + Rs3 \end{pmatrix}$ $\mathbf{RL} := \begin{pmatrix} RL1 + RL2 & RL2 \\ RL2 & RL2 + RL3 \end{pmatrix}$	The relationship between variables and matrices is defined in Section 6.1.9 'Transient Cross Coupling-Reflection Equations'
$\mathbf{Rn} := \mathbf{Ro} + \mathbf{Rs} \quad \mathbf{Rf} := \mathbf{Ro} + \mathbf{RL}$ $\mathbf{Gn} := \mathbf{Rn}^{-1} \quad \mathbf{Gf} := \mathbf{Rf}^{-1}$	The conductance matrices are the inverse of the resistance matrices
$Rrad := 100$	The value of the radiation resistance was a 'select-on-test' component
$\mathbf{Rs2} := \begin{pmatrix} Rrad + Rs2 & Rs2 \\ Rs2 & Rs2 + Rs3 \end{pmatrix}$	The resistance matrix for open-circuit conditions is defined by Figure 6.31(b)
$\mathbf{Gn2} := (\mathbf{Rs2} + \mathbf{Ro})^{-1}$	The conductance matrix for open circuit conditions
$T := 49 \cdot 10^{-9}$	The time taken for a pulse to traverse the length of the signal link. That is, for a single sweep
$n := 10$	The number of segments in the simulation
$dt := \frac{T}{n} = 4.9 \times 10^{-9}$	The time taken for current to flow into a segment
$N := 100 \cdot n \quad N \cdot dt = 4.9 \times 10^{-6}$	The number of sweeps in the simulation
$M := 82 \cdot n \quad M \cdot dt = 4.018 \times 10^{-6}$	The number of the sweep at which the transistor switch at the near end goes open-circuit
$i := 1..N$	Control variable
$t_i := (i - 1) \cdot dt$	The time vector
$\mathbf{Z} := \begin{pmatrix} 0 \\ 0 \end{pmatrix}$	A zero vector. Used in the main computation

Figure 6.32 First page of Mathcad worksheet: Assigning values to the variables

vector ***V2*** records the amplitude of the voltage at the near end of the victim loop at each time step.

6.4.8 Simulation

Figure 6.34 shows the waveforms of the output voltages *V1* and *V2* over the period of 5 μs. Comparing them with the waveforms of Figure 6.29 shows that there is a clear correlation between the response of the hardware and that of the simulation.

The main discrepancy is the fact that the amplitudes of the simulation are about twice the value of that measured by the oscilloscope. This is probably due to the fact that there are two paths available for the output current; via the screens of the coaxial cables to the conducting surfaces of the scope and signal generator. The voltage probe can only monitor the action of current departing via the oscilloscope.

```
ON(Ini, Vg) := | Ina ← Gn·(2·Ro·Ini + Vg)
               | Inr ← Ina − Ini
               | Vn ← Rs·Ina − Vg
               | (Inr  Ina  Vn)
```

Calculating the values of the currents and voltages at the near end after each time step, during the period when transistor switch is short-circuit

```
OFF(Ini) := | Ina ← Gn2·(2·Ro·Ini)
            | Inr ← Ina − Ini
            | Vn ← Rs2·Ina
            | (Inr  Ina  Vn)
```

Calculating the values of the currents and voltages at the near end after each time step, during the period when transistor switch is open-circuit

```
forward(F, Inr) := | F_1 ← Inr
                   | for i ∈ n + 1 .. 1
                   |   F_{i+1} ← F_i
                   | Ifi ← F_{n+2}
                   | (F  Ifi)
```

Action of shift register in moving every charge one step forward. Takes charges from the near end and delivers them to the far end

```
back(B, Ifr) := | B_{n+1} ← Ifr
                | for i ∈ 2 .. n + 1
                |   B_{i−1} ← B_i
                | Ini ← B_1
                | (B  Ini)
```

Action of shift register in moving every charge one step back. Takes charges from the far end and delivers them to the near end

```
far (Ifi) := | Ifa ← Gf·(2·Ro·Ifi)
             | Ifr ← Ifa − Ifi
             | (Ifr  Ifa)
```

Calculating the values of the currents and voltages at the far end after each time step

```
(V1  V2) := | for i ∈ 1 .. n + 2
            |   | F_i ← Z                                   Defining the shift registers
            |   | B_i ← Z
            | Ini ← Z                                       Defining initial values for currents and
            | Vg ← Z                                        voltages at near end
            | for i ∈ 1 .. N                                Calculations at each time step.
            |   | Vg_1 ← 4  if i > 2                        Switching input voltage ON
            |   | (Inr  Ina  Vn) ← ON(Ini, Vg)  if i < M    Responses when switch is ON
            |   | (Inr  Ina  Vn) ← OFF(Ini)  if i ≥ M       and OFF
            |   | (F  Ifi) ← forward(F, Inr)                One step forward
            |   | (Ifr  Ifa) ← far (Ifi)                    Response at the far end
            |   | (B  Ini) ← back(B, Ifr)                   One step back
            |   | V1_i ← Vn_1                               Voltages at the near end after
            |   | V2_i ← Vn_2                               every step
            | (V1  V2)                                      Output vectors
```

Figure 6.33 Second page of the worksheet: The functions and the main program

6.4.9 Illustration

Many systems on aircraft use the structure as the return conductor for power supplies. It is reasonable to assume that the characteristic resistance of the power line

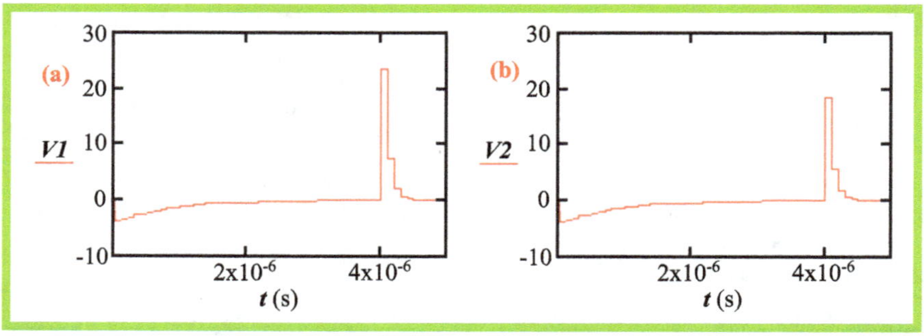

Figure 6.34 Simulation of the transient voltages at the near end of the rig: (a) voltage V1 developed at the near end of the culprit loop and (b) voltage V2 developed at the near end of the victim loop

would be 200 Ω. If the supply current *Is* is 5 A, then at switch OFF, the dynamic voltage stored in the cable will be

$$Vs = Ro \cdot Is = 1000 \text{ V} \tag{6.48}$$

If the time *T* for a pulse to propagate along the cable is 100 ns, the energy stored in the line would be

$$W = 2 \cdot T \cdot Vs \cdot Is = 10^{-3} \text{ J} \tag{6.49}$$

This could just be enough to ignite a flammable mixture of fuel and air, such as exists in the fuel tank.

If the wiring of the fuel quantity indication system includes a single-point reference, a mechanism exists for this energy to be coupled into the wiring for the fuel level sensors in the tank. Since the fuel pumps provide an inductive load, switching a pump OFF would inevitably create a much higher voltage than 1 kV and this would last for much longer than 100 ns. So, it is entirely possible for such a system to deliver enough energy to ignite the fuel.

A single fault condition in the tank wiring such as an intermittent open circuit would provide a spark gap. If such a fault coincided with a fuel pump being switched off, the fuel in the tank would ignite.

6.4.10 Conclusion

Ground bounce is an ever-present phenomenon in vehicles and aircraft because the structure is used to carry the return current for power supplies. When any supply is switched OFF, the energy stored in the power link will depart explosively into the local environment via the conducting structure.

Voltage spikes will appear in series with the signal carried by every cable which happens to be routed along that structure. Some would be puzzling, some would cause temporary upset, some could cause emergency repair action on an aircraft sitting on the ground, and some would be catastrophic.

The modelling technique described here can be applied by any circuit designer to analyse such an event. The preceding sections in this book identify ways of eliminating this hazard condition.

Chapter 7
Further EMC/EMI research

7.1 Dipole transients

7.1.1 Introduction

A transient model of a dipole antenna is developed from an earlier model which used an LCR circuit to simulate the frequency response of a single conductor. A transformer at the centre provided the source voltage. The resistive component of the model was that defined by electromagnetic theory as the 'radiation resistance'.

The new model uses a waveform generator to deliver current to the central section and shift registers to simulate the propagation of charges along the conductor. When the waveform is sinusoidal, the response is shown to be exactly the same as that of the LCR circuit.

It is reasoned that the radiation resistance provides a measure of the power used in converting the radial flow of charges in the conductor to that of photons emanating from the surface.

7.1.2 Frequency analysis

Figure 7.1(a) shows a general circuit model of a length of conductor acting as a dipole transmitter. During the first half cycle of a sinusoidal signal, the voltage source at the centre draws current from the left-hand monopole and delivers it to the right-hand monopole. During the second half cycle the process is reversed. When the frequency of the source corresponds to the half-wavelength of the conductor, the current reaches a maximum value, as does the radiated energy.

Invoking the relationships of transmission line theory allows the model of Figure 7.1(a) to be transformed into the distributed parameter model of Figure 7.1(b). Frequency analysis of this model provides a simulation which correlates with that of the actual hardware.

Figure 7.1(c) shows the setup used to carry out measurements of the response of a 15 m length of conductor. Recording the test results allows a graph to be created of the frequency response of the assembly. Analysis of the response of the model of Figure 7.1(b) results in a curve which is very similar to that of the hardware. Running the program several times whilst adjusting the values of the circuit parameters provides a curve which replicates that of the equipment under test. When these values are assigned to the variables of Figure 7.1(a), the result is the definitive circuit model of Figure 7.1(d).

Section 4.1 describes this process.

7.1.3 Transient analysis

The processing involved in simulating the transient behaviour of a transmitting dipole is much the same as that in previous transient models. Section 6.4 provides an example. Figure 7.2 shows the process in block diagram form. Charges propagate out from the central section, flow along the right monopole, arrive at the far end, and are immediately reflected back to the source. The time delays are simulated by shift registers and the terminations at the far end are each simulated as an open circuit. The same happens with the left-hand monopole.

Figure 7.3 is a circuit diagram which simulates the action of the central section of Figure 7.2. The voltage source is a sine wave generator of peak amplitude *Vg*.

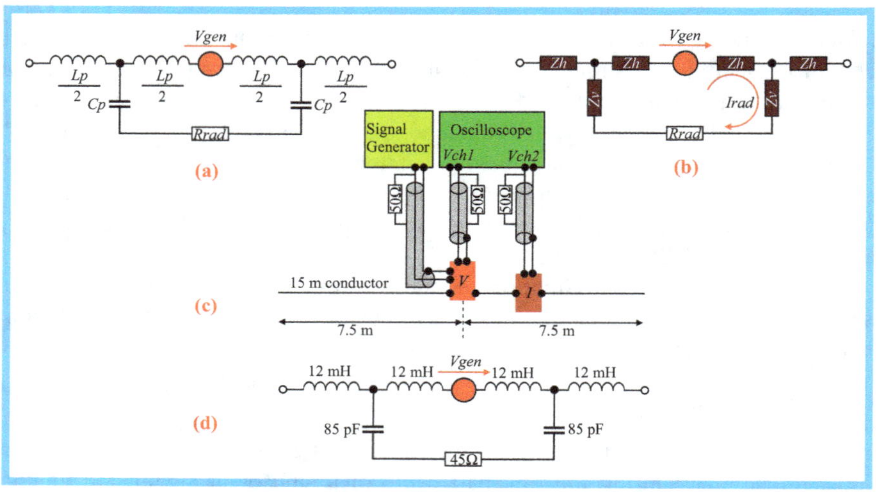

Figure 7.1 Modelling the response of a dipole emitter: (a) lumped parameter circuit model; (b) distributed parameter circuit model; (c) test setup for a 15 m conductor; and (d) definitive circuit model of the configuration under review

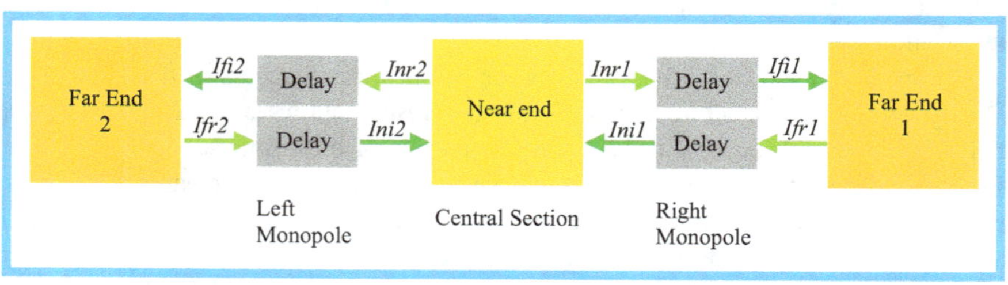

Figure 7.2 Signal processing block diagram for transient analysis

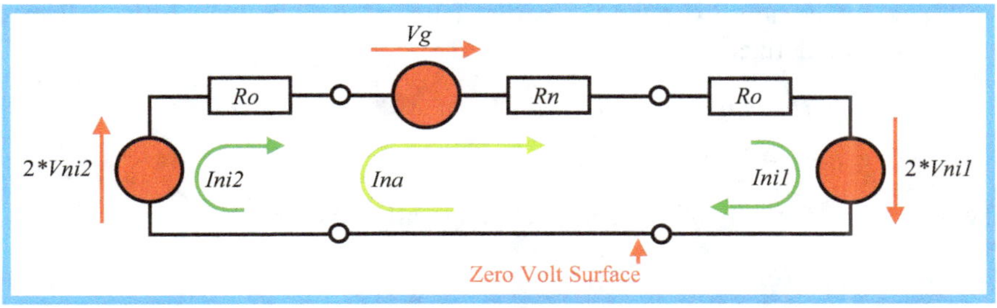

Figure 7.3 Circuit model of the central section of the dipole

The value of the series resistor is *Rn*. The incoming signal from the right monopole is represented as the voltage $2 \cdot Vni1$ applied via the characteristic resistance *Ro*,

$$Ro = \sqrt{\frac{Lp}{Cp}} \tag{7.1}$$

The amplitudes of the incident voltages *Vni1* and *Vni2* are

$$Vni1 = Ro \cdot Ini1 \tag{7.2}$$

$$Vni2 = Ro \cdot Ini2 \tag{7.3}$$

Although there is only one circuit model, it is necessary to set up three equations which enable the reflected currents *Inr1* and *Inr2* to be defined, as well as the total current *Ina*. The only way to do this is to invoke the principle of superposition.

$$2 \cdot Vni1 = (2 \cdot Ro + Rn) \cdot Ina1 \tag{7.4}$$

$$2 \cdot Vni2 = (2 \cdot Ro + Rn) \cdot Ina2 \tag{7.5}$$

$$Vg = (2 \cdot Ro + Rn) \cdot Ina3 \tag{7.6}$$

This gives

$$Ina1 = \frac{2 \cdot Ro \cdot Ini1}{2 \cdot Ro + Rn} \tag{7.7}$$

$$Ina2 = \frac{2 \cdot Ro \cdot Ini2}{2 \cdot Ro + Rn} \tag{7.8}$$

$$Ina3 = \frac{Vg}{2 \cdot Ro + Rn} \tag{7.9}$$

The total current in the central section is the sum of the partial currents:

$$Ina = Ina1 + Ina2 + Ina3 = \frac{2 \cdot Ro \cdot Ini1 + 2 \cdot Ro \cdot Ini2 + Vg}{2 \cdot Ro + Rn} \tag{7.10}$$

Since values now exist for the partial currents *Ina1* and *Ina2*, the reflected currents can be defined:

$$Inr1 = Ina1 - Ini1 \tag{7.11}$$

$$Inr2 = Ina2 - Ini2 \tag{7.12}$$

The time taken for a signal to propagate along either monopole is

$$T = \sqrt{Lp \cdot Cp} \tag{7.13}$$

The frequency of quarter-wave resonance is

$$Fq = \frac{1}{4 \cdot T} \tag{7.14}$$

7.1.4 Worksheet: first page

Now that all the basic equations have been defined, there is enough data to compile the first page of the worksheet. This is shown on Figure 7.4.

The values of *Lp* and *Cp* are those recorded in the definitive circuit model of this particular dipole resistance *Ro*. This is an encouraging result, since the intrinsic impedance of free space is 377 Ω, and since the value defined in Figure 7.1 was derived from a simple test setup using general-purpose test equipment: Figure 7.1(d). Invoking (7.1) leads to a value of 376 Ω for the characteristic equipment.

The time *T* for a charge to traverse the length of either monopole is 31.93 ns, and this correlates with a quarter-wave resonant frequency of 7.83 MHz.

The peak value of the voltage generator is defined as 10 V. It did not take long to assign a value of 45 Ω to *Rn*: the same value as the radiation resistor of Figure 7.1(d). The number of sections in the right monopole, *n1,* is defined as 100. So, the time increment *dt* is 319.3 picoseconds. The number of sections in the left monopole, *n2*, is the same.

The function *near(Ini1,Ini2,Vg)* is compiled using (7.7), (7.8), (7.10), (7.11), and (7.12). The other functions are copies of those used in previous analyses. Each function defines the processing involved in the associated block in Figure 7.2.

7.1.5 Worksheet: second page

The second page of the worksheet is duplicated by Figure 7.5.

The number of traverses *N* of the signal is set at 300. This allows the time vector ***t*** to be defined and for the simulation time of the scan to be set at just under 10 μs. The frequency *f* is set at the same value as the quarter-wave frequency of a monopole.

The main program defines the series of functions involved in the process. Figure 7.2 illustrates this in the form of a block diagram. The process carried out by each block relates to one of the functions defined in the first page of the worksheet. The arrowed lines indicate the signal which is delivered from the output of one block to the input of the next.

The output of this program is stored in the vector ***Ina.***

$$Lp := 12\cdot 10^{-6} \qquad Cp := 85\cdot 10^{-12}$$

$$Ro := \sqrt{\frac{Lp}{Cp}} = 375.845 \qquad T := \sqrt{Lp\cdot Cp} = 3.194\times 10^{-8} \qquad Fq := \frac{1}{4\cdot T} = 7.83\times 10^{6}$$

$$Vg := 10 \qquad Rn := 45 \qquad RL := 10^{7}$$

$$n1 := 100 \qquad n2 := n1 \qquad dt := \frac{T}{n1} = 3.193\times 10^{-10}$$

$$near(Ini1, Ini2, Vg) := \left| \begin{array}{l} Ina1 \leftarrow \dfrac{2\cdot Ro\cdot Ini1}{2\cdot Ro + Rn} \\ Ina2 \leftarrow \dfrac{2\cdot Ro\cdot Ini2}{2\cdot Ro + Rn} \\ Ina \leftarrow Ina1 + Ina2 + \dfrac{Vg}{2\cdot Ro + Rn} \\ Inr1 \leftarrow Ina - Ini1 \\ Inr2 \leftarrow Ina - Ini2 \\ (Inr1\ \ Inr2\ \ Ina) \end{array} \right.$$

$$forward(\boldsymbol{F}, Inr, n) := \left| \begin{array}{l} \boldsymbol{F}_1 \leftarrow Inr \\ for\ \ i \in n+1 .. 1 \\ \quad \boldsymbol{F}_{i+1} \leftarrow \boldsymbol{F}_i \\ Ind \leftarrow \boldsymbol{F}_{n+2} \\ (\boldsymbol{F}\ \ Ind) \end{array} \right.$$

$$far\ (Ifi) := \left| \begin{array}{l} Ifa \leftarrow \dfrac{2\cdot Ro\cdot Ifi}{Ro + RL} \\ Ifr \leftarrow Ifa - Ifi \\ (Ifr\ \ Ifa) \end{array} \right.$$

$$back(\boldsymbol{B}, Ifr, n) := \left| \begin{array}{l} \boldsymbol{B}_{n+1} \leftarrow Ifr \\ for\ \ i \in 2 .. n+1 \\ \quad \boldsymbol{B}_{i-1} \leftarrow \boldsymbol{B}_i \\ Ini \leftarrow \boldsymbol{B}_1 \\ (\boldsymbol{B}\ \ Ini) \end{array} \right.$$

Figure 7.4 First page of worksheet: defining constants and functions

Since there are too many transitions in the scan to provide a coherent picture, the waveform depicted in the bottom left-hand corner of Figure 7.5 is limited to a period of 1 μs. It is obvious that the current is basically a sine wave and that the amplitude increases exponentially.

The final function on the page is effectively a snapshot of the waveform over the final microsecond of the simulation. This is displayed on the graph on the right-hand side. It is clear that the waveform has settled down to a sinusoidal waveform of peak amplitude of 222 mA. This is the amplitude of the current in a 45 Ω resistor when a 10 V signal is applied.

$N := 300 \cdot n1 \qquad n := 1..N \qquad t_n := n \cdot dt \qquad N \cdot dt = 9.581 \times 10^{-6} \qquad f := Fq$

```
Ina := | F1[n1+2] ← 0
       | F2[n2+2] ← 0
       | B1[n1+1] ← 0
       | B2[n2+1] ← 0
       | for i ∈ 1..N
       |    | Vgen ← Vg·sin(2·π·f·t[i])
       |    | (Inr1 Inr2 Ina) ← near(Ini1, Ini2, Vgen)
       |    | (F1 Ifi1) ← forward(F1, Inr1, n1)
       |    | (F2 Ifi2) ← forward(F2, Inr2, n2)
       |    | (Ifr1 Ifa1) ← far (Ifi1)
       |    | (Ifr2 Ifa2) ← far (Ifi2)
       |    | (B1 Ini1) ← back(B1, Ifr1, n1)
       |    | (B2 Ini2) ← back(B2, Ifr2, n2)
       |    | Iout[i] ← Ina
       | Iout

m := 30·n1     S := | for j ∈ 1..m
                    |    | p ← N − m + j
                    |    | S[j] ← Ina[p]
                    | S
```

$f = 7.828 \times 10^{6} \qquad max(\boldsymbol{S}) = 0.222 \qquad min(\boldsymbol{S}) = -0.222$

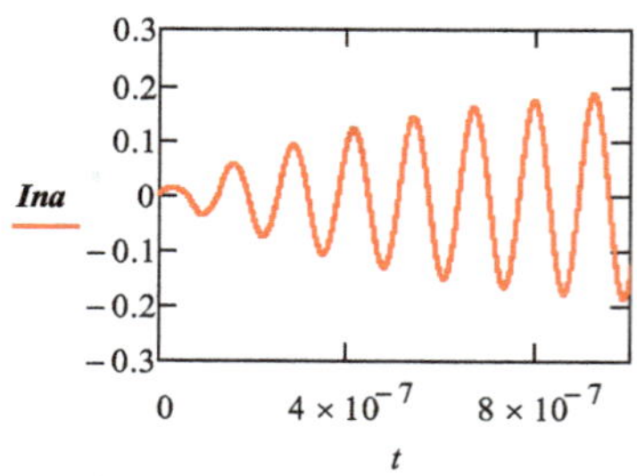

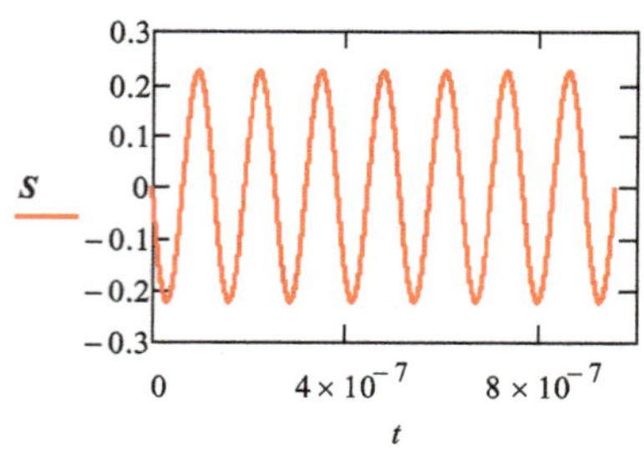

Figure 7.5 Second page of Mathcad worksheet

7.1.6 Other frequencies

By selecting different values for the frequency, it was possible to simulate the transient response of the dipole over the range 1–20 MHz. The left-hand graph of Figure 7.6 shows the initial response to the application of a sine wave, whilst the right hand graph shows the response after a time-lag of 10 μs. The function *max*(***S***) defines the peak value at this time; 2.7 mA.

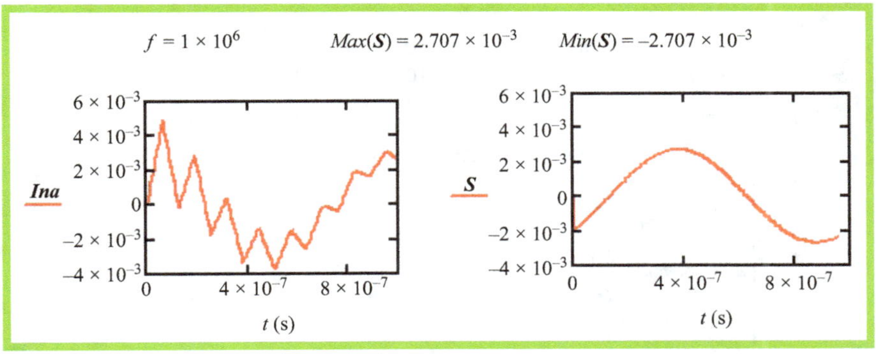

Figure 7.6 Response of the model to a 1 MHz source

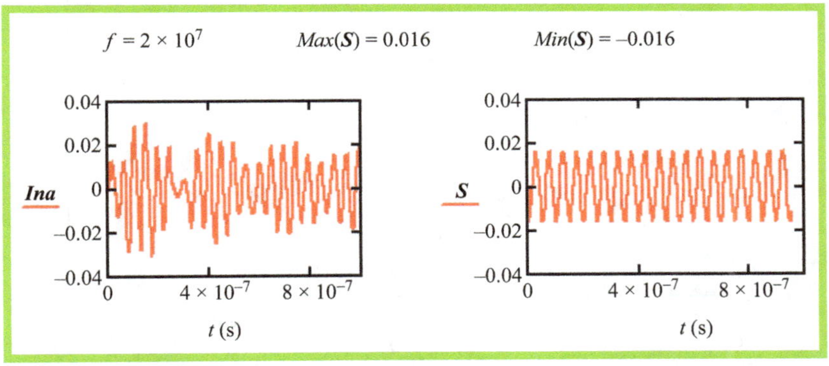

Figure 7.7 Response of the model to a 20 MHz source

The response of the model to a source operating at 20 MHz is shown on Figure 7.7. Again, the time delay between the initial and final snapshots is about 10 μs; but the duration of each snapshot is 100 ns.

7.1.7 Correlating the responses

Carrying out this exercise over a number of spot frequencies allowed a table to be compiled relating the value of *Ina* to frequency. These values are recorded as the blue circles on the graph of Figure 7.8. This figure also includes a copy of the worksheet used to generate the red curve.

7.1.8 Assessment

Using shift registers to simulate the propagation of charges backward and forward along the length of any conductor necessarily assumes that it is lossless. This is catered for by placing resistors at the terminations to simulate the energy lost. With

$Rp := 0.2 \quad Gp := 0 \quad Fq := 7.83 \cdot 10^{6} \quad Vg := 10$

$Rrad := 45 \quad Lp := 12 \cdot 10^{-6} \quad Cp := 85 \cdot 10^{-12} \quad i := 1..2000 \quad f_i := i \cdot 10^{4}$

$$Im_i := \begin{vmatrix} \omega \leftarrow 2 \cdot \pi \cdot f_i \\ \theta \leftarrow \sqrt{(Rp + j \cdot \omega \cdot Lp) \cdot (Gp + j \cdot \omega \cdot Cp)} \\ Zo \leftarrow \sqrt{\dfrac{Rp + j \cdot \omega \cdot Lp}{Gp + j \cdot \omega \cdot Cp}} \\ Zh \leftarrow Zo \cdot tanh\left(\dfrac{\theta}{2}\right) \\ Zv \leftarrow Zo \cdot csch(\theta) \\ Z3 \leftarrow 2 \cdot Zh + 2 \cdot Zv + Rrad \\ \dfrac{Vg}{|Z3|} \end{vmatrix}$$

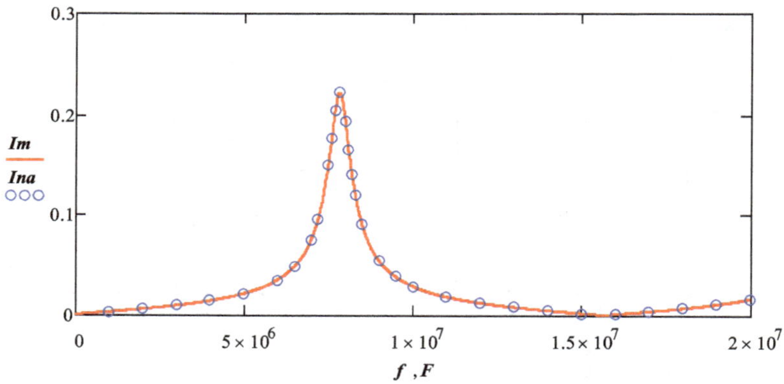

Red curve: Output of frequency response model
Blue circles: Responses derived from transient simulation

Figure 7.8 Comparing the responses of the two models to a sinusoidal source of 10 V

two-conductor and three-conductor signal links, this loss is due to the resistivity of the material.

However, the need to assign a value of 45 Ω to *Rn* is a clear indication that the power losses involved in this particular configuration are not due to the resistivity of the material. So, they must be due to the power loss in creating the electromagnetic field.

By calculating the power density at every point on the surface of a sphere which encloses the dipole and integrating this value over the entire surface, electromagnetic theory shows that the total power delivered to that surface is

$$P = Rrad \cdot Irms^2 \tag{7.15}$$

Where

$$Irms = \sqrt{2} \cdot Ina \tag{7.16}$$

That is, the power delivered to the 45 Ω resistor in the frequency response model of Figure 7.1(d) is a measure of the power delivered to the electromagnetic field. But the analysis of electromagnetic theory indicates that the radiation resistance only exists when the waveform along the dipole is half a sine-wave. That is, at the frequency of resonance.

The analysis of the transient model of Figure 7.2 produces the same relationship between frequency and current at the resonant frequency as the model of Figure 7.1. It also shows that the assignment of a value of 45 Ω to *Rn* will create an identical frequency response. Since the value of *Rn* in Figure 7.3 is identical to that of *Rrad* in Figure 7.1(d), this means that the radiation resistance is not a function of frequency. It has a fixed value, dependent on the dimensions of the conductor.

Electromagnetic theory also identifies the existence of the displacement current; a current which flows in the radial direction, rather than axially along the conductor. Since it flows out radially, it must arrive at the conductor surface within picoseconds. At the surface, the energy manifests itself as an electromagnetic field.

Since the photon is the *quantum* of the *electromagnetic field*, the charges delivered to the surface and held at that surface must cause photons to be released.

Since the power delivered to the electromagnetic field is the same as the voltage developed across the radiation resistor, the inevitable conclusion must be that the energy lost in this resistor is due to that expended at the surface of the conductor in converting the radial flow of charges to the emission of photons.

At frequencies other than resonance, the radiated emission is much less because the signal becomes less and less coherent.

With cross-coupling analysis, the radiation resistance does not feature in the equations. This indicates that the energy lost due to the emission of photons from the send conductor is delivered immediately to the return conductors.

7.1.9 Conclusion

The radiation resistance is an essential parameter in any circuit model used to relate the currents in the system to the level of emission from that system or to define the susceptibility of the system to an external field.

A simple test such as that summarised in the section on frequency analysis can be used to measure the value of this parameter.

7.2 Coaxial coupling model

7.2.1 Introduction

A method of characterising the electromagnetic coupling between a coaxial cable and the conducting structure of an electronic system is described. Frequency response tests are carried out on a rig and the data so collected is used to develop a circuit model which replicates that response. Assessment of the model provides a clear insight into the mechanisms involved in cross coupling. This form of analysis can be integrated into the system design process.

7.2.2 Setup

Figure 7.9 shows the setup. The rig itself is a 15 mm copper pipe routed round three walls of a room. A wooden batten fixed along the top of the pipe acts as a spacer to separate the cable under test from the pipe. This setup allows different cable assemblies to be characterised. In this case it is a 50 Ω coaxial cable. One end is terminated at an interface module which allows a signal to be applied via a low impedance source and for the amplitude of that signal to be monitored by channel 1 of an oscilloscope. Both terminals at the far end are short-circuited to the copper pipe. A current transformer is used to measure the current delivered to the cable generator.

Such a setup can be correlated with a coaxial cable routed along the structure of a vehicle, aircraft, spacecraft, or ship. The copper pipe replicates the action of the conducting material adjacent to the cable in creating the electromagnetic field which defines the cross coupling.

Figure 7.10 provides details of the cross-section of the test rig. The length of the cable was 11.4 m.

7.2.3 Test method

The signal generator was set to deliver a sinusoidal signal and the amplitudes of the voltage and current delivered to the cable were monitored at channels 1 and 2 of the

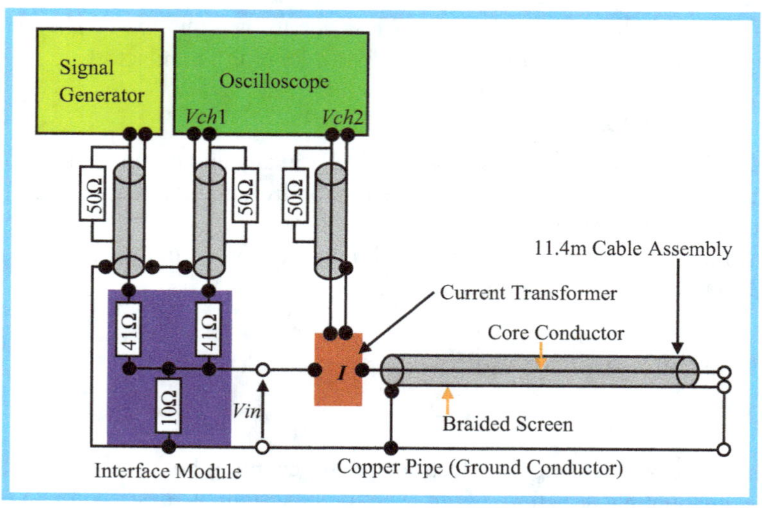

Figure 7.9 Test setup for characterising the coaxial cable

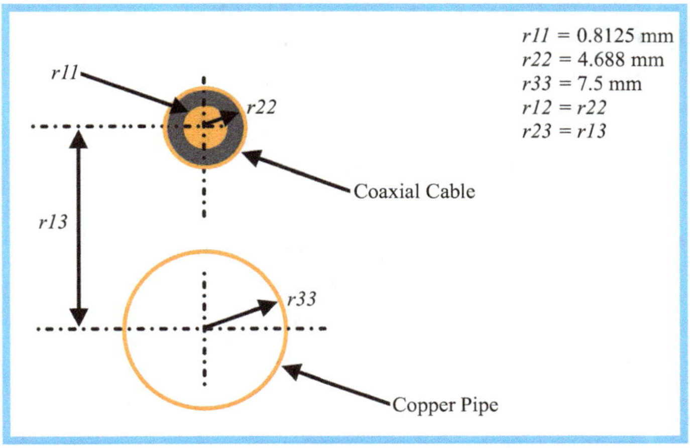

Figure 7.10 Cross-section of the test rig

oscilloscope. These data were recorded at a set of spot frequencies and a graph of the transfer admittance is created by the Mathcad worksheet shown by Figure 7.11.

Details of the current transformer are provided in Section 2.5. Section 2.6 relates the function ***Yt1*** to the circuit model of the transformer. This function calculates the amplitude of the differential-mode current that would flow when the voltage applied between the core and the screen was 1 V.

The setup of Figure 7.9 was then changed to route the entire cable through the current transformer. This allowed the common-mode current to be measured over the same range of frequencies. Figure 7.12 is a copy of page 2 of the Mathcad worksheet. This defines the frequency response of the transfer admittance ***Yt2***, the current that would flow in the common-mode loop when the voltage applied between the cable input terminals was 1 V.

The graph at the bottom of Figure 7.11 defines the differential-mode response of the cable under test. The graph at the bottom of Figure 7.12 defines the frequency response of the cross coupling between the differential and common-mode loops.

7.2.4 Developing the model

Having determined the actual response of the configuration, the next step is to create a model which replicates that response.

Previous sections have defined the relationship between Electromagnetic Theory and Circuit Theory and have invoked this relationship to derive formulae for the inductance and capacitance of a three-conductor assembly.

When these formulae are used to define the reactive parameters for the assembly of Figure 7.9, it transpires that the value of the inductance of the screen is zero and the capacitance is infinite. This results in the circuit model

$$data1 := \begin{pmatrix} 1 & 700 & 110 \\ 2 & 740 & 50 \\ 3 & 760 & 38 \\ 4 & 760 & 8 \\ 5 & 760 & 13 \\ 6 & 760 & 38 \\ 7 & 760 & 86 \\ 8 & 480 & 240 \\ 8.33 & 180 & 300 \\ 8.5 & 400 & 250 \\ 9 & 700 & 160 \\ 10 & 780 & 70 \\ 11 & 800 & 39 \\ 12 & 500 & 17 \\ 13 & 800 & 3 \\ 14 & 800 & 19 \\ 15 & 790 & 50 \\ 16 & 660 & 120 \\ 17 & 620 & 170 \\ 18 & 800 & 74 \\ 19 & 800 & 40 \\ 19.8 & 800 & 28 \end{pmatrix}$$

$$R1 := 100$$
$$R2 := 51$$
$$R3 := 50$$
$$R4 := 850$$
$$L1 := 200\cdot 10^{-6}$$
$$C1 := 60\cdot 10^{-12}$$
$$Turns := 10$$
$$K := \frac{50+41}{50} = 1.82$$

$$s := 1..rows(data1) \qquad F1_s := data1_{s,1}\cdot 10^6$$

$$Yt1_s := \left| \begin{array}{l} Vch1 \leftarrow data1_{s,2}\cdot 10^{-3} \\ Vin \leftarrow K\cdot Vch1 \\ Vch2 \leftarrow data1_{s,3}\cdot 10^{-3} \\ \omega \leftarrow 2\cdot\pi\cdot F1_s \\ Z1 \leftarrow R4 + \dfrac{1}{j\cdot\omega\cdot C1} \\ Y2 \leftarrow \dfrac{1}{R1} + \dfrac{1}{R2} + \dfrac{1}{R3} + \dfrac{1}{j\cdot\omega\cdot L1} + \dfrac{1}{Z1} \\ Isec \leftarrow |Y2|\cdot Vch2 \\ Iprim \leftarrow Isec\cdot Turns \\ \dfrac{Iprim}{Vin} \end{array} \right.$$

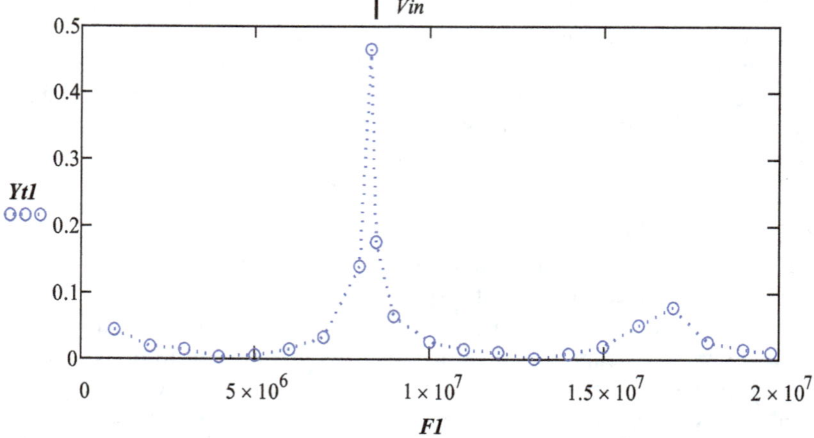

Figure 7.11 Worksheet page 1: Frequency response of the cable admittance

$$\mathbf{data2} := \begin{pmatrix} 1 & 800 & 0 \\ 2 & 800 & 0 \\ 3 & 800 & 0 \\ 4 & 800 & 0.8 \\ 5 & 800 & 1 \\ 6 & 800 & 0.8 \\ 7 & 800 & 0.8 \\ 8 & 600 & 4.8 \\ 8.5 & 680 & 12 \\ 9 & 680 & 12 \\ 10 & 840 & 15 \\ 10.5 & 840 & 36 \\ 10.6 & 840 & 38 \\ 11 & 840 & 9 \\ 11.8 & 840 & 1 \\ 12 & 840 & 1.2 \\ 13 & 860 & 3 \\ 14 & 840 & 5.6 \\ 15 & 840 & 5 \\ 16 & 760 & 3 \\ 17 & 460 & 8 \\ 18 & 840 & 9 \\ 19 & 860 & 9.8 \\ 19.8 & 860 & 11 \end{pmatrix}$$

$$s := 1..rows(\mathbf{data2}) \qquad \mathbf{F2}_s := \mathbf{data2}_{s,1} \cdot 10^6$$

$$\mathbf{Yt2}_s := \begin{array}{|l} Vch1 \leftarrow \mathbf{data2}_{s,2} \cdot 10^{-3} \\ Vin \leftarrow K \cdot Vch1 \\ Vch2 \leftarrow \mathbf{data2}_{s,3} \cdot 10^{-3} \\ \omega \leftarrow 2 \cdot \pi \cdot \mathbf{F2}_s \\ Z1 \leftarrow R4 + \dfrac{1}{j \cdot \omega \cdot C1} \\ Y2 \leftarrow \dfrac{1}{R1} + \dfrac{1}{R2} + \dfrac{1}{R3} + \dfrac{1}{j \cdot \omega \cdot L1} + \dfrac{1}{Z1} \\ Isec \leftarrow |Y2| \cdot Vch2 \\ Iprim \leftarrow Isec \cdot Turns \\ \dfrac{Iprim}{Vin} \end{array}$$

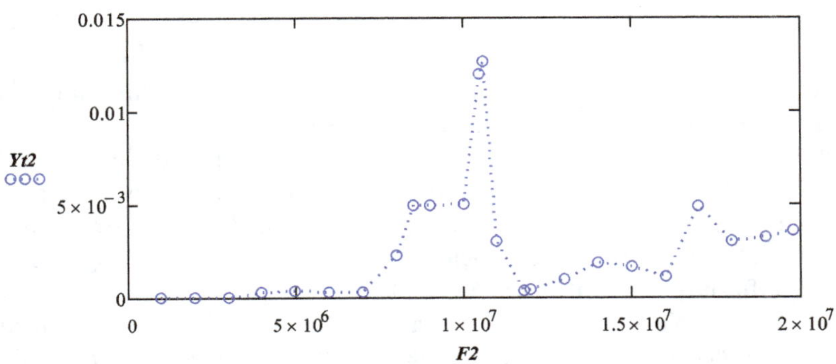

Figure 7.12 Worksheet page 2: Frequency response of the transfer admittance

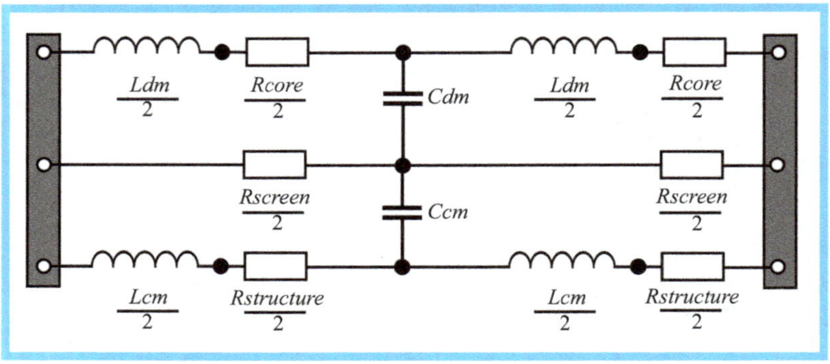

Figure 7.13 Basic model of coaxial cable routed over structure

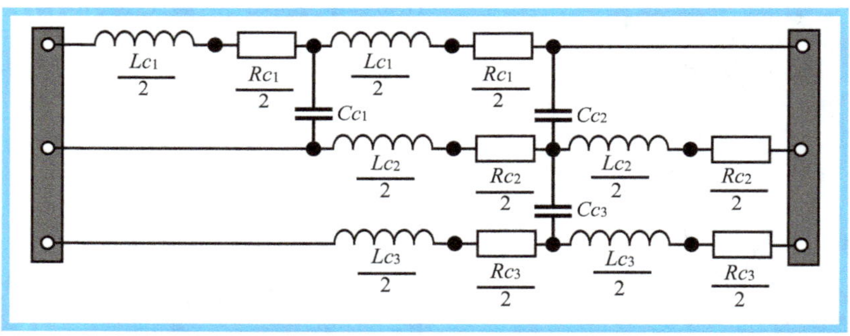

Figure 7.14 Lumped parameter model of braided-screen cable routed over structure

of Figure 7.13, where the only component coupling the differential-mode and common-mode loops is the resistance of the screen.

High-frequency coupling exists between the two loops, as illustrated by Figure 7.12. This is because the strands which form the screen are longer than the core conductor. If a length of coax cable is dissembled and a single strand of the screen is laid alongside the core conductor, it will be found that the length of that strand is significantly longer than the core.

That is, the cable can be simulated as two sections in series; one length representing perfect shielding and a shorter length simulating the extra length of the unshielded portion. This is illustrated by Figure 7.14.

The final step in the simulation is to transform every T-network of lumped parameters into the equivalent network of distributed parameters. This is shown in Figure 7.15.

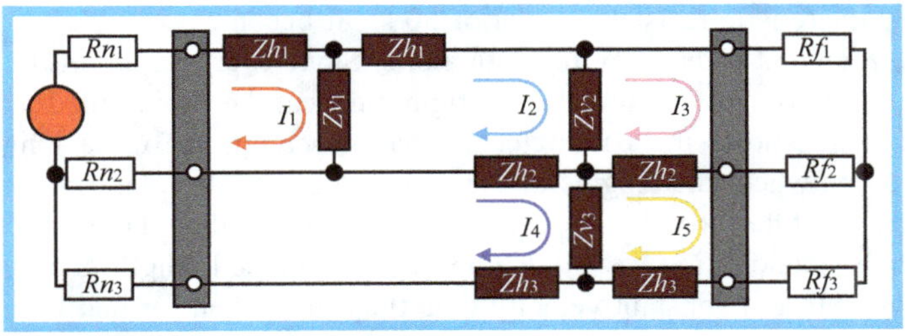

Figure 7.15 Distributed-parameter model of braided-screen cable routed over structure

$$\mu_o := 4\cdot\pi\cdot 10^{-7} \quad len := 11.4 \quad r11 := 0.325\cdot 10^{-3} \quad r22 := 1.875\cdot 10^{-3} \quad r12 := r22$$

$$r33 := 7.5\cdot 10^{-3} \quad r23 := 23\cdot 10^{-3} \quad r13 := r23$$

$$Lscreen := \frac{\mu_o\cdot len}{2\cdot\pi}\cdot ln\left(\frac{r12\cdot r23}{r22\cdot r13}\right) = 0 \qquad Ldm := \frac{\mu_o\cdot len}{2\cdot\pi}\cdot ln\left(\frac{r12\cdot r13}{r11\cdot r23}\right) = 3.996\times 10^{-6}$$

$$Lcm := \frac{\mu_o\cdot len}{2\cdot\pi}\cdot ln\left(\frac{r13\cdot r23}{r22\cdot r33}\right) = 8.271\times 10^{-6} \quad Fdm := 8.33\cdot 10^{6} \quad Tdm := \frac{1}{2\cdot Fdm} = 6.002\times 10^{-8}$$

$$Rdm := 50 \quad k := 0.01 \qquad T1 := (1-k)\cdot Tdm = 5.942\times 10^{-8} \qquad T2 := k\cdot Tdm = 6.002\times 10^{-10}$$

$$Lc_1 := Rdm\cdot T1 \qquad Lc_2 := Rdm\cdot T2 \qquad Lc_3 := Lcm$$

$$Cc_1 := \frac{T1}{Rdm} \qquad Cc_2 := \frac{T2}{Rdm} \qquad Cc_3 := \frac{Tcm^2}{Lcm} = 2.69\times 10^{-10}$$

$$Fcm := 10.6\cdot 10^{6} \qquad Tcm := \frac{1}{2\cdot Fcm} = 4.717\times 10^{-8}$$

$$Rdc := \begin{pmatrix} 0.2 \\ 0.02 \\ 0.005 \end{pmatrix} \quad Rn := \begin{pmatrix} 0 \\ 0 \\ 0 \end{pmatrix} \quad Rf := \begin{pmatrix} 0 \\ 0 \\ 0 \end{pmatrix} \quad Rins := \begin{pmatrix} 10^7 \\ 10^7 \\ 6000 \end{pmatrix} \quad Gins := \frac{1}{Rins} = \begin{pmatrix} 1\times 10^{-7} \\ 1\times 10^{-7} \\ 1.667\times 10^{-4} \end{pmatrix}$$

Figure 7.16 Third page of worksheet: assigning parameter values to the circuit of Figure 7.14

7.2.5 Assigning parameter values

Figure 7.16 is a copy of the third page of the worksheet.

The first step on this page is to copy the geometrical data of Figure 7.10 onto the worksheet. Invoking the relationships derived in Section 2.3 allows theoretical values for the inductances of the differential-mode loop, *Ldm*, and the common-mode loop *Lcm* to be derived. It also demonstrates the theoretical value of the inductor representing the screen *Lscreen* is zero.

A review of tabulated results recorded in Figure 7.11 shows that the admittance of the differential-mode loop peaks at a frequency *Fdm* of 8.33 MHz. This is the frequency at which half-wave resonance occurs. So, it is possible to derive the time taken *Tdm* for a signal to propagate along to cable. Since the coax is specified as having a characteristic resistance *Rdm* of 50 Ω, it becomes possible to derive the inductance *Ldm* and *Cdm* from this data alone. Since every strand of the screen is longer than the core conductor, the configuration can be represented as two sections; one representing the parameters one representing the extra length of the strands and one where the lengths of the core and screen are identical.

If the ratio of the extra length to the total length is *k*, then the length where the core is totally shielded is $(1 - k) \cdot len$ and the unshielded length is $k \cdot len$. Since, the signal travels at a constant velocity then the times taken for that to happen, *T1* and *T2*, can be deduced.

By invoking the relationships:

$$L = Ro \cdot T \tag{7.17}$$

And

$$C = \frac{T}{Ro} \tag{7.18}$$

Values can be assigned to Lc_1, Lc_2, Cc_1, and Cc_2 of Figure 7.14.

The tabulated results of Figure 7.12 identify the half-wave frequency *Fcm* of the current in the common-mode loop as 10.6 MHz. This allows the time taken *Tcm* for a pulse to propagate along that loop.

The inductance of the common-mode loop *Lcm* has already been derived, from geometrical data.

Invoking the relationship,

$$C = \frac{T^2}{Lcm} \tag{7.19}$$

Allows values to be assigned to Lc_3 and Cc_3.

Having assigned values for the reactive components of Figure 7.14, the remaining task is to define the series resistors. These are recorded in the vector ***Rdc***. The values of the resistors at the near end ***Rn*** and at the far end ***Rf*** are all zero. Since each capacitor is associated with a conductance, this parameter is defined in terms of a resistance. The vector ***Rins*** records the value of each.

7.2.6 Defining the functions

The basic functions are defined on Figure 7.17, the fourth page of the worksheet.

The function *Zbranch(f)* calculates the value of each impedance of Figure 7.15 at a specified frequency *f*, using the values of the lumped parameter components of Figure 7.14. The transformation equations are derived in Section 2.2. The output of this function is two vectors. The vector ***Zh*** defines the values of Zh_1, Zh_2, and Zh_3, whilst ***Zv*** defines the values of Zv_1, Zv_2, and Zv_3.

The computation assumes that the series resistance of the core conductor is constant up to a frequency *Fx* and then increases in proportion to the square root of the frequency. This takes into account the increase in resistance due to skin effect.

The computation also causes the conductance to increase with frequency to account for the fact that radiated emission increases with frequency.

$$Fx := 25 \cdot 10^{3}$$

$$Zbranch(f) := \left|\begin{array}{l} \omega \leftarrow 2 \cdot \pi \cdot f \\ \textit{for } i \in 1..3 \\ \quad \left|\begin{array}{l} \boldsymbol{Rc}_i \leftarrow \boldsymbol{Rdc}_1 \\ \boldsymbol{Rc}_1 \leftarrow \boldsymbol{Rdc}_1 \cdot \sqrt{1 + \dfrac{f}{Fx}} \\ \boldsymbol{Gc}_i \leftarrow \boldsymbol{Gins}_i \cdot \dfrac{f}{Fdm} \\ \theta \leftarrow \sqrt{\left(\boldsymbol{Rc}_i + j \cdot \omega \cdot \boldsymbol{Lc}_i\right) \cdot \left(\boldsymbol{Gc}_i + j \cdot \omega \cdot \boldsymbol{Cc}_i\right)} \\ Zo \leftarrow \sqrt{\dfrac{\boldsymbol{Rc}_i + j \cdot \omega \cdot \boldsymbol{Lc}_i}{\boldsymbol{Gc}_i + j \cdot \omega \cdot \boldsymbol{Cc}_i}} \\ \boldsymbol{Zh}_i \leftarrow Zo \cdot tanh\left(\dfrac{\theta}{2}\right) \\ \boldsymbol{Zv}_i \leftarrow Zo \cdot csch(\theta) \end{array}\right. \\ (\boldsymbol{Zh} \quad \boldsymbol{Zv}) \end{array}\right.$$

$$Zloop(f) := \left|\begin{array}{l} (\boldsymbol{Zh} \quad \boldsymbol{Zv}) \leftarrow Zbranch(f) \\ Z11 \leftarrow \boldsymbol{Rn}_1 + \boldsymbol{Zh}_1 + \boldsymbol{Zv}_1 + \boldsymbol{Rn}_2 \\ Z12 \leftarrow -\boldsymbol{Zv}_1 \\ Z13 \leftarrow 0 \\ Z14 \leftarrow -\boldsymbol{Rn}_2 \\ Z15 \leftarrow 0 \\ Z22 \leftarrow \boldsymbol{Zv}_1 + \boldsymbol{Zh}_1 + \boldsymbol{Zv}_2 + \boldsymbol{Zh}_2 \\ Z23 \leftarrow -\boldsymbol{Zv}_2 \\ Z24 \leftarrow -\boldsymbol{Zh}_2 \\ Z25 \leftarrow 0 \\ Z33 \leftarrow \boldsymbol{Zv}_2 + \boldsymbol{Rf}_1 + \boldsymbol{Rf}_2 + \boldsymbol{Zh}_2 \\ Z34 \leftarrow 0 \\ Z35 \leftarrow \boldsymbol{Rf}_2 + \boldsymbol{Zh}_2 \\ Z44 \leftarrow \boldsymbol{Rn}_1 + \boldsymbol{Zh}_2 + \boldsymbol{Zv}_3 + \boldsymbol{Zh}_3 + \boldsymbol{Rn}_3 \\ Z45 \leftarrow -\boldsymbol{Zv}_3 \\ Z55 \leftarrow \boldsymbol{Zv}_3 + \boldsymbol{Zh}_2 + \boldsymbol{Rf}_2 + \boldsymbol{Rf}_3 + \boldsymbol{Zh}_3 \\ \begin{pmatrix} Z11 & Z12 & Z13 & Z14 & Z15 \\ Z12 & Z22 & Z23 & Z24 & Z25 \\ Z13 & Z23 & Z33 & Z34 & Z35 \\ Z14 & Z24 & Z34 & Z44 & Z45 \\ Z15 & Z25 & Z35 & Z45 & Z55 \end{pmatrix} \end{array}\right.$$

$$\boldsymbol{V} := \begin{pmatrix} 1 \\ 0 \\ 0 \\ 0 \\ 0 \end{pmatrix}$$

Figure 7.17 Fourth page of worksheet: defining the functions

The function *Zloop(f)* takes the output of *Zbranch(f)* and uses that data to calculate the loop impedance of every circuit loop of Figure 7.12 and the value of every impedance which forms part of a pair of loops. These parameters are collated into the five-by-five array which constitutes the output of the function. Creating this list is simply a matter of inspecting Figure 7.12. For example, *Z22* is the sum of all the impedances through which I_2 flows, whilst *Z23* is the impedance which shares I_2 and I_3.

Finally, the voltage vector V is defined. This defines the source voltage in each of the five loops of Figure 7.15. In this case, the source is in loop 1 and the amplitude is 1 V.

7.2.7 Main program

Figure 7.18 is a copy of the fifth page of the worksheet.

The first action is to create a vector $\boldsymbol{F}$ which defines all the frequencies used in the main calculation. This is the range from 100 kHz to 20 MHz in 100 kHz steps.

The function $\boldsymbol{Ym}_i$ collects the elements of the five-by five-array created by *Zloop(f)*. Then the built-in function *lsolve*(**Z**,*V*) calculates the value of every current in the circuit of Figure 7.15 at the frequency *f*. The output is the amplitude of the current $\boldsymbol{I}_1$, that is the differential-mode current at the near end.

The function ***Ym2*** repeats the process, this time providing a vector which records the amplitudes of the current in loop 2 over the same range of frequencies, that is the common-mode current at the near end.

Enough data are now available to create a pair of graphs which compare the test results from the setup of Figure 7.9 with the results of the model of Figure 7.15.

The top graph shows the variation of differential-mode current ***Ym1*** with frequency *F* as a red curve. It also records the results calculated in page 1 of the worksheet, ***Yt1*** as a set of circles connected by blue dotted lines. That is, it compares the response of the simulation with that of the hardware.

The bottom graph shows the variation of the common-mode current over the same range of frequencies when the voltage at the differential-mode input is 1 V; ***Ym2*** versus ***F***. This is shown as a red curve. The circles on the bottom graph define the actual common-mode current.

It was necessary for the program to be run several times to allow the values of the circuit components of Figure 7.14 to be varied. Each run allowed the red curves to move into closer correlation with the circles. The graphs on Figure 7.18 are the eventual result.

Most of the input parameters to the computation are dependent variables. Examples are as follows: Lc_1, Lc_2, Cc_1, and Cc_2 are functions of *k*, *Rdm*, and *Fdm*, whilst *Rdm* is defined as 50 Ω and *Fdm* is defined by the peak frequency 8.33 MHz recorded in the test results of Figure 7.11.

This means that the only variables which needed to be adjusted were as follows:

k: The extra proportion of the length assigned to the shield.
Rc1: The resistance of the core conductor.
Gc3: The conductance of the common-mode loop.

$i := 1..200 \qquad F_i := i \cdot 10^5$

$$Ym1_i := \left| \begin{array}{l} f \leftarrow F_i \\ Z \leftarrow Zloop(f) \\ I \leftarrow lsolve(Z, V) \\ |I_1| \end{array} \right. \qquad Ym2_i := \left| \begin{array}{l} f \leftarrow F_i \\ Z \leftarrow Zloop(f) \\ I \leftarrow lsolve(Z, V) \\ |I_4| \end{array} \right.$$

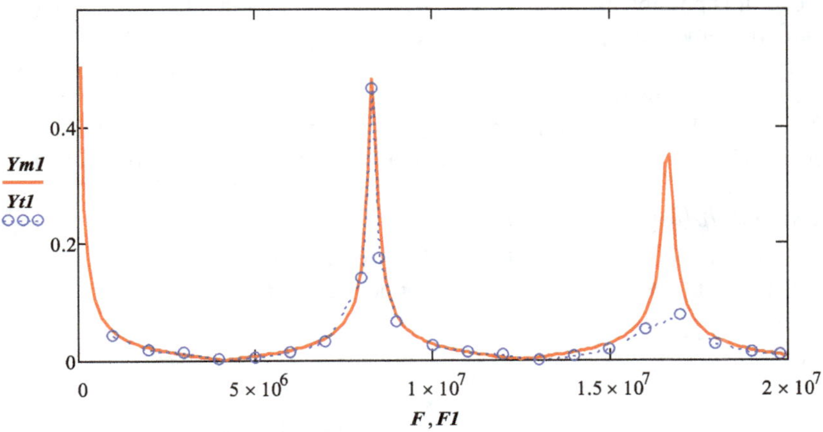

(a)

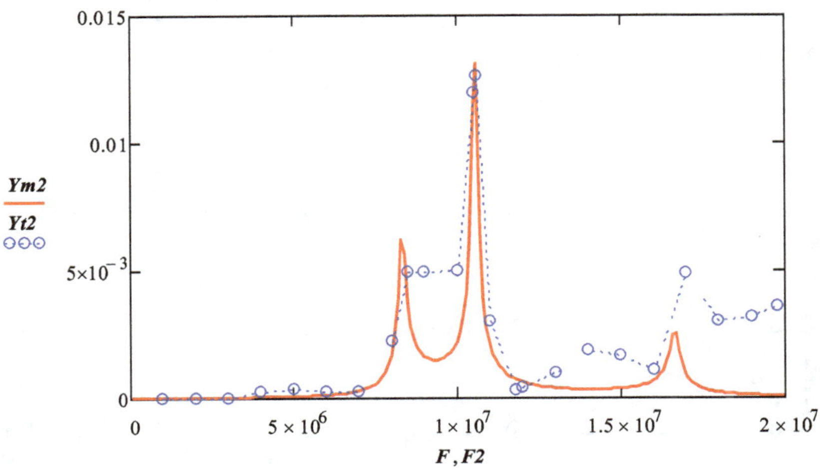

(b)

Blue circles: test results   Red curves: response of the model

Figure 7.18 Fifth page of worksheet: correlating the response of the model with the test results: (a) differential-mode current and (b) common-mode current

7.2.8 Definitive circuit model

Figure 7.19 is a printout of the values assigned to the model of Figure 7.14 which create the simulation recorded on the graphs of Figure 7.18. When these values are used to replace the variables of Figure 7.14, the result is the definitive circuit model of Figure 7.20. This defines the characteristic of the signal link used in the setup of Figure 7.9.

This model does not include the components of the interface circuitry of the equipment at the near and far ends of the signal link. When these values are included, it is possible to simulate the cross coupling between the differential-mode and common-mode loops over all the frequencies at which EMI would be critical.

The final model could simulate the common-mode interference created by any signal carried by the link.

It could also be transformed into a model which can analyse the transient response of the link-under review. Section 6.1 describes the transformation process.

7.2.9 Coupling mechanism

The circuit models of Figures 7.13 and 7.14 indicate that there is no capacitive coupling between the core conductor and the structure, but there is!

$$\frac{\boldsymbol{Lc}}{2} = \begin{pmatrix} 1.486\times 10^{-6} \\ 1.501\times 10^{-8} \\ 4.135\times 10^{-6} \end{pmatrix} \qquad \frac{\boldsymbol{Rdc}}{2} = \begin{pmatrix} 0.1 \\ 0.01 \\ 2.5\times 10^{-3} \end{pmatrix}$$

$$\boldsymbol{Cc} = \begin{pmatrix} 1.188\times 10^{-9} \\ 1.2\times 10^{-11} \\ 2.69\times 10^{-10} \end{pmatrix} \qquad \boldsymbol{Rins} = \begin{pmatrix} 1\times 10^{7} \\ 1\times 10^{7} \\ 6\times 10^{3} \end{pmatrix}$$

Figure 7.19 Sixth page of worksheet: Values assigned to the circuit model of Figure 7.14

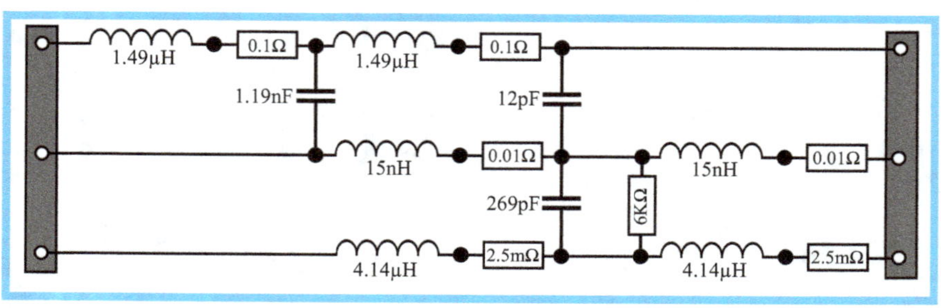

Figure 7.20 Definitive circuit model of assembly under review

Section 3.3 shows that there is coupling between these two conductors. This creates current in the structure; current which flows in the opposite direction. In turn, this creates current in the shield which flows in the same direction as that in the core. The core conductor also creates current in the shield, flowing in the opposite direction. So there are two currents in the shield, of equal amplitude, flowing in the opposite direction. The net current in the shield is zero.

Figure 7.14 shows that the extra length of the strands in the shield links with the ground conductor in exactly the same way as a normal return conductor.

7.2.10 System considerations

Analysis of the definitive circuit model clearly identifies the range of frequencies at which electromagnetic interference could cause problems.

If the link-under-review is carrying power, then an interface such as described in Section 5.4 would minimise conducted emission and radiated emission.

If the link-under-review is carrying a firing signal to an electro-explosive device, then the same filter would minimise conducted susceptibility and radiation susceptibility.

If the link is carrying a digital signal, then fitting a 50 Ω resistor at the receiving end would minimise cross-coupling hazards. There would be no reflections to create resonances.

7.2.11 Conclusion

Frequency response tests of the coupling between the differential-mode signal in a coaxial cable and the common-mode signal in the structure can be used to compile sufficient data to create a definitive circuit model of the assembly. Such a model can be used during the development stage of a project to estimate the electromagnetic compatibility of such a signal link.

Assessment of the model provides a clear insight into the mechanisms involved in cross coupling.

7.3 Relating photons to charges

7.3.1 Summary

A definitive relationship is established between photons and charges. Charges flow axially along the send conductor and radially towards the surface. At the surface, they are stopped by atomic forces. Photons are emitted. The rate of change of the number of departing photons is proportional to the number of charges arriving at the surface. The energy lost to the conductor is measured by the voltage developed along the inductance. The energy stored in the conductor is measured by the number of charges held at the surface. This assessment was obtained by successive refinements of the transmission line model of the differential-mode current waveform in a twin-conductor cable which is open circuit at the far end.

7.3.2 Introduction

A test is described where a step voltage is applied to the near end of a twin-conductor cable which is open circuit at the far end and a photographic record is taken of the differential-mode current. Then, a circuit model is created which replicates the recorded waveform. Analysis of the model provides an insight into the mechanisms involved.

Details are provided of the setup, the test method, and the test result; a waveform with multiple discontinuities. Then a circuit model is created to replicate that waveform. Details are provided of the reasoning used to create that model. A Mathcad worksheet is developed to simulate the recorded waveform. Details are provided of the values assigned to the constants and all the variables are defined. The purpose of each function (subroutine) is described and a block diagram provided of the computation process.

The end result is a waveform which correlates quite closely with that created by the test equipment. Assessment of the results provides a definitive relationship between the movement of charges along the conductor and the propagation of photons to and fro between the conductors.

Since the terminals at the far end are open circuit, the eventual state is a constant voltage between the conductors due to charges trapped on the surfaces. But this is sustained by photons which continue to propagate to and fro between the conductors.

7.3.3 Setup

In the setup illustrated by Figure 7.21, the signal generator was set to provide a square wave voltage to a potential divider which allows the resistance of the source

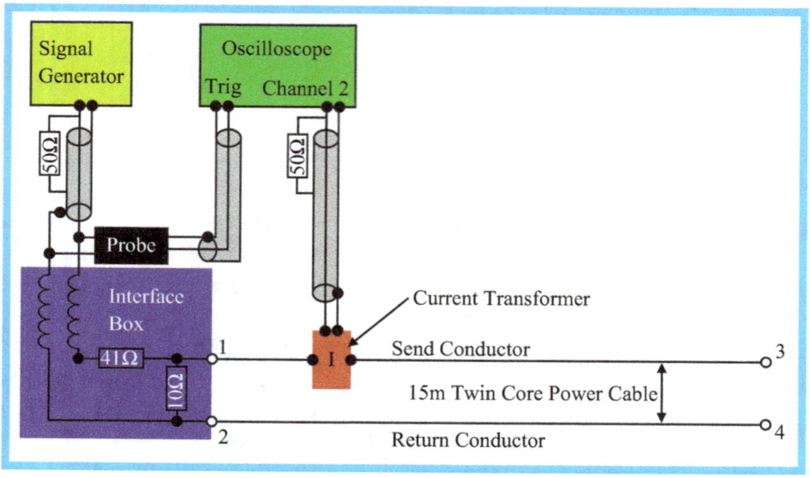

Figure 7.21 Measuring the response of twin-conductor cable

applied at the near end terminals of the cable to be approximately 9 Ω. This means that every transient step will cause a step current to propagate along the cable, be reflected at the open-circuit terminals at the far end of the line, and be reflected again by the low resistance at the near end. Textbook theory indicates that the waveform of the current in the line would be a square wave which gradually reduces in amplitude.

The differential-mode current was measured by an oscilloscope connected to a current transformer. Details of this transformer are provided in Section 2.5.

A common-mode choke is used to minimise the common-mode current flowing between the cable and the signal generator.

7.3.4 Test method

By selecting the operating frequency of the signal generator to be much lower than the frequency of the line transients, the response of the line to a single voltage step can be measured. Enough time between the rising and falling edges of the signal generator output is provided to allow the transients in the cable to decay to an imperceptible level.

Once the frequency of the square wave output of the signal generator had been selected, a voltage probe was used to measure the peak-to-peak amplitude of the voltage applied to the input terminals of the cable.

Then the waveform of the differential waveform was displayed on the oscilloscope. The use of an input trigger was used to lock the scan time to the source waveform.

Photographs were taken of the current waveform, and one of these is shown by Figure 7.22.

Vertical scale = 10 mV/div
Horizontal scale = 100 ns/div

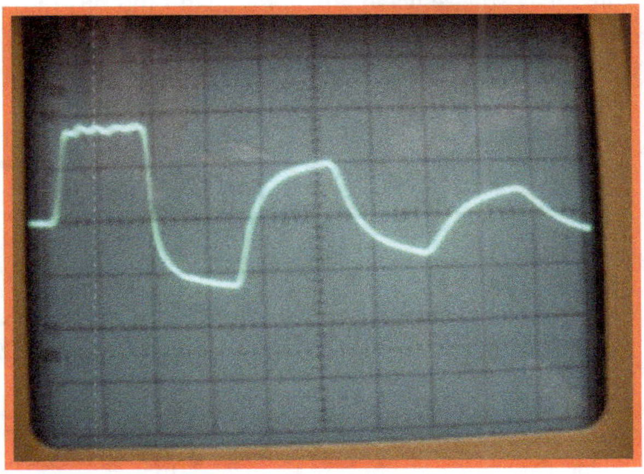

Figure 7.22 Waveform of the differential-mode current

7.3.5 Basic model

The model illustrated by Figure 7.23 allows the basic parameters to be related to currents and voltages at the line terminations.

Vgen is the source voltage.

Rg is the source resistance at the near end.

RL is the load resistance at the far end.

T is the time taken for a signal to propagate from the near end to the far end.

Ro is the characteristic resistance of the differential-mode loop.

Ina is the loop current flowing at the near end.

Vna is the voltage induced at the near end by the transients in the line.

Ini and *Inr* are the incident and reflected currents at the near end.

Ifa is the loop current flowing at the far end.

Vfa is the voltage induced at the far end by the transients in the line.

Ifi and *Ifr* are the incident and reflected currents at the far end.

The relationships between these parameters are derived in Section 5.2. A simulation of this model would result in a square wave of gradually decreasing amplitude.

The waveform of Figure 7.22 identifies features of the signal that are not accounted for in these basic relationships. But the nature of these features provides significant clues as to what is happening as the front edge of the step as it propagates forward and backward along the cable.

During the first transit forward and the first transit back the current delivered to the line is constant. The ripples are probably due to discontinuities in the layout of the cable. The finite time of the leading edge is due to the reaction time of the current transformer and the frequency response of the oscilloscope.

Since the first trailing edge of the waveform resembles that of an R–C network which goes open circuit, such a circuit was incorporated into the model. The net result was a square wave which gradually morphs into a sinusoidal waveform. But it did not replicate the discontinuities which occur during each half cycle of Figure 7.23.

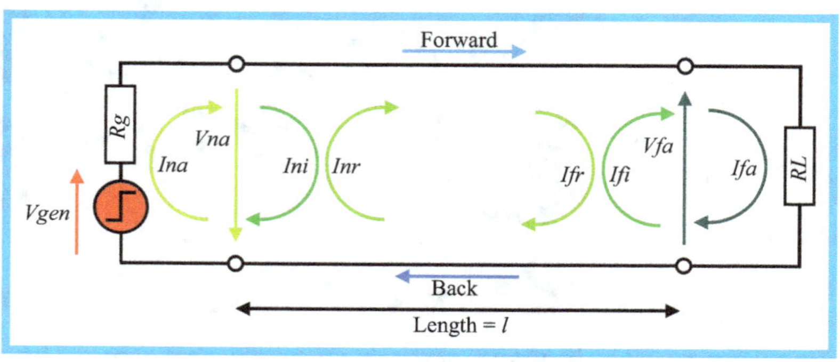

Figure 7.23 Basic model of transmission line: defining the parameters

7.3.6 Antenna-mode current

In the setup of Figure 7.21 it is clear that the current monitored by the current transformer is the differential-mode current. Any current which flows along both conductors in parallel would not be detected. That is, antenna-mode current which flows forward along the cable and out into the environment would have no effect on the waveform displayed by Figure 7.22.

The differential-mode current is that which flows along the send conductor and back along the return conductor. If some of the field emanating from the send conductor during forward propagation creates a current which flows forward along the return conductor, then this current would not be detected either.

It was reasoned that the exponential decay could also be due to inductive effects. The current at the far end rises exponentially with time. The missing portion of the original step function could still be due to antenna-mode emission. This reasoning led to the replacement of the R-C circuit with the L–R circuit illustrated by Figure 7.24(a). This model provides the loop equation:

$$Ro \cdot Inr = Ro \cdot Int + Lrad \cdot \frac{dInt}{dt} \tag{7.20}$$

where *Int* is the amplitude of the current arriving at the far end of the send conductor and *Lrad* is the radiation inductance. If *Int* is the current transmitted to the far end, then the current lost to the differential-mode loop during the transit from near end to far end must be

$$Ins = Inr - Int \tag{7.21}$$

The current lost to the differential-mode loop is stored in the antenna-mode loop and this can also be calculated using the model of Figure 7.24(b). The loop equation here is

$$Ro \cdot Inr = Ro \cdot Ins + \frac{Qns}{Crad} \tag{7.22}$$

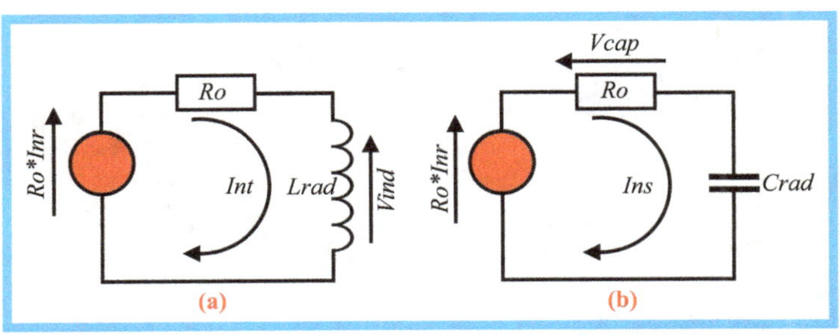

Figure 7.24 Modelling antenna-mode current (a) calculating transmitted current (b) calculating stored current

where *Crad* is the radiation capacitance and

$$Qns = Ins \cdot dt \tag{7.23}$$

It is possible to simulate the exponential fall in the antenna-mode current at the far end using either Figure 7.24(a) or (b). The computer program used to simulate this effect is based on the model of Figure 7.24(a); that is, on (7.20) and (7.21). The use of an inductor to simulate the rise time at the far end allows it to be reasoned that some of the missing current could be stored in the return conductor.

7.3.7 Antenna-mode reflection

Photons departing from the surface of the send conductor and arriving at the surface of the return conductor will induce charges which can flow in both directions. Some will be reflected back to the send conductor and will enhance the forward flow in the send conductor.

Charges which flow forward along the return conductor are propagating in the same direction as those in the send conductor. This creates antenna-mode current, and photons created by this current propagate out into the environment.

When the leading edge of the step arrives at the far end of the cable, the return conductor is carrying approximately half of the antenna-mode current. Conversely, the current arriving at the end of the send conductor is zero. Current arriving at an open-circuit terminal immediately reverses in direction. So now it is the return conductor which is delivering antenna-mode current to the send conductor.

The differential-mode current arriving back at the near end consists of two components; that which has been reflected back from the far end, and the proportion *k* of the antenna-mode current delivered back from the return conductor.

This mechanism can be simulated by the worksheet functions defined by the block diagram of Figure 7.25.

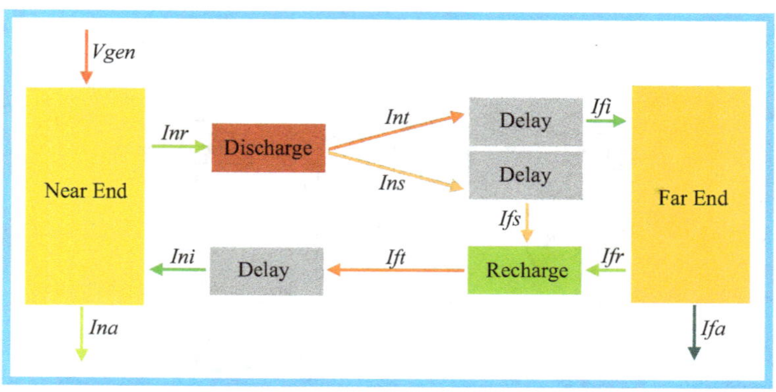

Figure 7.25 Signal processing functions for worksheet

7.3.8 Worksheet constants

A copy of the first page of the worksheet is provided by Figure 7.26. The values of the input parameters need some explanation.

The amplitude of the voltage step between terminals 1 and 2 of Figure 7.21, as monitored by the voltage probe, was

$$Vin = 0.84\ \text{V} \tag{7.24}$$

$$Vg := 0.91 \qquad Ro := 112 \qquad Rg := 9 \qquad RL := 100\cdot 10^{6} \qquad RT := 2.27$$

$$T := 82.5\cdot 10^{-9} \qquad n := 100 \qquad dt := \frac{T}{n} \qquad Trad := 32\cdot 10^{-9} \quad k := 0.5$$

$$Lrad := Trad\cdot Ro = 3.584\times 10^{-6}$$

$$T1 := 20\cdot 10^{-9} \qquad N1 := floor\left(\frac{T1}{dt}\right) = 24$$

$$T2 := 1000\cdot 10^{-9} \qquad N2 := floor\left(\frac{T2}{dt}\right) = 1.212\times 10^{3}$$

$$point(i) := \left|\begin{array}{l} p1 \leftarrow mod(i,n) \\ p1 \leftarrow n \ \ if\ \ p1 = 0 \\ p2 \leftarrow n - p1 + 1 \\ (p1 \ \ p2) \end{array}\right.$$

$$delay(\mathbf{D}, Iin, p) := \left|\begin{array}{l} Iout \leftarrow \mathbf{D}_p \\ \mathbf{D}_p \leftarrow Iin \\ (\mathbf{D} \ \ Iout) \end{array}\right.$$

$$near(Ini, Vg) := \left|\begin{array}{l} Ina \leftarrow \dfrac{2\cdot Ro\cdot Ini + Vg}{Ro + Rg} \\ Inr \leftarrow Ina - Ini \\ (Inr \ \ Ina) \end{array}\right.$$

$$charge(Inr, Int) := \left|\begin{array}{l} dInt \leftarrow \dfrac{dt}{Lrad}\cdot Ro\cdot (Inr - Int) \\ Int \leftarrow Int + dInt \\ Ins \leftarrow Inr - Int \\ (Int \ \ Ins) \end{array}\right.$$

$$far(Ifi) := \left|\begin{array}{l} Ifa \leftarrow \dfrac{2\cdot Ro\cdot Ifi}{Ro + RL} \\ Ifr \leftarrow Ifa - Ifi \\ (Ifr \ \ Ifa) \end{array}\right.$$

$$recharge(Ifs, Ifr) := \left|\begin{array}{l} Ift \leftarrow Ifr - k\cdot Ifs \\ Ift \end{array}\right.$$

Figure 7.26 First page of Mathcad worksheet. Input variables and processor functions.

Examination of the amplitude of the first step in the waveform of Figure 7.22 gave a value for *Vch2* of 17 mV. The relationship between this voltage and the amplitude of the current monitored by the current transformer is *RT*, the transfer resistance. The value of this parameter has been measured to be 2.27 Ω. Details of the calibration of this transformer can be found in Section 2.5.

Hence the value of the differential-mode current flowing in the cable was

$$Idiff = \frac{Vch2}{RT} = 7.49 \text{ mA} \tag{7.25}$$

So, the characteristic resistance *Ro* of the cable was

$$Ro = \frac{Vin}{Idiff} = 112 \ \Omega \tag{7.26}$$

The value of the source resistance *Rg* was determined by examination of the circuitry of the Interface Box of Figure 7.21.

$$Rg = \frac{10 \cdot (41 + 50)}{10 + 41 + 50} = 9 \ \Omega \tag{7.27}$$

The voltage *Vg* necessary to deliver *Vin* to the terminals at the near end must be

$$Vg = Vin \cdot \frac{Rg + Ro}{Ro} = 0.91 \text{ V} \tag{7.28}$$

The terminals at the far end of the cable are open circuit. Setting the value of the load resistance *RL* at 100 MΩ allows this interface to be simulated.

Figure 7.22 can be used to define the time *T* taken for the step to propagate along the cable. The time between the leading and trailing edges of the first pulse is that taken for two traverses; forward and back. Inspection of Figure 7.22 gives

$$T = \frac{185 - 20}{2} = 82.5 \text{ ns}$$

In this simulation, the number of segments *n* into which the line is divided is 100. So the time *dt* for the edge to propagate along one segment is

$$dt = \frac{T}{n} = 825 \text{ ps} \tag{7.29}$$

An approximate value for the time constant *Trad* of the exponential decay of the trailing edge can be deduced from Figure 7.22. Assigning an approximate value to this parameter allows the inductance *Lrad* of the antenna-mode loop to be calculated.

$$Lrad = Ro \cdot Trad \tag{7.30}$$

Successive iterations of the program allow the value of the time constant to be refined (*Trad* is the only select-on-test parameter).

T1 is the time delay between the start of the scan and the first leading edge. *T2* is the duration of the scan.

7.3.9 Worksheet functions

In previous simulations, a shift register was used to simulate the propagation of charges from one segment of the line to the next. So the entire contents of the register changed at each step. In this simulation, the only change is to the value of a single segment. Pointers are used to identify that segment. If *i* is the number of the iteration, the number *n* of segments in the register is 10, then the relationship between iteration and pointer is defined by Table 7.1.

The pointer *p1* identifies the segment of the ***Forward*** register which is to be updated and *p2* identifies the associated segment of the ***Back*** register.

The function *point* (*i*) calculates the values of *p1* and *p2* for any value of *i*. The Mathcad function *mod* (*i, n*) returns the remainder on dividing *n* by *i*.

The function *delay* (***D***, *In, p*) records the value of the element *p* of the register ***D*** as *Iout.* Then it updates that element with the value of *Iin.* The output of the function is the value *Iout.* This value is that which was entered during the previous scan. Since a complete scan represents a time delay of *T* seconds, the input is the value of *Iin* at one end of the line, and the output *Iout* is the value at the other end.

The functions *near* (*Ini, Vg*) and *far* (*Ifi*) have been explained in previous sections.

The function *charge*(*Inr*,*Int*) uses (7.20) and (7.21) to calculate the amplitudes of the current *Int* transmitted to the far end and the current *Ins* stored in the antenna-mode loop.

During the first half cycle, the current *Ins* simulates the total current radiated away from the send conductor. Half radiates away from the outward-facing surface in the form of photons departing at the velocity of light. Photons departing from the surface facing the return conductor create antenna-mode current in that conductor.

The antenna-mode current arriving at the far end of the return conductor must be half the value of *Ifs*. The antenna-mode current arriving at the far end of the send conductor must be zero, since only half can be collected by the return conductor. Any current arriving at an open-circuit terminal is immediately reflected back. So this reflected current reappears as differential-mode current.

Hence, the total amplitude of the differential-mode current reflected back from the far end is

$$Ift = Ifr - k \cdot Ifs \tag{7.31}$$

where $k = 0.5$.

Table 7.1 Relating the segments of the registers to the number of time steps

i	1	2	3	4	5	6	7	8	9	10
i	11	12	13	14	15	16	17	18	19	20
i	21	22	23	24	25	26	27	27	29	30
i	31	32	33	...	...	...				
p1	1	2	3	4	5	6	7	8	9	10
p2	10	9	8	7	6	5	4	3	2	1

7.3.10 Main program

The main program is shown on Figure 7.27.

The control variable i is set to step between unity and $N2$; the number of steps needed to simulate a single sweep of the oscilloscope.

```
i := 1..N2                      t_i := i·dt

Out := | Forward_n ← 0
       | Back_n ← 0
       | Antenna_n ← 0
       | for i ∈ 1..N2
       |    | (p1  p2) ← point(i)
       |    | Vgen ← Vg  if  i > N1
       |    | (Inr  Ina) ← near(Ini, Vgen)
       |    | (Int  Ins) ← charge(Inr, Int)
       |    | (Forward  Ifi) ← delay(Forward, Int, p1)
       |    | (Antenna  Ifs) ← delay(Antenna, Ins, p1)
       |    | (Ifr  Ifa) ← far(Ifi)
       |    | Ift ← recharge(Ifs, Ifr)
       |    | (Back  Ini) ← delay(Back, Ift, p2)
       |    | Out_i ← Ina·RT
       | Out
```

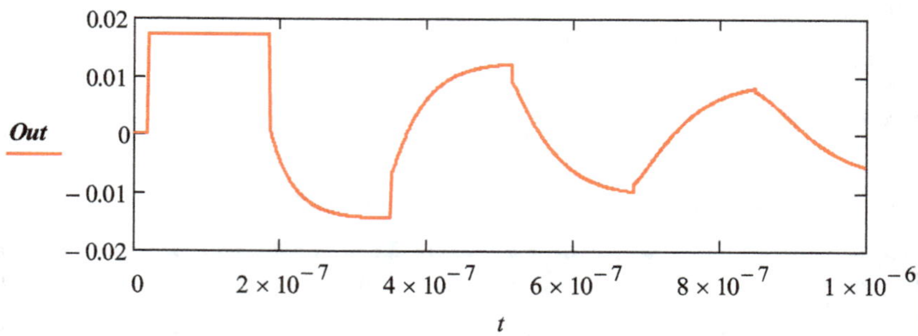

Figure 7.27 Second page of worksheet. Definition of time steps, and main program.

The number of elements in each register is defined as *n*, the number of segments into which the cable is divided. The register ***Antenna*** holds the values of the parameter *Ins*, the current 'stored' in the antenna-mode loop.

The function *point* (*p1,p2*) defines the value of each pointer at each instant. The amplitude of the voltage source *Vgen* is held at zero until the time *T1* corresponds to that of the leading edge of the recorded waveform. Then the sequence of operations follows that of the block diagram of Figure 7.25.

The output vector ***Out*** can collect the value of any parameter in Figure 7.25 at every step of the program. In the worksheet of Figure 7.27, this is the value corresponding to the amplitude of the waveform displayed on the scope: the current at the near end *Ina*, as monitored by the current transformer, multiplied by the transfer resistance *RT*.

7.3.11 Assessment

The waveform of Figure 7.27 now bears a close resemblance to that of the configuration-under-review. Since the frequency response of the oscilloscope is about 20 MHz, the spot on the screen cannot move as fast as the rate of change of the current being monitored. This is evident in the finite rise time on the leading edge and the finite fall time of the first trailing edge of Figure 7.22.

The significant feature of the simulated waveform is that it replicates the kinks in the actual response. These discontinuities are visible at every peak of Figure 7.27.

7.3.12 Alternative method

It is possible to create exactly the same simulation by replacing the function *charge*(*Inr*, *Int*) on the first page of the worksheet with that shown on Figure 7.28, where

$$Crad = \frac{Trad}{Ro} \tag{7.32}$$

This means that the circuit models of Figure 7.24(a) and (b) are equally valid.

$$charge(Inr, Qns) := \left| \begin{array}{l} Ins \leftarrow Inr - \dfrac{Qns}{Ro \cdot Crad} \\ Int \leftarrow Inr - Ins \\ Qns \leftarrow Qns + Ins \cdot dt \\ (Int \quad Ins \quad Qns) \end{array} \right.$$

Figure 7.28 Alternative method of calculating components of antenna-mode current

7.3.13 Charges and photons

The classical analyses of the transient responses of Figure 7.24(a) and (b) are

$$Int = \frac{Vnr}{Ro} \cdot \left(1 - e^{-\theta}\right) \tag{7.33}$$

And

$$Ins = \frac{Vnr}{Ro} \cdot e^{-\theta} \tag{7.34}$$

where the time constant *Trad* is

$$Trad = \frac{Lrad}{Ro} = Crad \cdot Ro \tag{7.35}$$

And the angle θ is the ratio

$$\theta = \frac{t}{Trad} \tag{7.36}$$

Differentiating (7.33)

$$\frac{dInt}{dt} = \frac{Vnr}{Ro} \cdot e^{-\theta} = Ins = \frac{dQns}{dt} \tag{7.37}$$

This relationship indicates that *Inr* and *Ins* are flowing in opposite direction and that *Ins* is equal to the rate of change of the charge *Qns*. It correlates with the assumption that the rate at which photons are emitted from the surface is proportional to the sudden deceleration of charges as they arrive at the surface of the conductor.

It can also be inferred that the rate at which photons arrive at the surface of a conductor is equal to the amplitude of the current delivered to that surface.

7.3.14 Magnetic field

The work done in ejecting photons from the surfaces of the conductors is the voltage *Vind* depicted in Figure 7.24(a). The work done in driving charges to the surfaces of the conductors is *Vcap* as depicted in Figure 7.24(b).

This means that the energy *Vcap* needed to hold charges at the surfaces of the conductors is the same as the energy *Vind* involved in transmitting photons into the environment; that is, the same as the energy of the magnetic field.

7.3.15 Electric field

The differential-mode voltage at the output terminals *Vout* can be calculated using the relationship

$$Vout = RL \cdot Ifa \tag{7.38}$$

The waveform settles down into 'steady-state' conditions after about 5 μs, as shown by Figure 7.29. Even so, the photons continue to propagate to and fro between the

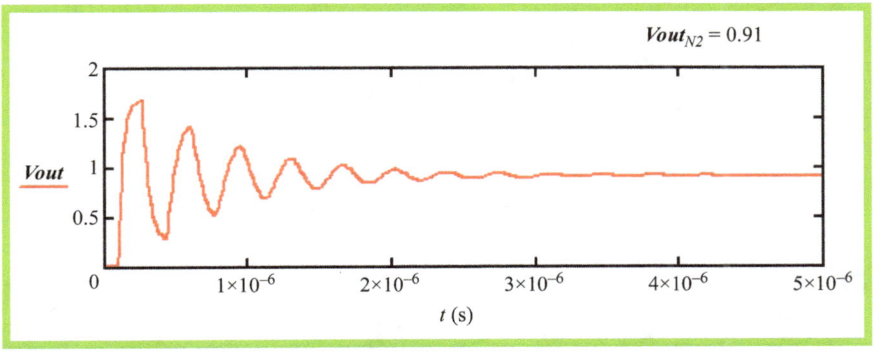

Figure 7.29 Waveform of the differential-mode voltage at the far end terminals

conductors. This is sustained by charges on the surfaces of both conductors. That is, the energy is now trapped at each surface by atomic forces. The atoms must be vibrating at an extremely high frequency. The actual state of the system is anything but stationary. But the number of photons in motion is orders of magnitude less than that associated with the magnetic field.

It is obvious that the voltage at the far end is equal to the open-circuit voltage *Vg*. So there is no voltage drop in the source resistor *Rg*. This can only happen if the differential-mode current is infinitesimal. If the amplitude of the current is minute, then the number of photons it ejects must also be tiny.

7.3.16 Conclusion

Cross-coupling is caused by imbalance in the antenna-mode flow of current. This flows along both conductors and emits photons which depart into the environment at the velocity of light. The amplitude of current in one conductor is greater than that in the other. The conductor with the highest current acts as a transmitter, the other as a receiver. The role of each conductor changes after each reflection of the leading edge of the waveform at the open-circuit termination. This imbalance is rapidly damped down by energy absorbed by the resistance at the near end of the line.

Invoking classical theory of transients in inductors and capacitors, a definitive relationship is established between photons and charges. The rate at which photons are emitted is equal to the energy lost in decelerating charges which arrive radially at the surface of a circular-section conductor. The radial flow of charges into a circular section conductor is proportional to the rate at which photons arrive at the surface from the environment.

Appendix A

Inductance and capacitance primitives

A1 Analysis of the integration process

A1.1 First differential and integral

As with the approach used in textbooks on engineering mathematics, the technique used here is to differentiate selected expressions and use the result of each differentiation to define an integral. That is,

if

$$\frac{dy}{dx} = u,$$

then

$$\int u\,dx = y$$

Let

$$y = a + \sqrt{a^2 + x^2}$$

And let

$$u = a^2 + x^2,$$

giving

$$\frac{du}{dx} = 2 \cdot x$$

Then

$$y = a + \sqrt{u}$$

And

$$\begin{aligned} \frac{dy}{du} &= \frac{1}{2 \cdot \sqrt{u}} \\ \frac{dy}{dx} &= \frac{dy}{du} \cdot \frac{du}{dx} = \frac{1}{2 \cdot \sqrt{u}} \cdot 2 \cdot x = \frac{x}{\sqrt{a^2 + x^2}} \end{aligned} \tag{A.1}$$

Giving

$$\frac{d}{dx}\left(a + \sqrt{a^2 + x^2}\right) = \frac{x}{\sqrt{a^2 + x^2}} \tag{A.2}$$

And

$$\int \frac{x}{\sqrt{a^2 + x^2}} dx = a + \sqrt{a^2 + x^2} \tag{A.3}$$

A1.2 Second differential

Let

$$y = \frac{a + \sqrt{a^2 + x^2}}{x}. \tag{A.4}$$

The differential of this function of x can be obtained by letting $y = u \cdot v$

Where

$$u = a + \sqrt{a^2 + x^2}$$

And

$$v = \frac{1}{x}$$

Since

$$\frac{d}{dx}(u \cdot v) = v \cdot \frac{du}{dx} + u \cdot \frac{dv}{dx}$$

And from (A.2)

$$\frac{du}{dx} = \frac{d}{dx}\left(a + \sqrt{a^2 + x^2}\right) = \frac{x}{\sqrt{a^2 + x^2}}$$

Then:

$$\begin{aligned}
\frac{dy}{dx} &= \frac{1}{x} \cdot \frac{x}{\sqrt{a^2 + x^2}} - \frac{1}{x^2} \cdot (a + \sqrt{a^2 + x^2}) \\
&= \frac{1}{\sqrt{a^2 + x^2}} - \frac{a + \sqrt{a^2 + x^2}}{x^2} \\
&= \frac{x^2 - a \cdot \sqrt{a^2 + x^2} - (a^2 + x^2)}{x^2 \cdot \sqrt{a^2 + x^2}} \\
&= \frac{-a \cdot \sqrt{a^2 + x^2} - a^2}{x^2 \cdot \sqrt{a^2 + x^2}}
\end{aligned}$$

Giving

$$\frac{d}{dx}\left(\frac{a+\sqrt{a^2+x^2}}{x}\right) = -\frac{a\cdot\left(a+\sqrt{a^2+x^2}\right)}{x^2\cdot\sqrt{a^2+x^2}} \tag{A.5}$$

A1.3 Second integral

From (A.1)

$$\frac{d}{dx}[\ln(y)] = \frac{f^{/}(x)}{f(x)} \tag{A.6}$$

Thus

$$\int\frac{f^{/}(x)}{f(x)}dx = \ln(y) \tag{A.7}$$

If

$$y = f(x) = \frac{a+\sqrt{a^2+x^2}}{x},$$

then, from (A.5) and (A.4)

$$\frac{f^{/}(x)}{f(x)} = -\frac{a\cdot\left(a+\sqrt{a^2+x^2}\right)}{x^2\cdot\sqrt{a^2+x^2}}\cdot\frac{x}{a+\sqrt{a^2+x^2}} = -\frac{a}{x\cdot\sqrt{a^2+x^2}}$$

Invoking (A.7):

$$\int\frac{a}{x\cdot\sqrt{a^2+x^2}}dx = -\ln\left(\frac{a+\sqrt{a^2+x^2}}{x}\right) \tag{A.8}$$

A1.4 Combining first and second integrals

Let

$$y = \int\frac{\sqrt{a^2+x^2}-x}{x}dx \tag{A.9}$$

That is:

$$\begin{aligned} y &= \int\left(\frac{\sqrt{a^2+x^2}}{x}-1\right)dx \\ &= \int\left(\frac{a^2+x^2}{x\cdot\sqrt{a^2+x^2}}-1\right)dx \\ &= \int\left(\frac{a^2}{x\cdot\sqrt{a^2+x^2}}+\frac{x}{\sqrt{a^2+x^2}}-1\right)dx \end{aligned}$$

Invoking (A.8) and (A.3)

$$y = -a \cdot \ln\left(\frac{a + \sqrt{a^2 + x^2}}{x}\right) + \left(a + \sqrt{a^2 + x^2}\right) - x$$

Rearranging:

$$\int \frac{\sqrt{a^2 + x^2} - x}{x} dx = a \cdot \left[\frac{\sqrt{a^2 + x^2} - x}{a} + 1 - \ln\left(\frac{a + \sqrt{a^2 + x^2}}{x}\right)\right] \tag{A.10}$$

A1.5 Third integral

Let

$$y = \sqrt{(a - x)^2 + b^2} \tag{A.11}$$

And let

$$\mu = (a - x)^2 + b^2.$$

Then

$$y = \sqrt{u}$$

And

$$\frac{dy}{du} = \frac{1}{2 \cdot \sqrt{u}}$$

Since

$$u = a^2 - 2 \cdot a \cdot x + x^2 + b^2, \frac{du}{dx} = -2 \cdot a + 2 \cdot x = 2 \cdot (x - a)$$

Since

$$\frac{dy}{dx} = \frac{dy}{du} \cdot \frac{du}{dx},$$

Then

$$\frac{dy}{dx} = \frac{1}{2 \cdot \sqrt{u}} \cdot 2 \cdot (x - a) = \frac{x - a}{\sqrt{(a - x)^2 + b^2}} \tag{A.12}$$

Since

$$\int \frac{dy}{dx} dx = y,$$

then

$$\int \frac{a-x}{\sqrt{(a-x)^2+b^2}}dx = -\sqrt{(a-x)^2+b^2} \tag{A.13}$$

A1.6 Definite integrals

Substituting l for a and r for b in (A.13)

$$\int \frac{l-x}{\sqrt{(l-x)^2+r^2}}dx = -\sqrt{(l-x)^2+r^2} \tag{A.14}$$

Finding the definite integral of (A.14) for the range $x = 0$ to l

$$\int_0^l \frac{l-x}{\sqrt{(l-x)^2+r^2}}dx = -\left[\sqrt{(l-l)^2+r^2}\right] + \left[\sqrt{l^2+r^2}\right] = \sqrt{l^2+r^2} - r \tag{A.15}$$

Substituting r for a in (A.3)

$$\int \frac{x}{\sqrt{r^2+x^2}}dx = r + \sqrt{r^2+x^2} \tag{A.16}$$

Finding the definite integral of (A.16) for the range $x = 0$ to l

$$\int_0^l \frac{x}{\sqrt{r^2+x^2}}dx = \left[r+\sqrt{r^2+l^2}\right] - \left[r+\sqrt{r^2}\right] = \sqrt{r^2+l^2} - r \tag{A.17}$$

Adding (A.15) and (A.17)

$$\int_0^l \left(\frac{l-x}{\sqrt{(l-x)^2+r^2}} + \frac{x}{\sqrt{x^2+r^2}}\right)dx = 2\cdot\left(\sqrt{r^2+l^2}-r\right)$$

Hence

$$\frac{1}{2}\cdot\int_0^l \frac{1}{r}\cdot\left(\frac{l-x}{\sqrt{(l-x)^2+r^2}} + \frac{x}{\sqrt{x^2+r^2}}\right)dx = \left(\frac{\sqrt{l^2+r^2}-r}{r}\right) \tag{A.18}$$

And

$$\frac{1}{2}\cdot\int_{r_{1,1}}^{\infty}\int_0^l \frac{1}{r}\cdot\left(\frac{l-x}{\sqrt{(l-x)^2+r^2}} + \frac{x}{\sqrt{x^2+r^2}}\right)dx\,dr = \int_{r_{1,1}}^{\infty}\left(\frac{\sqrt{l^2+r^2}-r}{r}\right)dr \tag{A.19}$$

Substituting l for a and r for x in (A.10)

$$\int \frac{\sqrt{l^2+r^2}-r}{r}dr = l\cdot\left[\frac{\sqrt{l^2+r^2}-r}{l}+1-\ln\left(\frac{l+\sqrt{l^2+r^2}}{r}\right)\right] \tag{A.20}$$

Finding the definite integral of (A.20) for the range $r = r_{1,1}$ to ∞

$$\int_{r_{1,1}}^{\infty} \frac{\sqrt{l^2+r^2}-r}{r}dr = l\cdot\left[\frac{r-r}{r}+1-\ln\frac{r}{r}\right] - l\cdot\left[\frac{\sqrt{l^2+r_{1,1}{}^2}-r_{1,1}}{l}+1-\ln\left(\frac{l+\sqrt{l^2+r_{1,1}{}^2}}{r_{1,1}}\right)\right]$$

$$\int_{r_{1,1}}^{\infty} \frac{\sqrt{l^2+r^2}-r}{r}dr = l\cdot\left[\ln\left(\frac{l+\sqrt{l^2+r_{1,1}{}^2}}{r_{1,1}}\right)+\frac{r_{1,1}-\sqrt{l^2+r_{1,1}{}^2}}{l}\right] \tag{A.21}$$

Substituting $l/2$ for a and r for x in (A.8)

$$\int \frac{l/2}{r\cdot\sqrt{(l/2)^2+r^2}}dr = -\ln\left(\frac{l/2+\sqrt{(l/2)^2+r^2}}{r}\right) \tag{A.22}$$

Finding the definite integral of (A.22) for the range $r = r_{1,1}$ to ∞

$$\int_{r_{1,1}}^{\infty} \frac{l/2}{r\cdot\sqrt{(l/2)^2+r^2}}dr = -\ln\left(\frac{r}{r}\right)+\ln\left(\frac{l/2+\sqrt{(l/2)^2+r_{1,1}{}^2}}{r_{1,1}}\right) \tag{A.23}$$

A2 Primitive inductance

Figure A.1 illustrates a short section of conductor of radius $r_{1,1}$ with a constant current Ip flowing in the z-direction.

The magnetic field strength H at point $P(r,z)$ is

$$H = \frac{Ip}{4\cdot\pi\cdot r}\cdot[\sin(\alpha_1)+\sin(\alpha_2)] \tag{A.24}$$

Converting this equation to one which uses the co-ordinates z and r gives

$$H = \frac{Ip}{4\cdot\pi\cdot r}\cdot\left[\frac{\sqrt{l^2+r^2}-r}{l}-\ln\left(\frac{l+\sqrt{l^2+r^2}}{r}\right)\right] \tag{A.25}$$

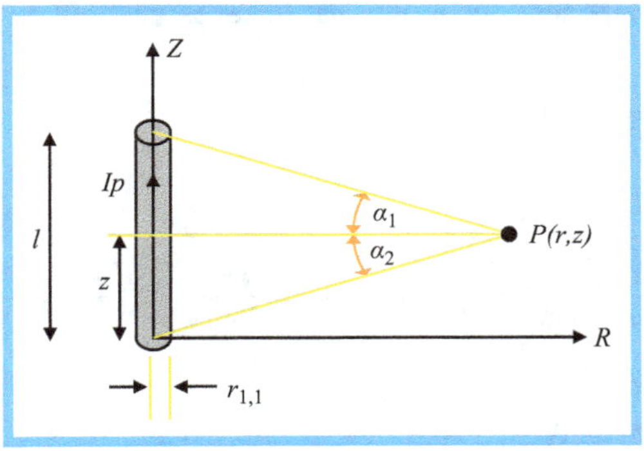

Figure A.1 Magnetic field at a point

The flux density B at this point is

$$B = \mu \cdot H \tag{A.26}$$

Where μ is the permeability of the medium. The total flux which passes through a rectangular strip extending from the surface of the conductor to a very large distance is

$$\varphi = \int B\, ds \tag{A.27}$$

where ds is the surface area of the rectangular strip. That is

$$\varphi = \frac{\mu \cdot Ip}{4 \cdot \pi} \cdot \int_{r_{1,1}}^{\infty} \int_{0}^{l} \frac{1}{r} \cdot \left(\frac{l-z}{\sqrt{(l-z)^2 + r^2}} + \frac{z}{\sqrt{z^2 + r^2}} \right) dz\, dr \tag{A.28}$$

Invoking (A.19)

$$\varphi = \frac{\mu \cdot Ip}{2 \cdot \pi} \cdot \int_{r_{1,1}}^{\infty} \cdot \left(\frac{\sqrt{l^2 - r^2} - r}{r} \right) dr \tag{A.29}$$

Invoking (A.20)

$$\varphi = \frac{\mu \cdot Ip \cdot l}{2 \cdot \pi} \cdot \left[\ln\left(\frac{l + \sqrt{l^2 + {r_{1,1}}^2}}{r_{1,1}} \right) + \frac{r_{1,1} - \sqrt{l^2 + {r_{1,1}}^2}}{l} \right] \tag{A.30}$$

Inductance L is defined as the ratio of the magnetic flux to the current creating that flux:

$$L = \frac{\mu \cdot l}{2 \cdot \pi} \cdot \left[\ln\left(\frac{l + \sqrt{l^2 + {r_{1,1}}^2}}{r_{1,1}} \right) + \frac{r_{1,1} - \sqrt{l^2 + {r_{1,1}}^2}}{l} \right] \tag{A.31}$$

Adding the component due to internal linkages:

$$Ltotal = \frac{\mu \cdot l}{2 \cdot \pi} \cdot \left[\ln\left(\frac{l + \sqrt{l^2 + r_{1,1}{}^2}}{r_{1,1}} \right) + \frac{r_{1,1} - \sqrt{l^2 + r_{1,1}{}^2}}{l} + \frac{1}{4} \right] \tag{A.32}$$

If $l >> r$, then

$$Ltotal = \frac{\mu \cdot l}{2 \cdot \pi} \cdot \left[\ln\left(\frac{2 \cdot l}{r_{1,1}} \right) - 1 + \frac{1}{4} \right] = \frac{\mu \cdot l}{2 \cdot \pi} \cdot \left[\ln\left(\frac{l}{r_{1,1}} \right) + \ln(2) - 1 + \frac{1}{4} \right]$$

$$= \frac{\mu \cdot l}{2 \cdot \pi} \cdot \ln\left(\frac{l}{r_{1,1}} - 0.057 \right)$$

To a first approximation

$$Ltotal \approx \frac{\mu \cdot l}{2 \cdot \pi} \cdot \ln\left(\frac{l}{r_{1,1}} \right) \tag{A.33}$$

This is the inductance of a length l of circular section conductor. It is the simplest possible parameter which has the dimensions of inductance. Since it is one of the fundamental building blocks of every circuit model, it can be described as a 'primitive'.

A3 Primitive capacitance

Figure A.2 illustrates a short length l of a conductor of radius $r_{1,1}$ charged up to a voltage Vp.

If the charge density is ρ, then the intensity of the electric field at point $P(r,l)$ is

$$E_r = \frac{\rho}{4 \cdot \pi \cdot \varepsilon} \cdot \frac{1}{r} (\sin(\alpha_1) + \sin(\alpha_2)) \tag{A.34}$$

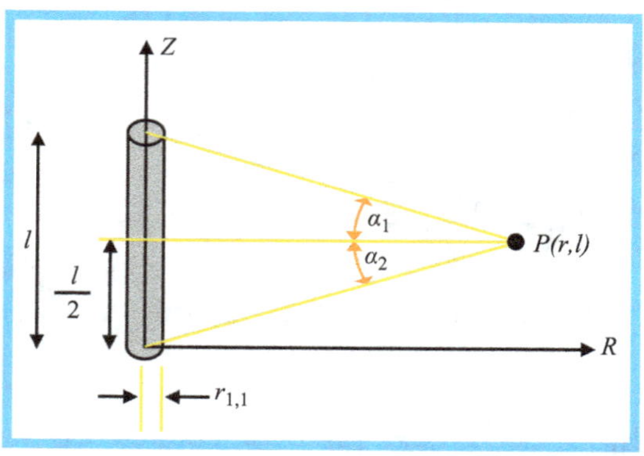

Figure A.2 Electric field at a point

In rectangular co-ordinates:

$$E_r = \frac{\rho}{2 \cdot \pi \cdot \varepsilon} \cdot \frac{\frac{l}{2}}{r \cdot \sqrt{\left(\frac{l}{2}\right)^2 + r^2}} \tag{A.35}$$

At all points along the surface defined by $z = l/2$, the E_z component of the electric field is zero. If unit charge were to follow a path along this surface from a point in the far distance to the surface of the conductor, then the work done would be:

$$Vp = \frac{\rho}{2 \cdot \pi \cdot \varepsilon} \cdot \int_{r_{1,1}}^{\infty} \frac{l/2}{r \cdot \sqrt{(l/2)^2 + r^2}} dr \tag{A.36}$$

Invoking (A.23),

$$Vp = \frac{\rho}{2 \cdot \pi \cdot \varepsilon} \cdot In \left(\frac{l/2\sqrt{(l/2)^2 + {r_{1,1}}^2}}{r_{1,1}} \right) \tag{A.37}$$

If $r_{1,1} << l$, then

$$Vp = \frac{\rho}{2 \cdot \pi \cdot \varepsilon} \cdot \ln\left(\frac{l}{r_{1,1}}\right) \tag{A.38}$$

The capacitance Cp of the conductor is the ratio of charge $\rho \cdot l$ to voltage Vp. That is

$$Cp = \frac{\rho \cdot l}{Vp} \tag{A.39}$$

Using (A.38) to substitute for the voltage in (A.39):

$$Cp \approx \frac{2 \cdot \pi \cdot \varepsilon \cdot l}{\ln\left(\frac{l}{r_{1,1}}\right)} \tag{A.40}$$

This is the capacitance of a length l of circular section conductor. It is the simplest possible parameter which has the dimensions of capacitance. Since it is one of the fundamental building blocks of every circuit model, it can be described as a 'primitive'.

Appendix B
Mathcad worksheets

As far as circuit modelling is concerned, Mathcad software eliminates most of the tedium involved in developing programs. It can handle tasks that are well beyond the scope of SPICE software.

In a programming language, equations look something like:

$$\mathrm{x} = (-\mathrm{B} + \mathrm{SQRT}(\mathrm{B} * *2 - 4 * \mathrm{A} * \mathrm{C}))/2 * \mathrm{A}$$

With Mathcad software, the same equation looks the same way as it does in a reference book:

$$x = \frac{-b + \sqrt{b^2 - 4 \cdot a \cdot c}}{2 \cdot a}$$

This makes the programs much easier to read and understand. Mathcad software also avoids the need to use a rigorous procedure to write a program. If the programmer gets the syntax wrong, the software refuses to accept the entry and provides a message indicating what is wrong.

On a worksheet, the equations can be set out from left to right, top to bottom. Text can be included anywhere on the page.

Doubtless there are other software packages which provide the same sort of facility. However, for consistency in presentation, Mathcad is the one used here.

Since this software has been available for over a decade, there is a fair probability that the reader is already using it. If not, it should be fairly easy task to modify the programs in the worksheets of this document to convert them into programs which work with other software packages.

A few symbols in Mathcad have a special meaning:

$a := b + c$ 'a' is defined as the sum of b and c

$a = 2$ the value of 'a' is 2

If the 'equals' sign is in bold typeface, the software interprets this as the 'Boolean equal' and returns a zero or a one.

The programming operator looks like that shown in Figure B.1.

The programming operator behaves like a function, taking input variables and returning an output. This output is the last variable to be declared. In the example above it is the value of the ratio of 'num' divided by '*denom*'. The program can return a single variable, a vector, or an array.

```
root(a,b,c) :=  | discr ← b^2 − 4·a·c
                | num ← −b + √discr
                | denom ← 2·a
                | num
                | -----
                | denom
```

Figure B.1 Simple Mathcad program

Local variables defined in the function are not visible outside. However, variables declared in the worksheet above the program function are visible within the function.

Included in the software are a number of built-in functions. As far as circuit modelling is concerned, the most important is 'lsolve(*M*,**v**)'. The argument '*M*' is a square matrix, whilst '**v**' is a vector. '*M*' contains as many rows as there are elements in '**v**'. This function returns a solution vector '**x**' such that $M\mathbf{x} = \mathbf{v}$.

Another useful characteristic of the software is that it can distinguish between the '*j*' operator as used with complex numbers and the variable '*j*'. It can be defined as an imaginary number by typing the characters '1*j*'. It appears on the worksheet as '*j*'.

Armed with this information, there should be no real problems in understanding the programs contained in the Mathcad worksheets of this book. Since the figures in the book which depict Mathcad worksheets have been hand-copied from copies of the actual worksheets, it is possible that errors have been introduced in the transcription process. Further errors could be introduced by the reader hand-copying the text onto his or her own computer. To prevent the chance of such errors creeping in, copies of the original Mathcad files are available in a zip folder which can be downloaded from www.designemc.info. These files can be run by any computer which has Mathcad version 15.0 (or higher) software installed.

The techniques described in the previous pages can be used to simulate the interference-coupling characteristics of any signal link. Every electrical or electronic system will have its own particular set of interference problems. It should be possible for individual designers to carry out tests and create circuit models of their own critical links. Such information can be shared. Mathcad users have access to PlanetPTC, a mix of dynamic channels that enable PTC customers and product development professionals to actively participate in exchanging ideas.

However, it does not really matter which software is used to carry out the calculations. The key feature of any circuit model is the fact that it is a shorthand method of defining a set of equations. It becomes possible to describe the hardware, the tests, the model, and to illustrate the results in a single report. There are many forums and many communities which can be used to publish such reports.

Appendix C

Links to Internet resources and references

Updating Circuit Theory
https://www.electronicsworld.co.uk/?s=Updating+Circuit+Theory
Accident Report
https://www.ntsb.gov/investigations/AccidentReports/Reports/aar0003.pdf
elementary particle
https://en.wikipedia.org/wiki/Elementary_particle
quantum
https://en.wikipedia.org/wiki/Quantum
electromagnetic field
https://en.wikipedia.org/wiki/Electromagnetic_field
electromagnetic radiation
https://en.wikipedia.org/wiki/Electromagnetic_radiation
light
https://en.wikipedia.org/wiki/Light
radio waves
https://en.wikipedia.org/wiki/Radio_wave
force carrier
https://en.wikipedia.org/wiki/Force_carrier
electromagnetic force
https://en.wikipedia.org/wiki/Electromagnetic_force
massless
https://en.wikipedia.org/wiki/Massless_particle
speed of light in vacuum
https://en.wikipedia.org/wiki/Speed_of_light
photon
https://en.wikipedia.org/wiki/Photon
www.designemc.info
http://www.designemc.info/

Appendix D

Definitions

Aerial-mode current	Unidirectional current in the conductors of a cable. That which flows between the cable and the environment.
Circuit equations	A set of equations which relate currents to voltages in a circuit model.
Circuit model	A model which obeys the rules of circuit theory and simulates the behaviour of the assembly under review.
Circuit parameter	A parameter used in a circuit model.
Common-mode current	Current in the loop formed by cable and structure.
Common-mode gain	The ratio of the output voltage to the common-mode input voltage.
Common-mode rejection	The ratio of the differential-mode gain to the common-mode gain. Usually quoted in terms of decibels.
Composite conductor	A set of elemental conductors, aligned in parallel, which enables the distribution of currents or voltages in the actual conductor to be simulated.
Conducted emission	The current induced in the common-mode loop by a voltage source in the differential-mode loop.
Conducted susceptibility	The current induced in the differential-mode loop by a voltage source in the common-mode loop.
Culprit loop	The loop in which the interfering voltage source is located.
Current transformer	A transducer which monitors the magnetic field surrounding a group of conductors and generates a voltage proportional to the sum of the currents in those conductors.
Definitive circuit model	A circuit model which simulates the interference coupling mechanisms of a specific signal link.
Delay-line model	A circuit model which simulates the differential-mode behaviour of a transmission line.
Differential-mode current	Current which flows in the loop formed by the signal and return conductors.
Differential-mode gain	The ratio of the output voltage to the differential-mode input voltage of an amplifier.
Distributed parameters	Parameters which cater for the fact that voltage and current both vary along the length of any conductor.
Earth	The conductor(s) designated to carry fault current in an AC distribution network.
Elemental conductor	A conductor which represents a small segment of the surface of a composite conductor.

EMC	Electromagnetic compatibility. The ability of a device, unit of equipment, or system to function satisfactorily in its electromagnetic environment without introducing intolerable electromagnetic disturbances to anything in that environment.
EMI	Electromagnetic interference.
EUT	Equipment Under Test.
General circuit model	A model which includes interface circuit components as well as cable coupling components, but which does not specify component values.
Floating configuration	A wiring configuration where one end of the signal link is isolated from the structure.
Ground	Another name for the structure.
Grounded configuration	A wiring configuration where the return conductor of the signal link is bonded to local structure at each end.
Loop parameters	Parameters which have been derived from groups of primitive parameters. Loop parameters can be measured by electrical test equipment.
Loop equations	A set of equations, derived from the primitive equations, which relate loop voltages to loop currents.
Lumped parameters	Resistors, capacitors, and inductors, as used in textbooks on circuit theory.
Partial parameters	Those parameters associated with the behaviour of composite conductors such as aerials. The term is also used to distinguish between incident and reflected currents in transmission lines.
Partial inductance	The inductance of a composite conductor, or the inductance of a loop segment.
Partial capacitance	The capacitance of a composite conductor, or the capacitance of a loop segment.
Partial current	A portion of the current flowing in a conductor.
Partial voltage	A portion of the voltage in a circuit loop.
Partial inductor	Inductor associated with partial current and partial voltage.
Primitive current	Unidirectional current flow in a single conductor of a multi-conductor assembly.
Primitive voltage	Voltage along a conductor due to electromagnetic coupling.
Primitive parameters	Parameters associated with conductors which are treated as aerials.
Primitive equations	A set of equations relating primitive voltages to primitive currents.
Radiated emission	The current induced in the aerial-mode loop by a voltage source in the differential-mode loop.
Radiation resistance	The apparent resistance of an aerial when it is carrying maximum current during resonance conditions.
Radiation susceptibility	The current induced in the differential-mode loop by a voltage source in the aerial-mode loop.
Return conductor	The conductor designated to complete the loop carrying signal current.
Signal conductor	The conductor which is defined as carrying the signal under review.

Signal link	The conductors and interface circuitry involved in transmitting a signal from one location to another in a particular system.
SPICE	Simulation Program with Integrated Circuit Emphasis.
Structure	The conducting elements of the structure which act as a shield to electromagnetic radiation.
Threat environment	The frequency response of the power density of the worst-case external radiation to which the equipment can be subjected. It can also be defined in terms of the waveform of a transient current or voltage.
Threat voltage	The amplitude of the voltage in the culprit loop.
Time-step analysis	A method used to predict currents and voltages in the circuit model a discrete time after a step change has occurred in any current or voltage in that model.
Transfer admittance	The ratio of the current in the victim loop to the source voltage in the culprit loop, when there are no other voltage sources.
Victim loop	The loop carrying the signal which is regarded as being susceptible to interference.
Virtual conductor	An imaginary conductor which enables the coupling between cable and environment to be simulated.
Voltage transformer	A transducer which uses magnetic field coupling to induce a voltage in the loop-under-test and which allows the induced voltage to be measured.

Index

www.ingramcontent.com/pod-product-compliance
Lightning Source LLC
LaVergne TN
LVHW011848200926
848484LV00002B/422
* 9 7 8 1 8 3 9 5 3 9 5 3 4 *